Invading Nature - Springer Series in Invasion Ecology

Volume 11

Series Editor
Daniel Simberloff
Ecology and Evolutionary Biology
Knoxville
Tennessee
USA

Biological Invasions represent one of those rare themes that cut across the disciplines of academic biology, while having profound environmental, philosophical, socioeconomic, and legislative implications at a global scale. There can be no doubt that biological invasions represent the single greatest threat to biodiversity past the activities of humankind itself. The implications are far reaching. Novel ecological and evolutionary forces are now directing the future expression of life itself, as native species and the communities that they comprise contend with invading species. The rules of the game have been suddenly and irrevocably changed.

Invading Nature - Springer Series in Invasion Ecology is a new book series topically spanning the breadth of invasion biology. The series is of singular importance as an integrative venue focusing on the broader ecological and evolutionary issues arising from non-native species, the impacts such species have in particular environs, trends patterns and processes, as well as causes and correctives. The series seeks novel and synthetic approaches to invasions including experimental, theoretical, systematic and conceptual treatments.

Prospective authors and/or editors should consult the **Series Editor Daniel Simberloff** for more details:

e-mail: tebo@utk.edu

More information about this series at http://www.springer.com/series/7228

Fanghao Wan • Mingxing Jiang • Aibin Zhan
Editors

Biological Invasions and Its Management in China

Volume 1

Springer

Editors
Fanghao Wan
Institute of Plant Protection
State Key Lab for Biology of Plant Diseases & Insect Pests
Chinese Academy of Agricultural Sciences
Beijing, China

Mingxing Jiang
Institute of Insect Sciences
Zhejiang University
Hangzhou, China

Aibin Zhan
Research Center for Eco-Environmental Sciences (RCEES)
Chinese Academy of Sciences (CAS)
Beijing, China

Invading Nature - Springer Series in Invasion Ecology
ISBN 978-94-024-0946-8 ISBN 978-94-024-0948-2 (eBook)
DOI 10.1007/978-94-024-0948-2

Library of Congress Control Number: 2017930663

Printed on acid-free paper

This Springer imprint is published by Springer Nature
The registered company is Springer Science+Business Media B.V.
The registered company address is: Van Godewijckstraat 30, 3311 GX Dordrecht, The Netherlands

Foreword

China is one of the countries severely affected by biological invasions. By the end of 2016, at least 610 alien invasive species (AIS) had been identified in China, 50 of which are on the list of the world's 100 worst invasive species (IUCN). AIS have caused huge economic losses, estimated to be 17 billion US dollars per year. In addition, AIS negatively influence multiple ecosystems and biodiversity, leading to large-scale degradation of invaded ecosystems. Thus, biological invasions are considered as one of the biggest biosecurity issues in China.

To cope with biological invasions, a great number of research projects have been awarded by central and local governments in China since the beginning of the twenty-first century. These projects cover all aspects of biological invasions, from surveys of species distribution and damage to conducting advanced research on the mechanisms of invasion success and developing control techniques/strategies. Great progress has been made in several fields, including identifying which species have the highest possibility of being introduced into China, why AIS are successful in invasions in different regions/environments, and where they would have a high risk of causing serious damage. Moreover, Chinese scientists have successfully developed techniques for rapid early detection and field monitoring, and most importantly, effective control methods have been developed for highly invasive species. So far, we have clarified the mechanisms underlying successful invasions for several AIS, such as the asymmetric mating interactions in the whitefly *Bemisia tabaci*, and communication between pinewood nematodes, insect vectors, and associated microbes.

As an outcome of extensive researches in this field, Chinese scientists have published more than 1400 papers in ISI-indexed journals. There are also a large number of papers published in Chinese journals. Furthermore, numerous field examples of successful control activities have been implemented but not reported. To present a comprehensive view of these results, the publication of an English monograph that summarizes the major findings/experiences in China is timely. The book *Biological Invasions and Its Management in China* edited by Fanghao Wan, Mingxing Jiang, and Aibin Zhan and their colleagues covers topics of current interest and research progress in the field of biological invasions in China.

The book contains several important themes, such as what invasion problems have occurred in Chinese major ecosystems, what has been done to solve current problems, and what are the major research directions in China. In addition, the book includes findings from other parts of the world to provide comprehensive information to readers. Therefore, readers will find many areas of interesting research in this book. We expect this book to appeal widely to scientists and staff who work in the field of biological invasions.

The arrival of alien species in China and their impacts will undoubtedly continue, mainly owing to increased international trade and travel. The invasions may escalate as some of the plans to open up the country are implemented in China, such as the "One Belt and One Road" (OBOR) policy. As a result, the issues of biological invasions faced by China will also affect other countries, such as those along the OBOR. It is, therefore, crucial to promote exchanges and collaborations between Chinese scientists/governments and those overseas in the field of AIS. From this point of view, this book is expected to become a very valuable shared source of information.

Academician of the Chinese Academy of Engineering, Kongming Wu
Institute of Plant Protection
Chinese Academy of Agricultural Sciences
Beijing, China

Preface

During the past decades, China has been greatly challenged by a dramatic increase in alien invasive species (AIS), leading to significant negative impacts on the economy, ecology, and even social development. Many ecosystems in China have been seriously affected by AIS, including agricultural and forest ecosystems. As a consequence, biological invasions have become a major focus for scientific research and administrative management in China, in particular since the beginning of the twenty-first century.

Our purpose in editing this book stems from the research advances that have been achieved in the last two decades in the field of biological invasions in China. These advances relate not only to scientific research but also to the management of invasive species. Our book will provide readers with information on what we have already done and what we propose to do in future studies, by reviewing a large volume of research findings and management experiences in China.

Exchanges between China and other countries are increasing. China has a very diverse range of ecosystems and is developing extensive trade in multiple categories of products. These trends are predicted to persist for the next few decades, particularly after the implementation of "One Belt and One Road" strategy. Human-mediated introductions of AIS create some features of biological invasions specific to China. Clearly, studying the trend of new types of invasions, as well as the possible mechanisms underlying invasions, will contribute to the control and management of AIS and will also enhance international collaborations to mitigate the negative impacts of AIS.

Despite the enormous efforts dedicated to control, the level and rate of invasions are continuing to increase due to the dramatic growth in international trade and travel, as well as in the nationwide transport of multiple categories of products and materials. Central and local governments need to exert more efforts on regulatory and administrative activities. The public will need to increase their awareness of biological invasions and to be more extensively involved in the control of AIS. Overall, some serious invasive species have not been effectively managed and are still causing serious damage to China, while new ones are being introduced that are not yet recognized because their populations are small or they have not caused environmental problems. There is little doubt that many fundamental topics remain to be answered in invasion biology. All these problems are examined in our book, which will benefit readers who want a comprehensive understanding of the position on biological invasions and relevant research in China.

Although we aim to provide a thorough coverage of the topic to readers, this book cannot cover all aspects of biological invasions in China. Articles by Chinese scientists are being published at a rapid rate, and thus readers may find that some valuable results are not included in our book. Moreover, as biological invasion itself is a fast-growing field, readers will find that some issues in the book have not yet been solved or even examined in detail. We hope that this book will promote active discussions in the field and draw attention to the problem of invasive species in China.

This book consists of two major parts: first, Chaps. 2, 3, 4, 5 and 6 that represent biological invasions in different types of ecosystems and, second, Chaps. 8, 9, 10, 11, 12, 13, 14, 15, 16, 17, 18, 19, 20, 21, 22, 23, 24, 25, 26, 27, 28, 29, 30, 31, 32 and 33 that address the invasion and management of representative invasive species. For each chapter, we include the results of the most relevant studies and management strategies/techniques. An index at the end of the book will help readers to find topics of interest to them. The book will be of interest to researchers, regulatory administrators, environmental managers, and the public.

Beijing, China Fanghao Wan
Hangzhou, China Mingxing Jiang
Beijing, China Aibin Zhan

Acknowledgments

We would like to thank foremost all authors for their huge efforts dedicated to this book, which finally makes the book a reality. We are very grateful to the scientists who generously provided the information, figures, photos, and articles we requested. We wish to appreciate the reviewers who kindly read the drafts and provided invaluable suggestions and comments for our chapters. We also offer special thanks to the series editor, Abbey Huang at Springer, for her invaluable assistance in preparing this volume and to Atma Biswal and his colleagues for their assistance during the production process.

Numerous scientific findings described in our book are generated from the research projects financially supported since 2002 by the Ministry of Science and Technology (MOST), Ministry of Agriculture (MOA), National Natural Science Foundation of China (NSFC), etc., particularly the national projects of "National Key Research and Development Programs" (2016YFC1200600, 2016YFC1202100, 2016YFC1201200) and the "948 Program" (2016-X48).

Beijing, China — Fanghao Wan
Hangzhou, China — Mingxing Jiang
Beijing, China — Aibin Zhan

Contents of Volume 1

Part I Biological Invasions in China: Outline

1 **Biological Invasion and Its Research in China: An Overview** 3
Ying Yan, Xiaoqing Xian, Mingxing Jiang, and Fanghao Wan

2 **Biological Invasions in Agricultural Ecosystems in China**.................. 21
Mingxing Jiang, Yunshan Huang, and Fanghao Wan

3 **Biological Invasions in Forest Ecosystem in China**............................. 53
Min Lu and Jianghua Sun

4 **Biological Invasions in Aquatic Ecosystems in China** 67
Aibin Zhan, Ping Ni, Wei Xiong, Yiyong Chen, Yaping Lin, Xuena Huang, Yuzhan Yang, and Yangchun Gao

5 **Biological Invasions in Desert Green-Islands and Grasslands** 97
Amanulla Eminniyaz, Juan Qiu, Carol C. Baskin, Jerry M. Baskin, and Dunyan Tan

6 **Biological Invasions in Nature Reserves in China**............................... 125
Hui Guo, Susan J. Mazer, Xinyu Xu, Xi Luo, Kailing Huang, and Xiaohong Xu

7 **Roles of Chinese Government on Prevention and Management of Invasive Alien Species**.. 149
Rui Wang, Fanghao Wan, and Bo Li

Part II Invasion and Management of Major Alien Insects in China

8 **The Whitefly *Bemisia tabaci* (Gennadius)**... 159
Xiaowei Wang and Nianwan Yang

9 Rice Water Weevil *Lissorhoptrus oryzophilus* Kuschel 183
Yunshan Huang, Michael O. Way, and Mingxing Jiang

10 Colorado Potato Beetle *Leptinotarsa decemlineata* (Say)..................... 195
Wenchao Guo, Chao Li, Tuerxun Ahemaiti, Weihua Jiang, Guoqing Li, Jiahe Wu, and Kaiyun Fu

11 Red Turpentine Beetle *Dendroctonus valens* LeConte......................... 219
Min Lu and Jianghua Sun

12 Coconut Leaf Beetle *Brontispa longissima* Gestro 229
Baoqian Lu and Zhengqiang Peng

13 Red Palm Weevil *Rhynchophorus ferrugineus* (Olivier) 245
Lu Peng and Youming Hou

14 Nipa Palm Hispid Beetle *Octodonta nipae* (Maulik) 257
Baozhen Tang and Youming Hou

15 Oriental Fruit Fly *Bactrocera dorsalis* (Hendel).................................. 267
Dong Wei, Wei Dou, Mingxing Jiang, and Jinjun Wang

16 Codling Moth *Cydia pomonella* (L.) 285
Maohua Chen, Xinle Duan, Yuting Li, Qiulei Men, and Fanghao Wan

17 Red Imported Fire Ant *Solenopsis invicta* Buren 299
Lei Wang and Yongyue Lu

Appendices.. 317

Index 361

Contents of Volume 2

Part III Invasion and Management of Major Alien Non-insect Animals, Plants and Microorganisms in China

18 Pinewood Nematode *Bursaphelenchus xylophilus* (Steiner and Buhrer) Nickle 3
Lilin Zhao and Jianghua Sun

19 Burrowing Nematode *Radopholus similis* (Cobb) 23
Birun Lin and Huifang Shen

20 Golden Apple Snails 33
Xiaoping Yu, Qianqian Yang, and Yipeng Xu

21 Red-Eared Slider *Trachemys scripta elegans* (Wied-Neuwied) 49
Kai Ma and Haitao Shi

22 Nile Tilapia *Oreochromis niloticus* (L.) 77
Dangen Gu, Yinchang Hu, Hui Wei, Yunjie Zhu, Xidong Mu, Du Luo, Meng Xu, and Yexin Yang

23 North African Catfish *Clarias gariepinus* (Burchell) 91
Dangen Gu, Yinchang Hu, Hui Wei, Yunjie Zhu, Xidong Mu, Du Luo, Meng Xu, and Yexin Yang

24 Common Ragweed *Ambrosia artemisiifolia* L. 99
Zhongshi Zhou, Fanghao Wan, and Jianying Guo

25 Crofton Weed *Ageratina adenophora* (Sprengel) 111
Guoqing Yang, Furong Gui, Wanxue Liu, and Fanghao Wan

26 Mile-a-Minute Weed *Mikania micrantha* Kunth 131
Shichou Han, Zhigang Li, Qiyun Xu, and Lingling Zhang

27 Canada Goldenrod *Solidago canadensis* L. 143
Fengjuan Zhang and Fanghao Wan

28 Coastal Plain Yellowtops *Flaveria bidentis* (L.) Kuntze 153
Fengjuan Zhang and Fanghao Wan

29 Alligator Weed *Alternanthera philoxeroides* (Mart.) Griseb. 163
Mengzhu Shi and Jianwei Fu

30 Water Hyacinth *Eichhornia crassipes* (Mart.) Solms 175
Jianwei Fu, Mengzhu Shi, and Jianyu Li

31 Saltmarsh Cordgrass *Spartina alterniflora* Loisel 187
Ruiting Ju, Hui Li, Lei Shang, Shiyun Qiu, Jing Li,
Ming Nie, and Bo Li

32 *Phytophthora sojae* 199
Xiaoren Chen and Yuanchao Wang

33 *Fusarium oxysporum* f. sp. *cubense* 225
Birun Lin and Huifang Shen

Part IV Perspectives and Future Research Directions

34 Research and Management of Biological Invasions in China: Future Perspectives 239
Mingxing Jiang, Aibin Zhan, Hui Guo, and Fanghao Wan

Index 249

Part I
Biological Invasions in China: Outline

Chapter 1
Biological Invasion and Its Research in China: An Overview

Ying Yan, Xiaoqing Xian, Mingxing Jiang, and Fanghao Wan

Abstract The rapid economic development in China significantly triggered the merchandise trade, international tourism, as well as species migration and exchange in the past 30 years. Under such circumstances, large-scale biological invasions happened intensively in agricultural lands, forests, grasslands, islands, fishery, wetlands, rivers, marine and natural reserves since the last decade of twentieth century. Invasive alien species (IAS) not only cause economic losses and ecosystem meltdown, but also pose threats to human health, food supply and national safety. Since the beginning of twenty-first century, China has considerably invested in basic and applied research of IAS to meet the urgent needs in terms of prevention and management. The theories and experiences of IAS research from other countries provide valuable references to China. On the other hand, due to its own geographic, economic, cultural and historic features, China has some different motivations, emphasizes and methods for IAS researches. In this chapter we give brief introduction about IAS research in China including: the occurrence and damage of IAS, the aspects that have been studied mostly, the achievements we already got, the role of government, and the international cooperation. We also illustrate the necessity, purpose and outline of this book. In a summary, the capacity of biological invasion research and IAS management in China significantly grow in the last 20 years, which has largely contributed to the social benefits, environmental protection and economic development of this country. For the coming decades, we expect the globalization, human activity and climate change will pose more threats in terms of biological invasions, and China will be ready and well prepared for such a challenge.

Keywords Biological invasions • Invasive alien species • Ecosystem • Biodiversity • Research

---

Y. Yan • X. Xian • F. Wan (✉)
State Key Lab for Biology of Plant Disease and Insect Pests, Institute of Plant Protection, Chinese Academy of Agricultural Sciences, Beijing 100193, China
e-mail: wanfanghao@caas.cn

M. Jiang (✉)
Ministry of Agriculture Key Laboratory of Agricultural Entomology, Institute of Insect Sciences, Zhejiang University, Hangzhou 310058, China
e-mail: mxjiang@zju.edu.cn

F. Wan et al. (eds.), *Biological Invasions and Its Management in China*, Invading Nature - Springer Series in Invasion Ecology 11,
DOI 10.1007/978-94-024-0948-2_1

1.1 Introduction

Human activities have extensively changed our world in the past century, and the interactions among people as well as countries are becoming more frequent than ever. As one of prominent outcomes, large numbers of species have been introduced unintentionally or intentionally to regions beyond their natural distributions (i.e. invasive alien species, IAS). Some of these species became invasive in the non-native range, causing damages to local economy, ecology, and/or other human's welfares.

As one of the fast globalizing countries, China is facing a high risk from biological invasions due to rapid increase in international trade, travel and transportation. In this country, concerns on IAS issues arose since the middle of 1990s, when a few scientists presented papers to introduce damage, biology and control of the IAS occurring in other countries (i.e., Yang and Liang 1996). Later, several papers addressed the distribution and damage of some serious IAS in China, such as fall webworm *Hyphantria cunea* (Drury) and pinewood nematode *Bursaphelenchus xylophilus*, to reveal the urgency of carrying out research and management over IAS (Lu 1999; Liu and Wen 2000; Xie et al. 2001). From then, concerns on IAS issues increased rapidly.

Parallel to scientists' response, China government began to take actions at the beginning of this century. In 2001, State Environmental Protection Administration funded a national three-year program for surveys on IAS status in China, and in the following year, Ministry of Science & Technology (MST) listed the research of IAS as a National Basic Research and Development Program of China which spanning from 2002 to 2013. At the same time, a great number of research programs were supported by Ministry of Agriculture (MOA) and National Natural Science Foundation of China (NSFC). In addition to financial supports, the government also founded several agencies to strengthen administrative management on IAS and related issues, such as Office for Alien Species Management.

After 20 years since the arising of concerns on biological invasions, now we are pleased to see that great progress has been achieved in both research and management of IAS. Most of serious IAS have been extensively investigated on their invasion biology, which generated a huge body of knowledge on their invasion process and the underlying mechanisms. Meanwhile, numerous practical control techniques have been developed, which greatly strengthen our capacity of dealing with biological invasions, from risk analysis and prevention to acquiring sustainable control strategy for certain IAS. Consequently, so far many serious IAS have been taken under control.

In this chapter, we give an overview on the past and present states of IAS in China, and the advance in research and management. We also show actions that the China governments have taken to strengthen the management and what China has done at the international level.

1.2 A Glance at IAS Occurrence and Damage in China

Before 2000, it was not clear how many IAS were in China. During 2001–2003, an extensive field investigation was carried out across the country to collect basic information of IAS, which included species composition, distribution, living habitat, life history, damage, etc. During 2007–2012, an additional field survey was conducted. So far we have been much clear about these aspects (Xu et al. 2006a, b, 2012b).

Based on available literature and field survey data, Xu et al. (2012b) compiled an inventory of IAS in China. This inventory lists a total of 488 species that occur in terrestrial habitats, inland waters and marine ecosystems, which included 171 animals, 265 plants, 26 fungi, 3 protists, 11 procaryots, and 12 viruses. Plants, insects and microorganisms are the major groups. The number of IAS must have been underestimated due to the overlooking of some species that have low density and thus can be rarely collected at early stages of biological invasions. Recently updated data show that there are 618 IAS in China (Center for Management of Invasive Alien Species, CMIAS, Ministry of Agriculture, China). Of them, plants are the major group, followed by insect and fish (Fig. 1.1).

About 70% of IAS in China were introduced after 1950. According to Xu et al. (2012b), only 33 IAS were recorded before 1850, and the recorded number increase slowly in the following 100 years, reaching about 180 specie by the end of 1950. Thereafter, however, the IAS number began to increase rapidly with at least 50 species being detected per 10 years (Xu et al. 2012b; and the data from CMIAS). It should be noted that a certain proportion of species probably had been present in China, but not identified before 1950, due to the lack of scientists working in this

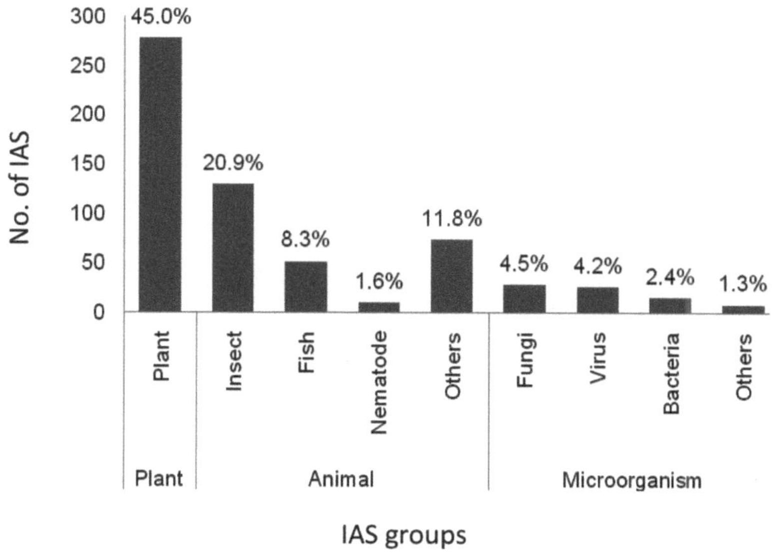

Fig. 1.1 The taxonomic composition of invasive alien species (IAS) in China

field during the war years. Thus, some IAS recorded in later years might have been introduced to China much earlier. Despite this, accumulating data strongly suggests that a large proportion of IAS in China were introduced in recent decades, due to the rapid development of economy, international trade, as well as frequent human activities (e.g., Lin et al. 2007; Ding et al. 2008; Xu et al. 2012b; Wan and Yang 2016).

In terms of invasion source, around two third of invasive plants and one third of invasive animals in China were derived from intentional introductions. According to Xu et al. (2012b), 51.1% of IAS have an origin from North and South Americas, 18.3% from Europe, 17.3% from other Asian countries, 7.2% from Africa, and 1.8% from Oceania.

The distribution of IAS in China has an evident geographic and ecosystematic bias. They are richest in eastern, southern coastal provinces and the Yunnan province in southwestern of China, where the economy is much developed or/and human activities are intensive. In comparison, much fewer IAS occur in inland provinces, and in some northwestern regions less than 100 IAS were recorded. Overall, the IAS richness in China shows a general pattern of declining gradually from Southeast to Northwest (Fig. 1.2). More than 60% of IAS have been recorded in farmlands, while only 14% and 16% occur in forests and water bodies (sea and inland waters), respectively.

IAS have caused serious damages to economy, ecology and other human welfares in China. Xu et al. (2006a) estimated that China suffered an economic loss ca. USD 14.5 billion annually, which accounted for approximately 1.36% of GDP (estimations for the year 2000). Of the diverse ecosystems, agroecosystems and forests

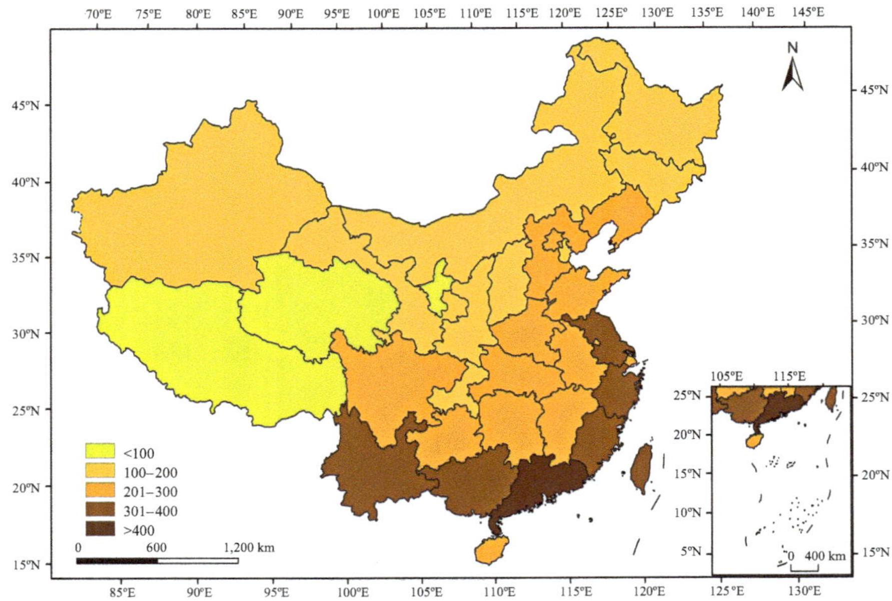

Fig. 1.2 Geographic distribution of invasive alien species in China

suffered the most serious losses (Xu et al. 2006a). In addition to huge direct economic loss, damages to ecosystems are also very severe in China. What to be most concerned is the loss of biodiversity. For instance, the smooth cordgrass *Spartina alterniflora* not only remarkably suppresses native annual plants and becomes dominant plant species, but also significantly decreases the diversity of rhizospheric microorganisms, insects and benthic animals. This plant can even reduce bird diversity due to the habit loss and food shortages. Some IAS may also impair ecosystem's structure and function. Studies showed that the crofton weed *Ageratina adenophora* could change nutrient cycles in soil and simplify plant community structure. In addition, some IAS such as common ragweed *Ambrosia artemisiifolia*, red imported fire ant *Solenopsis invicta* and South American ampullarid *Pomacea canaliculata*, can harm human health. Take *A. artemisiifolia* for an example, its pollen is a problematic aero-allergen which can induce severe health problems in human populations. In China, 2–3% of the population, ca. 14.5 million persons, in the *A. artemisiifolia* occurring regions ever suffered from allergy, which costs nearly 1.45 billion US dollars annually in China.

1.3 What Have Been Studied Mostly

Since the beginning of this century, the interest of studying IAS boomed in China, as indicated by the sudden increase of related papers. Scientists began to consider what were the most significant problems in China that had to be answered urgently. Below, we show what ecosystems and what kinds of IAS have been given much research attention, and what topics have been attracted much research interest.

1.3.1 Research Attention to Various Kinds of Ecosystems and IAS

China has diverse managed or natural ecosystems, including agricultural lands, forests, grasslands, deserts, wetlands, lakes, oceans, etc. Among them, agricultural ecosystem and forests are more extensively studied. This is because these two ecosystems are much larger, and as mentioned above they have been invaded by more IAS and suffered more damages than the other types of ecosystems (Xu et al. 2006a, 2012b). The other ecosystems, though much less studied, are being invaded by an increasing number of IAS, and some of them are suffering serious damages, such as the aquatic ecosystems.

Plant and insects are the major IAS groups so far extensively studied in China. This is consistent with the facts that they are the most diverse IAS in China and most of them occur in agroecosystems (Qiang et al. 2010; Wan and Yang 2016). Only a few invasive birds and mammals were studied in this country.

1.3.2 Major Research Topics

Our analysis on the papers published during 2001–2015 suggests that invasion process is the major research topic focused by China scientists in the past. That is, it had been the major task in the past to learn how the IAS was introduced to China, established, spread and then outbreak. Another major research topic is species invasiveness (invasive capacity) and underlying mechanisms. This aspect has been investigated in a great number of IAS regarding their reproduction, growth, development, adaptation to ambient stresses, and interspecific interactions. By contrast, ecosystem invisibility, one of core rationales in invasion biology, is much less investigated in China. The general pest control strategy in China might be a major reason for this, since in most cases people tend to target the IAS directly, rather than manipulate local systems to be resistant.

Specific aspects of IAS from a certain ecosystem were interested and often studied intensively by scientists. For example, for the invasive insect pests in agroecosystems, such aspects were often targeted: resistance to chemical insecticides, adaptation to ambient stresses, natural enemy release and symbiotic association (Wan and Yang 2016). For IAS in aquatic ecosystems, however, people generally want to know where the IAS come from (invasion pathways), what are the ecological roles of propagule pressure and colonization pressure, the impact on their native counterparts (competitiveness) and the ecosystem (invisibility).

Chinese scientists are advancing their research to the field of physiology and molecular biology with the aim of discovering the mechanisms of invasion success (e.g., Feng et al. 2009). They also have attempted to examine whether the species have genetically differentiated and evolved during invasions, which in turn increases their invasive capacity (Feng et al. 2009; Qin et al. 2013; Huang et al. 2015; Zheng et al. 2015).

To develop IAS management strategies, a large number of applied researches has been conducted for risk assessment, prevention, detection, monitor, eradication, and sustainable control methods. Consequently, effective control strategies have been developed for many serious IAS in China.

In deed, the topics selected for research priority differ with the type of species and ecosystems. For example, for invasive insects and plants in agroecosystems already extensively studied in the past decades, the development of sustainable control methods will be the first option in next decade. By contrast, for the IAS in nature reserves, grassland and desert oases, which were much less studied previously, their invasion process and invasion mechanisms have to be clarified first. Aquatic IAS, which may be introduced and spread via a diversity of human activity-related pathways, their invasion risk should be addressed and the sound strategies and control techniques are urgently needed.

1.4 What Achievements We Already Had

1.4.1 Current Status and Biological Background of IAS

As mentioned above, to date we have been much clear about the current status of IAS in China, including their taxonomy, invasion sources and pathways, geographic distribution, damages, etc. (Wan et al. 2005; Xu et al. 2006b; Ding et al. 2011; Xu and Qiang 2011). Moreover, a large number of IAS has been extensively investigated on their invasion biology features, in particular those associated with invasion process and invasion capacity. This has accumulated a huge body of information that is essential for further studies.

1.4.2 Novel Invasion Mechanisms

Several novel invasion mechanisms have been discovered in China. For example, the invasive MEAM1 of whitefly *B. tabaci* was found to be capable of interfering with indigenous whiteflies' reproduction through a mechanism "asymmetric mating interactions", which as a result leads to replacement of indigenous whiteflies (Liu et al. 2007). This invasive whitefly has also developed an indirect vector-virus mutualism with geminiviruses through which its invasive capacity is elevated (Jiu et al. 2007; Zhang et al. 2012; Luan et al. 2013). In the invasive plant *A. adenophora*, an evolutionary tradeoff was discovered in nitrogen allocation (to photosynthesis vs. cell walls), which allows the plant to grow more rapidly and thus to become more competitive in the non-native range (Feng et al. 2009). This plant can also undertake a quicker return energy-use strategy, which serves as a mechanism for the evolution of observed increased growth and competitive ability in invasive plants (Feng et al. 2011). In *C. odorata*, evolution of increased competitive ability, innate competitive advantages, and novel biochemical weapons were found to be able to act in concert with promoting its invasion (Qin et al. 2013; Zheng et al. 2015).The fungal pathogen *Phytophthora sojae*, which causes phytophthora root rot in soybeans, was found to be able to secrete a number of effector proteins that are capable of modulating or suppressing host plant immunity (Dong et al. 2011; Liu et al. 2014). The pinewood nematode *Bursaphelenchus xylophilus* (Steiner and Buhrer), a causative agent of pine wilt disease, was demonstrated to engage in a symbiotic partnership with multiple organisms (its insect vector as well as associated bacteria and fungi) to promote its pathogenicity and invasiveness (Zhao et al. 2014). At the ecological level, Wang et al. (2011b) suggested a 'combinatorial' invasion mechanism through which invasive plants can spread under heterogeneous ecogeographical environments at a rapid invasion rate.

1.4.3 Ecosystem Invasibility

Ecosystem invasibility have been examined by a number of studies for its relations with community structure, biodiversity, resource availability, human disturbance, soil biota, and physical stress imposed by environments (Xu et al. 2004; Li et al. 2005; Pan et al. 2006; Lu and Ma 2005; Wang et al. 2006; Jiang et al. 2007; Zhu et al. 2011; He et al. 2012; Liu et al. 2012; Gao et al. 2014; Zhu et al. 2015; Wu et al. 2016). Most of these studies focused on the IAS in agroecosystems. To date we have learned much about what kind of agroecosystems in China could be heavily invaded, such as for the species alligator weed *Alternanthera philoxeroides* and crofton weed *Eupatorium adenophorum*, two notorious invasive plants in China. The results give us the clues of regulating ecosystem's components and structure to reduce its resistance to biological invasions.

1.4.4 Risk Assessment and Early Warning

More than 100 invasive species in China have been quantitatively assessed for the risk of introduction, establishment, spread and outbreak in Chinese ecosystems. The species of high invasion risk has been identified, and the corresponding warning has been given to the regions that could be potentially invaded (Wan et al. 2010). Moreover, the platforms for remote, quantitative risk analyses of IAS are under construction.

1.4.5 Detection and Monitoring

DNA-based rapid detection techniques have been developed for the IAS which can not be reliably identified using conventional methods, such as those based on morphological features due to small body size, similarity with closely related species/strains/biotypes, or the species which have to be identified at juvenile stages with limited available diagnostic characters (Wan et al. 2011a). Moreover, some efficient kits have been developed and provided to personnel at entry & exit ports and quarantine stations. To increase detect efficiency, DNA barcoding technique has been used for more than 141 IAS in China, which include fruit flies, thrips, scale insects and whiteflies. Using this technique, most specimens can be identified within 1–2 days. Field-monitoring systems have been established for more than 80 invasive species, and accordingly a national monitoring network has been established. These detection techniques and monitoring systems have contributed greatly to the risk assessment, early warning, prevent/interception, and the development of management strategies.

1.4.6 Early Prevention and Efficient Interception

Based on the data generated from risk analyses, a large number of IAS has been successfully intercepted before introduction or spread to the local ecosystems or habitats. Effective prevention belts have been constructed for highly dangerous IAS that are occurring at a small a geographic range in China, such as Colorado potato beetle (CPB) *Leptinotarsa decemlineata* Say, red imported fire ant (RIFA) *Solenopsis invicta* Buren, codling moth *Cydia pomonella* (L.), and burrowing nematode *R. similis* (see Chap. 2). These pests have been successfully intercepted at the spreading frontier and prevented from spreading neighboring regions.

1.4.7 Sustainable Control Methods

A number of sustainable control techniques have been developed for the control of serious invasive species. For example, in the past four decades, more than 40 biological control agents were introduced from overseas, of which over ten are highly effective against invasive arthropods in agricultural and forest ecosystems. In addition, a great number of native natural enemies were also tested to be effective (Wan et al. 2008a; Yang et al. 2014). Through using biocontrol agents, some notorious invasive species have been successfully controlled in some regions, such as common ragweed *A. artemisiifolia* and alligator weed *Alternanthera philoxeroides* (Wan and Yang 2016).

Chinese scientists have also made efforts towards screening of native plant species that can be used for replacement control of invasive plants. In some regions, such a control method has been successfully employed for the control of *A. artemisiifolia*, crofton weed *A. adenophora* (Spreng.) King & Rob. and *Spartina alterniflora* Loisel. Great progresses have also been achieved in the screening of plant germplasms resistant to *P. sojae*, *R. similis*, and *Foc* race 4 of *Fusarium oxysporum* f. sp. *cubense* Schlechtend (Smith) Snyder & Hansen (a pathogenic fungus causing Fusarium wilt of banana) (see Chap. 2).

1.4.8 Database of Invasive Alien Species

A database of invasive alien species in China (http://www.chinaias.cn/wjPart/index.aspx) has been built at CMIAS. This database is composed of six systems, i.e., the basic information system of IAS in China, geographic distribution system of IAS, field-data collection system of IAS, biosafety evaluation system of IAS in agriculture and forestry ecosystems, DNA-barcoding based identification system of major invasive insects, and remote monitoring system of important invasive insects in China. It serves as an important information-exchange platform in the field of biological invasions in China, by providing comprehensive information for research institutions, administrative departments and individuals. This platform can also be used for releasing detection/monitoring information, querying for IAS data, and popularizing biological invasion knowledge (Xian et al. 2013).

1.4.9 Education to the Public

Since 2003, a series of education activities were conducted for the public to promote their awareness on biological invasions. To increase education efficiency, several books were published, which are excellent in both interpretation and pictures, easy to understand, and thus attractive to readers, such as *Biological Invasion: Illustrated Handbook for Invasion Plants in China* (Wan et al. 2015a), and *Biological Invasion: Illustrated Handbook for Invasion Animals in China* (Wan et al. 2016).

1.4.10 Framework of Biological-Invasion Research and Management

Through the development in the past four decades particularly the recent 20 years, a framework for research and management of biological invasions has been proposed in China (Wan et al. 2009a). In this framework, the invasion process is divided into five stages, i.e., introduction, establishment, lag phase, dispersal and outbreak; for each stage, the key scientific research topics the corresponding management actions to be developed are suggested. This framework was suggested to be referred when designing research and management programs for biological invasions.

Invasion biology has become one of important subjects in the science and technology realm of China. Owing to the development in theories and practices, invasion biology now possesses its own research framework (Wan et al. 2008a, b, 2009, 2010, 2011a, b). Moreover, it has been listed as a subject designated to be supported by NSFC. Some universities and colleges take it as a course of students using the recently published textbook *Invasion Biology* (Wan et al. 2015b). There is also a periodical "*Journal of Biosafety*" (issued in 2010) for publication of papers in the field of biological invasions.

1.4.11 Contributions to National Policy Making

The fruitful outcomes of biological invasion research and management in China are very suggestive for government when making decisions. So far CMIAS and the other research institutions have provided at least 15 consulting reports for the government. They also often offer suggestions to the government for the issues associated with biological invasions, of which 30 has been adopted so far.

1.5 Government's Role

Chinese government has played great roles in the management of IAS in China. In 2003 the State Council of China assigned Ministry of Agriculture (MOA) the leadership of IAS management in this country. MOA has established several organizations to ensure the policy making, rapid responses, and efficient activities towards biological invasion issues, including Office for Alien Species Management, Center for Management of Invasive Alien Species (CMIAS), and Rural Energy & Environment Agency. Each of these organizations is working efficiently. Since the beginning of this century, the government has also been a major financial supporter for researches in the field of biological invasions. For example, since 1999 more than 500 research programs have been supported by the Natural Science Foundation of China (NSFC), and during 2002–2013 two projects were supported as national key basic research programs by Ministry of Science and Technology (MST). NSFC has been the most important financial source, which provided 1.7–3.0 million US$ annually in recent years for researches in biological invasions.

More recently, IAS was considered as one of key biological threats to China social development and ecological security, and is now under research jointly implemented by more than research institutions across the country. Besides the administrative supervision and financial support, the Chinese government is also cooperating with other countries to promote management of the IAS that occurs at the regional level.

1.6 International Cooperation

Owing to the fact that invasive species are international and most of biological invasion issues have to be resolved at the international level, China has been collaborating closely with other countries in several fields. For example, in the past decade a number of IAS has been jointly studied by Chinese scientists and the scientists from the USA, Australia, CABI, Italy, Germany, Danmark, Israel, etc. In these programs, a number of IAS were extensively studied, such as whitefly *B. tabaci* (Liu et al. 2007; De Barro et al. 2011), the red turpentine beetle *Dendroctonus valens* LeConte (Sun et al. 2013), and *Phytophthora sojae* (Wang et al. 2011a; Jing et al. 2016), which largely increase our understating of some serious IAS' invasion biology. To solve the problems upcoming with initiation of novel opening-up policy (e.g., One Belt and One Road strategy as described above), China and relevant countries have set out to develop joint research/management programs aiming at reducing invasion risks. In long term, China is ready to create more opportunities for scientists of the country and overseas scientists to exchange their information and views and deepen collaborative research. As one might have seen, since 2004 a number of international congress/conferences have been held in China, such as APEC Symposium on the Prevention and Control of Biological Invasion (Beijing 2005), the 1st and 2nd International Congress of Biological Invasions (ICBC, Fuzhou 2009 and Qingdao 2013), and the 5th International Bemisia Congress (Guangzhou 2009), all of which facilitate our understanding the invasion issues over the world. More recently, to enhance collaborative research, several laboratories have been built jointly by China and other countries, such as the "Prevention & Control Center for Alien Pests" built in 2014 by Chinese Academy of Agricultural Sciences (CAAS) and North Carolina State University, and the "Eradication & Management Center for Alien Pests" built in 2014 by CAAS and Murdoch University.

1.7 About This Book

1.7.1 Why We Want to Publish This Book

Biological invasions have been one of prominent factors that deeply influence human society and ecology. In China, its significance lies in not only the resulted loss to economy and damage to ecosystem, but also in the potential threats to food safety and even national safety. Particularly, the threat to food safety is arising, as indicated by the fact that agriculture, the biggest industry and the crucial determinant of food-supplying capacity in China, is facing an increasing damage from IAS. Moreover, clearly, China will have to meet more severe challenges in coming decades, due to intensifying of the factors that favor biological invasions, such as human activities and global changes. Consequently, biological invasions in China are such issues imposing extensive, serious, long-term effects to the country and requiring utmost solutions. Under these circumstances, publishing a book focusing the biological invasions specifically in China would be of great value.

It is the time to publish such a book. As revealed above, we have accumulated huge knowledge from the large number of researches. Some ecosystems, such as farmlands and forests, have been intensively studied, from which a great of theoretical points as well as practical management tactics have been resulted. For the other ecosystems, including water bodies, nature reserves, grasslands and desert oases, great research progress has also been achieved, although the concerns and studies for these ecosystems were a bit later. In any cases, we think it is necessary to review in time what we have done in the past, so that we could extend our future work in more rational directions.

We expect this book will enhance the information exchange between China and other counties. Invasion biology is developing rapidly throughout the world, thereby novel findings in this field are being continually accumulated. Moreover, an increasing number of IAS are becoming common to different countries, which should be give concerns and studied at the international level. Therefore, international exchanges have become more necessary than ever. In this book, most of the ecosystems introduced also exist in other countries, such as farmlands, forests, water bodies and nature reserves, and most of the IAS descried also occur seriously elsewhere. Thus, we believe overseas readers will hopefully find what they are interested in.

1.7.2 What Are in This Book

The book is divided into four parts.

Part I details the biological invasions and their management in various ecosystems. Both the systems that are extensively managed (e.g., agroecosystem) or natural (e.g., deserts and nature conserves) were included, with the aim of giving readers a full picture of biological invasions in China. This part has five chapters, dealing with agroecosystems (Chap. 2), forest ecosystems (Chap. 3), aquatic ecosystems (Chap. 4), desert green-islands and grasslands (Chap. 5), and nature reserves (Chap. 6). In each chapter, we focused on topics including: what have contributed significantly to the biological invasions in the system, and what kinds of management strategies and methods have been implemented. At the end of this part, we addressed the roles of governments in the prevention and management of biological invasion in China (Chap. 7).

When introducing the ecosystems, we focused largely on these aspects: (i) diversity of IAS, (ii) history and pathways of IAS introductions to China, (iii) major research advances in species invasiveness, ecosystem invasibility, (iv) management strategies and techniques, and (v) suggested directions and perspectives for future research.

Parts II and III describe the invasion and management of major IAS in China, which include insects (Chaps. 8, 9, 10, 11, 12, 13, 14, 15, 16 and 17), nematodes (Chaps. 18 and 19), snails (Chap. 20), turtle (Chap. 21), fish (Chaps. 22 and 23), plants (Chaps. 24, 25, 26, 27, 28, 29, 30 and 31), and microbes (Chaps. 32 and 33). Most of these species have caused serious damages and received considerable attention in China. They are also largely the ones common to different countries, as described above. For each species, we focused on: (i) history and pathways of the species' introduction to China and its current distribution, (ii) invasion biology of the species, with the life features contributing its invasion in China be highlighted; (iii) management strategies and techniques, and (iv) suggested directions and perspectives for future research.

Part IV gives perspectives on future development and management of biological invasions in China (Chap. 34).

1.8 Conclusions

Before 1990s, we learn little about biological invasions, despite large number of IAS had been presented and caused serious damages in China. Yet, after 20 years, now we witness that great achievements were made in China both in basic research and practical management. A great number of serious IAS were extensively investigated for their invasion biology. At the same time, a great number of techniques have been developed for efficient prevention, interception and control of IAS. Accordingly, some serious invasive species has been under control with their damages being reduced substantially. During this process, each part of the country, including research institutions, government and the public, have concerned much to IAS issues and strived together to combat them. Based on these achievements and experiences, now a working framework for research/management of biological invasions has been established in China.

In conclusion, our research and management capacity towards biological invasion issues have increased greatly as compared with 20 years ago. The work in this field has largely contributed to the technological and economic development of this country. It also has greatly reduced the threats to ecological and society of this country. As biological invasions will likely become more severe in China in the future, which possibly impose more serious impacts on China economy, ecology and society (Liao et al. 2008; Weber et al. 2008; Xu et al. 2012a;Yuan et al. 2015), challenges to China will continue to increase. We hope reader will learn from this book the information about the past, current and future of biological invasions in China.

Acknowledgments We are grateful to Prof. Aibin Zhan from Chinese Academy of Sciences (CAS) China for his valuable comments on the manuscript. We also thank Prof. Daniel Simberloff from University of Tennessee USA for his helpful suggestions on the analysis of biological invasion research in this chapter, during/after the 3th National Congress of Biological Invasion, Haikou, China, 2010.

References

De Barro PJ, Liu SS, Boykin LM, Dinsdale AB (2011) *Bemisia tabaci*: a statement of species status. Annu Rev Entomol 56:1–19

Ding JQ, Mack RN, Lu P, Ren MX, Huang HW (2008) China's booming economy is sparking and accelerating biological invasions. Bioscience 58:317–324

Ding H, Xu HG, Qiang S, Meng L, Han ZM, Miu JL, Hu BS, Sun HY, Huang C, Lei JC, Le ZF (2011) Status and trends of biological invasion into China. J Ecol Rural Environ 27(3):35–41

Dong SM, Yin WX, Kong GH, Yang XY, Qutob D, Chen QH, Kale SD, Sui YY, Zhang ZG, Dou DL et al (2011) *Phytophthora sojae* avirulence effector Avr3b is a secreted NADH and ADP-ribose pyrophosphorylase that modulates plant immunity. PLoS Pathog 7:e1002353

Feng YL, Lei YB, Wang RF, Callaway RM, Valiente-Banuet A, Inderjit LYP, Zheng YL (2009) Evolutionary tradeoffs for nitrogen allocation to photosynthesis versus cell walls in an invasive plant. Proc Natl Acad Sci U S A 106:1853–1856

Feng YL, Li YP, Wang RF, Callaway RM, Valiente-Banuet A, Inderjit (2011) A quicker return energy-use strategy by populations of a subtropical invader in the non-native range: a potential mechanism for the evolution of increased competitive ability. J Ecol 99:1116–1123
Gao Y, Yu HW, He WM (2014) Soil space and nutrients differentially promote the growth and competitive advantages of two invasive plants. J Plant Ecol 7:396–402
He Q, Cui BS, An Y (2012) Physical stress, not biotic interactions, preclude an invasive grass from establishing in forb-dominated salt marshes. PLoS ONE 7:e33164
Huang FF, Peng SL, Chen BM, Liao HX, Huang QQ, Lin ZG, Liu G (2015) Rapid evolution of dispersal-related traits during range expansion of an invasive vine *Mikania micrantha*. Oikos 124:1023–1030
Jiang XL, Zhang WG, Wang G (2007) Biodiversity effects on biomass production and invasion resistance in annual versus perennial plant communities. Biodivers Conserv 16:1983–1994
Jing MF, Guo BD, Li HY, Yang B, Wang HN, Kong GH, Zhao Y, Xu HW, Wang Y, Ye WW et al (2016) A *Phytophthora sojae* effector suppresses endoplasmic reticulum stress-mediated immunity by stabilizing plant Binding immunoglobulin Proteins. Nat Commun 7:11685
Jiu M, Zhou XP, Tong L, Xu J, Yang X, Wan FH, Liu SS (2007) Vector-virus mutualism accelerates population increase of an invasive whitefly. PLoS ONE 2:e182
Li ZQ, Bogaert J, Nijs I (2005) Gap pattern and colonization opportunities in plant communities: effects of species richness, mortality, and spatial aggregation. Ecography 28:777–790
Liao CZ, Peng RH, Luo YQ, Zhou XH, Wu XW, Fang CM, Chen JK, Li B (2008) Altered ecosystem carbon and nitrogen cycles by plant invasion: a meta-analysis. New Phytol 177:706–714
Lin W, Zhou GF, Cheng XY, Xu RM (2007) Fast economic development accelerates biological invasions in China. PLoS ONE 2:e1208
Liu HX, Wen JB (2000) Pay attention to effects of biological invasion. World Agric 9:26–28
Liu MQ, Bjornlund L, Ronn R, Christensen S, Ekelund F (2012) Disturbance promotes non-indigenous bacterial invasion in soil microcosms: Analysis of the roles of resource availability and community structure. PLoS ONE 7:e45306
Liu SS, De Barro PJ, Xu J, Luan JB, Zang LS, Ruan YM, Wan FH (2007) Asymmetric mating interactions drive widespread invasion and displacement in a whitefly. Science 318:1769–1772
Liu TL, Song TQ, Zhang X, Yuan HB, Su LM, Li WL, Xu J, Liu SH, Chen LL, Chen TZ et al (2014) Unconventionally secreted effectors of two filamentous pathogens target plant salicylate biosynthesis. Nat Commun 5:4686
Lu QG (1999) Damage of biological invasion. World Agric 4:38–39
Lu ZJ, Ma KP (2005) Scale dependent relationships between native plant diversity and the invasion of croftonweed (*Eupatorium adenophorum*) in southwest China. Weed Sci 53:600–604
Luan JB, Yao DM, Zhang T, Walling LL, Yang M, Wang YJ, Liu SS (2013) Suppression of terpenoid synthesis in plants by a virus promotes its mutualism with vectors. Ecol Lett 16:390–398
Pan XY, Geng YP, Zhang WJ, Li B, Chen JK (2006) The influence of abiotic stress and phenotypic plasticity on the distribution of invasive *Alternanthera philoxeroides* along a riparian zone. Acta Oecologica Int J Ecol 30:333–341
Qiang S, Chen GQ, Li BP, Meng L (2010) Invasive alien species in Chinese agricultural ecosystems and their management. Biodivers Sci 18:647–659
Qin RM, Zheng YL, Valiente-Banuet A, Callaway RM, Barclay GF, Pereyra CS, Feng YL (2013) The evolution of increased competitive ability, innate competitive advantages, and novel biochemical weapons act in concert for a tropical invader. New Phytol 197:979–988
Sun JH, Lu M, Gillette NE, Wingfield MJ (2013) Red turpentine beetle: innocuous native becomes invasive tree killer in China. Annu Rev Entomol 58:293–311
Wan FH, Yang NW (2016) Invasion and management of agricultural alien insects in China. Annu Rev Entomol 61:77–98
Wan FH, Zheng XB, Guo JY (2005) Biology and management of invasive alien species in agriculture and forestry. Science Press, Beijing, 820 pp
Wan FH, Li BP, Guo JY (2008a) Biological invasions: biological control theory and practice. Science Press, Beijing, 596 pp
Wan FH, Xie BY, Chu D (2008b) Biological invasions: legislations and management strategies. Science Press, Beijing, 316 pp

Wan FH, Guo JY, Zhang F (2009) Research on biological invasions in China. Science Press, Beijing, 302 pp
Wan FH, Peng DL, Wang R (2010) Biological invasions: risk analysis and early prevention. Science Press, Beijing, 757 pp
Wan FH, Feng J, Xu J (2011a) Biological invasions: detection, surveillance and monitoring. Science Press, Beijing, 589 pp
Wan FH, Xie BY, Yang GQ (2011b) Invasion biology. Science Press, Beijing
Wan FH, Liu QR, Xie M (2015a) Biological invasion: illustrated handbook for invasion plants in China. Science Press, Beijing
Wan FH, Hou YM, Jiang MX (2015b) Invasion biology. Science Press, Beijing
Wan FH, Yu GY, Xie M (2016) Biological invasion: illustrated handbook for invasion animals in China. Science Press, Beijing
Wang Q, Wang CH, Zhao B, Ma ZJ, Luo YQ, Chen JK, Li B (2006) Effects of growing conditions on the growth of and interactions between salt marsh plants: implications for invasibility of habitats. Biol Invasions 8:1547–1560
Wang QQ, Han CZ, Ferreira AO, Yu XL, Ye WW, Tripathy S, Kale SD, Gu BA, Sheng YT, Sui YY et al (2011a) Transcriptional programming and functional interactions within the *Phytophthora sojae* RXLR effector repertoire. Plant Cell 23:2064–2086
Wang R, Wang JF, Qiu ZJ, Meng B, Wan FH, Wang YZ (2011b) Multiple mechanisms underlie rapid expansion of an invasive alien plant. New Phytol 191:828–839
Weber E, Sun SG, Li B (2008) Invasive alien plants in China: diversity and ecological insights. Biol Invasions 10:1411–1429
Wu H, Carrillo J, Ding JQ (2016) Invasion by alligator weed, *Alternanthera philoxeroides*, is associated with decreased species diversity across the latitudinal gradient in China. J Plant Ecol 9:311–319
Xian XQ, Chen H, Zhao J, Wan FH (2013) Introduction to the database of invasive alien species in China. Plant Prot 39(5):103–109
Xie Y, Li ZY, Gregg WP, Dianmo L (2001) Invasive species in China – an overview. Biodivers Conserv 10:1317–1341
Xu HG, Qiang S (2011) China's invasive alien species. Science Press, Beijing, 684 pp
Xu KY, Ye WH, Cao HL, Deng X, Yang QH, Zhang Y (2004) The role of diversity and functional traits of species in community invasibility. Bot Bull Acad Sinica 45:149–157
Xu HG, Ding H, Li MY, Qiang S, Guo JY, Han ZM, Huang ZG, Sun HY, He SP, Wu HR, Wan FH (2006a) The distribution and economic losses of alien species invasion to China. Biol Invasions 8:1495–1500
Xu HG, Qiang S, Han ZM, Guo JY, Huang ZG, Sun HY, He SP, Ding H, Wu HR, Wan FH (2006b) The status and causes of alien species invasion in China. Biodivers Conserv 15:2893–2904
Xu H, Chen K, Ouyang ZY, Pan XB, Zhu SF (2012a) Threats of invasive species for China caused by expanding international trade. Environ Sci Technol 46:7063–7064
Xu HG, Qiang S, Genovesi P, Ding H, Wu J, Meng L, Han ZM, Miao JL, Hu BS, Guo JY et al (2012b) An inventory of invasive alien species in China. NeoBiota 15:1–26
Yang PJ, Liang GQ (1996) Summary on ecological problems of biological invasions. Nat Enem Insects 18(1):91–97
Yang ZQ, Wang XY, Zhang YN (2014) Recent advances in biological control of important native and invasive forest pests in China. Biol Control 68:117–128
Yuan JJ, Ding WX, Liu DY, Kang H, Freeman C, Xiang J, Lin YX (2015) Exotic *Spartina alterniflora* invasion alters ecosystem-atmosphere exchange of CH_4 and N_2O and carbon sequestration in a coastal salt marsh in China. Glob Chang Biol 21:1567–1580
Zhang T, Luan JB, Qi JF, Huang CJ, Li M, Zhou XP, Liu SS (2012) Begomovirus-whitefly mutualism is achieved through repression of plant defences by a virus pathogenicity factor. Mol Ecol 21:1294–1304
Zhao LL, Mota M, Vieira P, Butcher RA, Sun JH (2014) Interspecific communication between pinewood nematode, its insect vector, and associated microbes. Trends Parasitol 30:299–308

Zheng YL, Feng YL, Zhang LK, Callaway RM, Valiente-Banuet A, Luo DQ, Liao ZY, Lei YB, Barclay GF, Pereyra CS (2015) Integrating novel chemical weapons and evolutionarily increased competitive ability in success of a tropical invader. New Phytol 205:1350–1359

Zhu XZ, Zhang JT, Ma KP (2011) Soil biota reduce allelopathic effects of the invasive *Eupatorium adenophorum*. PLoS ONE 6:e25393

Zhu DH, Wang P, Zhang WZ, Yuan Y, Li B, Wang J (2015) Sampling and complementarity effects of plant diversity on resource use increases the invasion resistance of communities. PLoS ONE 10:e0141559

Chapter 2
Biological Invasions in Agricultural Ecosystems in China

Mingxing Jiang, Yunshan Huang, and Fanghao Wan

Abstract Agriculture has served as a major contributor for the China's rapid economic development over past decades. However, agroecosystems in this coutry have been intensively invaded by invasive alien species (IAS), in particular since the beginning of twentieth century, which cause huge losses to economy and ecological service values each year. Moreover, now China agroecosystems are at the stage of being introduced with new IAS at a greater rate due to rapid growth in international trade and human activities. And, a number of dangerous IAS already introduced are continuing to expand rapidly. For these reasons, biological invasions in China agroecosystems have attracted great attention since the beginning of this century, both from governments and research institutions. In this chapter, we first depict the status of IAS in China agroecosystems regarding their diversity, source, distribution, spread and damage, and the factors perceived to have contributed to the IAS introduction and spreading in this country. Then, we show the invasibility of China agroecosystems. The major findings of researches regarding invasiveness of IAS in China are reviewed. Next, invasiveness and the underlying invasion mechanisms of some serious IAS were reviewed. Lastly, the major approaches employed for control of IAS in China agroecosystems were presented, including risk analysis, detection, monitoring, regional prevention, biological control, replacement control, breeding and use of resistant crop cultivars, as well as integrated use of invasive plants. The trend of introducing new IAS into China agroecosystems and the corresponding tactics to be taken for dealing with IAS in the future were discussed.

Keywords Agricultural ecosystems • Invasive alien species • Biological invasions • Invasion mechanisms • Invasibility • Invasion management

M. Jiang (✉) • Y. Huang
Ministry of Agriculture Key Laboratory of Agricultural Entomology, Institute of Insect Sciences, Zhejiang University, Hangzhou 310058, China
e-mail: mxjiang@zju.edu.cn

F. Wan (✉)
State Key Laboratory for Biology of Plant Disease and Insect Pests, Institute of Plant Protection, Chinese Academy of Agricultural Sciences, Beijing 100193, China
e-mail: wanfanghao@caas.cn

F. Wan et al. (eds.), *Biological Invasions and Its Management in China*,
Invading Nature - Springer Series in Invasion Ecology 11,
DOI 10.1007/978-94-024-0948-2_2

2.1 Introduction

China's agriculture, which has developed rapidly since the implementation of reform and opening-up policy, plays fundamental roles in the economic development of this country (Cao and Birchenall 2013). However, China agricultural ecosystems have been long the main recipient for invasive alien species (IAS), suffering huge losses each year in economy and ecological service values (Xu et al. 2004a, 2006; Wan et al. 2009a; Qiang et al. 2010; Wan and Yang 2016). Several reasons were suggested for such a case, including rapid economic development (Lin et al. 2007; Ding et al. 2008; Weber and Li 2008), intensified human activities (Liu et al. 2005; Wang et al. 2011b), and tremendous changes in land-use pattern (Wan and Yang 2016).

In the past several decades, in particular since the beginning of this century, biological invasions in agroecosystems have attracted much attention. A huge number of researches and management practices were undertaken (Wan et al. 2009a; Wan and Yang 2016). In this chapter, we first describe the overall status of biological invasions in China agroecosystem, including species composition, source, distribution, spread and damage. Then, IAS's invasiveness, major invasion mechanisms, and the agroecosystems' invasibility in China are reviewed. Next, we show several measures that have been largely undertaken in China for management of IAS. Lastly, we discuss the directions for future development and management of biological invasions in China agroecosystems.

2.2 Historical and Current Status of Biological Invasions in China Agroecosystem

2.2.1 Diversity and Source of IAS

According to documentary records, inventory (e.g., Xu and Qiang 2011; Xu et al. 2012a) and recent supplement (Xiaoqing Xian, pers. Commun.), there have been 618 IAS in China by the end of 2015. Of them, 382 species can occur in agroecosystem, including 254 plants, 88 animals, and 40 microorganisms which present 66.5, 23.0 and 10.5% of the total IAS in agroecosystem, respectively. Ten of them are members of the world's 100 worst invasive species defined by the International Union for the Conservation of Nature (Lowe et al. 2004), and 38 are listed among the first group of important IAS (52 species) in China that are being managed with priority (The Ministry of Agriculture of China 2013; Table 2.1). Some notorious species have been long present in China agroecosystem. For example, crofton weed *Ageratina adenophora* (Spreng.) King & Rob., alligator weed *Alternanthera philoxeroides* (Mart.) Griseb., common ragweed *Ambrosia artemisiifolia* L., Siam weed *Chromolaena odorata* (L.) King & Rob., water hyacinth *Eichhornia crassipes* (Martius) Solms-Laubach, and giant African snail *Achatina fulica* (Bowdich) have

Table 2.1 The most important invasive species in China agroecosystems that are being given prior management

IAS groups	Speciesa	Classification position	Year first recorded in China	Native range	Distribution
Plants	Crofton weed *Ageratina adenophora* (Spreng.) King & Rob.	Asterales: Asteraceae	1935	Mexico and Costa Rica	Nine provinces in southwest, south, mid and east.
	Common ragweed *Ambrosia artemisiifolia* L.	Asterales: Asteraceae	1935	North America	More than 20 provinces, in most parts of China except northwest.
	Giant ragweed *Ambrosia trifida* L.	Asterales: Asteraceae	1930s	North America	Twelve provinces in north, northeast, east and mid.
	*Siam weed *Chromolaena odorata* (L.) King & Rob.	Asterales: Asteraceae	1934	Mid and South America	Six provinces in southwest, south and east.
	*Mile-a-minute *Mikania micrantha* H.B.K.	Asterales: Asteraceae	1919 Hongkong 1984 Shenzhen	Central America	Guangdong, Hongkong, Macao, Taiwan, Yunnan.
	Yellowtop *Flaveria bidentis* (L.) Kuntze	Asterales: Asteraceae	1996–2001	South America and the West Indies	Tianjin, Hebei, Shandong, Henan.
	Whitetop weed *Parthenium hysterophorus* L.	Asterales: Asteraceae	1926 or 1980s	American tropics	Nine provinces in east, south and southwest.
	Praxelis *Eupatorium catarium* Veldkamp	Asterales: Asteraceae	1980s	South America	Six provinces in east, south and southwest.
	Canada goldenrod *Solidago canadensis* L.	Asterales: Asteraceae	1935	Northeastern and north-central North America	Sixteen provinces in most parts of China.

(continued)

Table 2.1 (continued)

IAS groups	Speciesa	Classification position	Year first recorded in China	Native range	Distribution
	Poison ryegrass *Lolium temulentum* L.	Graminales: Gramineae	1940s	Europe	More than 20 provinces, in most parts of China except the south.
	Johnson grass *Sorghum halepense* (L.) Pers.	Graminales: Gramineae	the early 20th century	Mediterranean region	Sixteen provinces in most parts of China except northwest.
	Tausch's goatgrass *Aegilops tauschii* Coss.	Graminales: Gramineae	Unclear	West Asia	Eight provinces in the north, northwest, southwest, mid and east.
	Field sandbur *Cenchrus incertus* M. A. Curtis	Graminales: Gramineae	The early 20th century	North America and tropical coastal region	Eleven provinces in northeast, in most parts of China expect northwest and mid.
	Alligator weed *Alternanthera philoxeroides* (Mart.) Griseb.	Chenopodiales: Amaranthaceae	1892	South America	More than 20 provinces; most parts of China except northeast and the majority of northwest.
	Carelessweed *Amaranthus palmeri* Watson	Chenopodiales: Amaranthaceae	1985	North America	Five provinces in north, northeast, and east.
	Spiny Amaranth *Amaranthus spinosus* L.	Chenopodiales: Amaranthaceae	1836 Macao 1857 Hongkong	Tropical Americas	Across the country except a few province in north, east and west.
	*Water hyacinth *Eichhornia crassipes* (Martius) Solms-Laubach	Commelinales: Pontederiaceae	1901	Tropical Americas	About 20 provinces/cities, in most of part of China excluding northwest and the majority of north and northeast.

	*Largeleaf lantana *Lantana camara* L.	Verbenales: Verbenaceae	1645	American tropics	Eleven provinces in southwest, south, east and mid.
	Buffalobur nightshade *Solanum rostratum* Dunal	Solanales: Solanaceae	1895	North America	Eight provinces in north, northeast, northwest, and east.
Animals	Western flower thrips *Frankliniella occidentalis* Pergande	Thysanoptera: Thripidae	2000 Taiwan 2003 Beijing	Southwestern United States	Eight provinces in north, east, south, and southwest.
	*Middle East-Asia Minor 1 (MEAM1) and Mediterranean (MED) of sweet potato whitefly *Bemisia tabaci* Gennadius	Hemiptera: Aleyrodidae	Mid-1990s (MEAM1) and 2003 (MED)	India Peninsula or North Africa to the Middle East	Across the country.
	Spiraling whitefly *Aleurodicus dispersus* Russell	Hemiptera: Aleyrodidae	1988 Taiwan 2006 Hainan	Central America and the Caribbean region	Taiwan and Hainan.
	Cotton mealybug *Phenacoccus solenopsis* Tinsley	Hemiptera: Pseudococcidae	2008	North America	Thirteen provinces in most parts of China except northeast.
	Colorado potato beetle *Leptinotarsa decemlineata* (Say)	Coleoptera: Chrysomelidae	1993, Xinjiang	Southwestern North America	Xinjiang.
	Rice water weevil *Lissorhoptrus oryzophilus* Kuschel	Coleoptera: Curculionidae	1988	North America	More than 20 provinces, in most parts of China except the majority of northwest.

(continued)

Table 2.1 (continued)

IAS groups	Speciesa	Classification position	Year first recorded in China	Native range	Distribution
	Oriental fruit fly *Bactrocera dorsalis* (Hendel)	Diptera: Tephritidae	1911 Taiwan 1934 Hainan	Japan	Fifteen provinces in east, south, southwest, and mid.
	Melon fruit fly *Bactrocera (zeugodacus) cucuribitae* (Coquillett)	Diptera: Tephritidae	Unclear	India and Sri Lanka	Ten provinces in east, south, southwest, and mid.
	Ber fruit fly *Carpomya vesuviana* Costa	Diptera: Tephritidae	2007	India	Xinjiang.
	Vegetable leafminer *Liriomyza sativae* Blanchard	Diptera: Agromyzidae	1993	South America	All provinces except Tibet.
	American serpentine leafminer, *Liriomyza trifolii* (Burgess)	Diptera: Agromyzidae	1988 Taiwan 2005 Guangdong	North America	Taiwan, Guangdong, Hainan.
	Codling moth *Cydia pomonella* (L.)	Lepidoptera: Tortricidae	1953	Europe	Seven provinces in north, northeast, and northwest.
	*Red imported fire ant *Solenopsis invicta* Buren	Hymenoptera: Formicidae	2003	South America	Ten provinces in south, southwest, east, and mid.
	*Golden apple snail *Pomacea canaliculata* (Lamarck)	Mollusca: Ampullariidae	1970s Taiwan 1981 Guangdong	South America	Thirteen provinces in south, east, southwest, and mid.
	*Giant African snail *Achatina fulica* (Bowdich)	Mollusca: Achatinidae	From the late 1920s to the early 1930s	East Africa	Seven provinces in the south, southwest, and east.
	Burrowing nematode *Radopholus similes* (Cobb) Thorne	Tylenchida: Pratylenchidae	Unclear	First reported in Fiji	Taiwan.

Microorganisms	*Fusarium oxysporum* f. sp. *cubense* Schlechtend (Smith) Snyder & Hansen Race 4	Sordariomycetes: Hypocreales	1967	Australia	Guangdong, Guangxi, Hainan, Fujian, Taiwan, and Yunnan.
	Phytophthora sojae Kaufmann & Gerdemann	Oomycetes: Pythiales	1991	USA	Six province in north, northeast, northwest, east, and mid.
	Clavibacter michiganensis subsp. *michiganensis* (Smith) Davis et al.	Firmibacteria: Actinomycetales	1981	USA	Twelve provinces in most parts of China except the south.

[a] The asterisk indicates the species is the member of the world's top 100 serious invasive species

been there since 1930s or even earlier. By contrast, some were introduced much later, such as Mediterranean of the sweet potato whitefly *Bemisia tabaci* Gennadius species complex, western flower thrips *Frankliniella occidentalis* Pergande, red imported fire ant (RIFA) *Solenopsis invicta* Buren, and cotton mealybug *Phenacoccus solenopsis* Tinsley, which were reported in the 2010s for the first time on China mainland.

Most of the IAS so far occurring in China agroecosystems were introduced after 1900. Before 1850 and in the latter half of the nineteenth century, respectively only 33 and 35 IAS were recorded from China farmland (not including the species uncertain with introduction years, the same below). However, the number of IAS introduced in twentieth century hit 223, with an average of 22 species (range 11–37 species) being found per decade (Fig. 2.1a). Moreover, from the beginning of twenty-first century, the newly introduced IAS is increasing at a rate much greater than ever: during the first decade and 2011–2015, there were 30 and 22 newly recorded IAS, respectively (Fig. 2.1a). Such a trend was particularly evident after 2005 (Fig. 2.1b).

About 45% of the IAS recorded in China agroecosystem have an origin of North and South America. Europe and Asia are the second (22.9%) and third (16.4%) important sources. More than 90% of animal IAS were unintentionally introduced, whereas at least 54% of plant IAS were intentionally introduced (Qiang et al. 2010).

2.2.2 Geographic Distribution and Spread of IAS

IAS distribute widely in China agroecosystems. Their richness is highest in southern and eastern provinces, while decreased towards northward and northwestward (Wu et al. 2006; Qiang et al. 2010). Some southern/eastern provinces (e.g., Guangdong, Fujian, Zhejiang, Jiangsu, Anhui and Taiwan) have been invaded by more than 100 IAS in local agroecosystem. Similar distribution pattern can also be observed for the major IAS (Fig. 2.2).

Distribution and spread of IAS in China are associated with a number of factors. Climatic condition was thought to have determined their overall distribution pattern in this country. For instance, tropical or subtropical regions generally have more IAS than other regions (Wan et al. 2009a; Qiang et al. 2010). Other kinds of physical factors, such as river, flooding, wind, *etc*., may have enhanced the regional dispersal of some plant IAS. For example, with the help of water stream and wind, seeds of crofton weed *A. adenophora*, which are very small, light (ca. 0.4 g per 1000 seeds) and have a specialized feather-like structure (pappus), can be easily carried and dispersed to neighboring regions (Wang et al. 2011b; Horvitz et al. 2014).

It is widely accepted that rapid economic development in China has induced the rapid introduction and spread of IAS (Liu et al. 2005; Wu et al. 2006; Lin et al. 2007; Ding et al. 2008; Weber and Li 2008; Qiang et al. 2010). Such an association can be illustrated by the fact that IAS-abundant regions normally have a higher

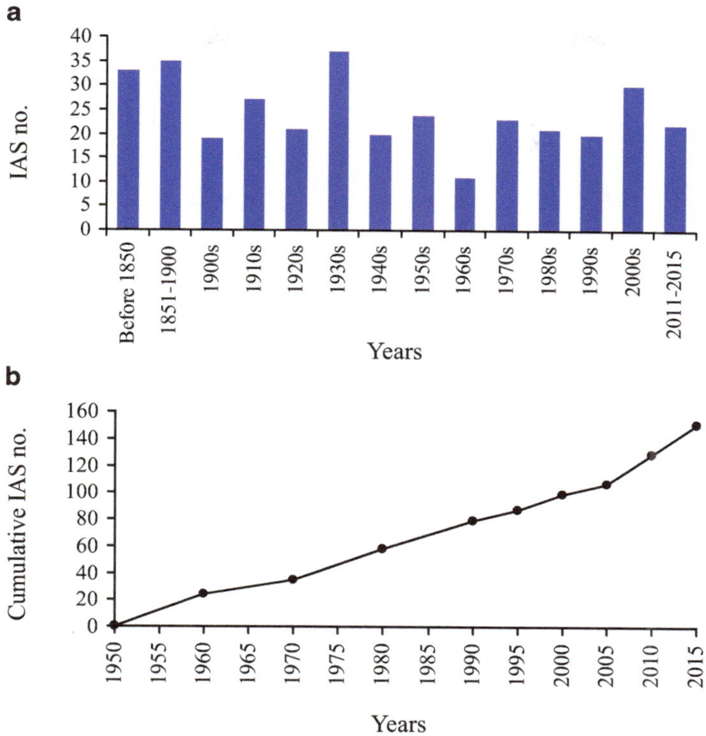

Fig. 2.1 Historical changes in the number of IAS introduced to China agroecosystems. (**a**) Number of IAS introduced at various historical stages; (**b**) Cumulative introduced IAS number since 1950. Summarized based on Xu et al. (2012a) and the data proviced by Xiaoqing Xian. Not including the species uncertain with introduction years

economic growth rate relative to other regions, such as the situation in southern and eastern China described above.

Intentional introductions of alien species among regions (e.g., for economic purpose) have proved causing wide spread of some serious IAS in China. One typical example comes from *A. philoxeroides*. This plant was introduced into Shanghai and Zhejiang (eastern China) in 1930s as horse fodder. In 1950s, it was further introduced to southern China as swine feedstuffs. From 1980s, its geographic range expanded rapidly, becoming one of the most notorious weeds in China agroecosystem (Pan et al. 2007).

Unintentional introductions by human activities were believed to have tremendously facilitated IAS dispersal in China (Liu et al. 2005; Huang et al. 2010; Wang et al. 2011b). When persons are traveling by cars, seeds of some invasive plants may be carried by cloth, cars and other items, thereby dispersed to far regions. For example, *A. adenophora* seeds can easily stick to clothing, footwear or vehicles by the pappus, thereby disperse along road and rail line for a long distance (Wang et al.

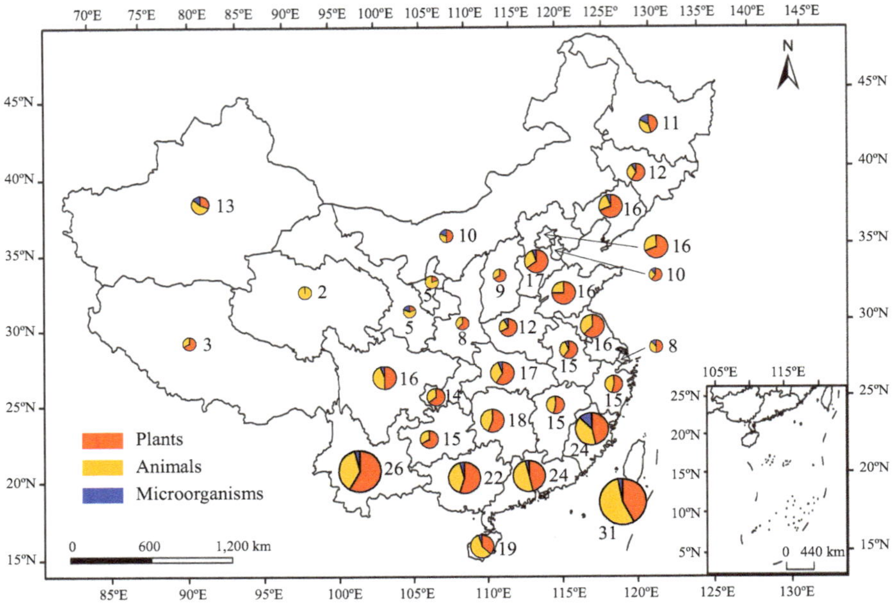

Fig. 2.2 Geographic distribution of the major IAS in China agroecosystems that have to be managed with priority. A total of 38 species were analyzed which include 19 plants, 16 animals and 3 microorganisms (Table 2.1). Number in parentheses indicates the number of IAS that have already invaded the corresponding province. The circles have been scaled with number of IAS

2011b). Ansong and Pickering (2013) estimated that nearly 1/3 of invasive weeds in China could be transported by cars. And, according to announcement of the Ministry of Public Security, there had been 172 million registered cares by the end of 2015. For these reasons, we can speculate that human activities, in particular those with using cars, will probably speed the IAS spread in China agroecosystems.

Other factors, such as urbanization of previous farming areas, land use pattern, and construction of new transport infrastructure, are also likely affect biological invasions in China. Promoted IAS dispersal due to rapid urbanization has been reported from Beijing (Wang et al. 2011a). Changes in land use pattern were thought to be affecting invasions of alien insect pests in China (Wan and Yang 2016). Establishing new sea, land and air links between China and other countries/regions or within the country will probably result introductions of novel IAS into China agroecosystems.

2.2.3 Damage

Biological invasions have caused huge economic loss in China. According to a loss assessment by Xu et al. (2006), in 2000 the IAS caused a direct economic loss of ca. USD 1.474 billion to China agriculture, most of which was resulted from crofton weed *A. adenophora*, common ragweed *A. artemisiifolia*, whitefly *B. tabaci* and *Trialeurodes vaporariorum* (Westwood), and vegetable leafminer *Liriomyza sativae* Blanchard. Potato, which has become the fourth most important food crop in China, can be reduced by 30–50% with the yield when damaged by Colorado potato beetle (CPB) *Leptinotarsa decemlineata* Say (Guo et al. 2016). In 2007 the reported economic loss from CPB damage reached USD 3.2 million (Liu et al. 2012). The direct economic loss caused by *A. artemisiifolia* and *A. philoxeroides* was estimated to be USD 4.4 billion annually (Zhou et al. 2011).

Meanwhile, biological invasions may also change agroecosystems' properties and service function. In some regions of southwestern China, for example, *A. adenophora* invasion was reported to have greatly reduced local soil nutrients (e.g., available nitrogen, phosphorus and potassium) (Wu et al. 2007). This plant has also reduced diversity of native plants in some invaded areas (Ding et al. 2007). Due to the damage by IAS, the overall of service function would be reduced. Xu et al. (2006) estimated that the total indirect economic loss to China agroecosystem service function was USD 1.405 billion annually.

Also, some IAS, such as RIFA and common ragweed, have impacted human health in the invaded regions. According to Xu et al. (2012b), more than 1/3 of the human population in RIFA-infested areas ever suffered stings of this ant, and nearly 10% of victims fevered, with some experiencing dizziness, generalized urticaria or anaphylactic shock. Hence, the allergic reactions to RIFA stings have been a serious public health problem in China.

2.3 Invasibility of China Agroecosystems

A number of biotic and abiotic factors can affect invasibility of ecosystems, such as biodiversity, interspecific relationships, disturbance and resource (Levine and D'Antonio 1999; Davis et al. 2000; Eschtruth and Battles 2009). In China, a number of studies have addressed the influence of these factors.

Associations of invasibility with biodiversity were investigated in *A. adenophora* and *A. philoxeroides*, two notorious invasive plants in China. It was found high richness of native plants along roads would increase ecosystem resistance to *A. adenophora* (Lu and Ma 2006). However, another study reported a reverse result, that when viewing the invasion at a larger (e.g., provincial) scale, increase in native plant richness would result a higher invasibility to this plant (Lu and Ma 2005). Thus, their actual associations may differ with the regions and geographic range under investigation. Invasibility to *A. philoxeroides* appears to be not much related with species diversity. However, the invasibility would be reduced at the presence of the native species that belong to the same functional group with *A. philoxeroides*, such as *Alternanthera sessilis* (Xu et al. 2004b). As *A. sessilis* can compete with *A. philoxeroides* for similar kinds of resources, the resource available to *A. philoxeroides* would be reduced and eventually its biomass reduced.

Human disturbance was suggested to be able to increase ecosystem invasibility for *A. adenophora* and Siam weed *C. odorata*. This is often observed in China that (i) these two plants tend to be abundant near roads (typical human-disturbing habitats) relative to the areas far from the roads (much less disturbed), and they also generally grow taller near roads (Liu et al. 2007a), and (ii) in forest ecosystems, they were present primarily in secondary forests (ever much disturbed) relative to native forests (Dong et al. 2011a).

Presence of high levels of resources in soil can increase invasibility. This has been found for a number of invasive plant species, including *A. philoxeroide* (Pan et al. 2006a), *A. adenophora* (Lei et al. 2012), mile-a-minute *Mikania micrantha* (Zhang et al. 2016b), *C. odorata* (Quan et al. 2015; Zhang et al. 2016b), praxelis *Eupatorium catarium* Veldkamp (Huang et al. 2016), and tropic ageratum *Ageratum conyzoides* L. (Huang et al. 2016). For example, high invasibility of abandoned fields and swamps for *A. philoxeroides* can be largely attributed to the presence of rich nutrients in soil, relative to the marsh dunes and grave dunes that are poor in nutrients and much less invaded by this plant (Pan et al. 2006a). Accordingly, some argued that increased use of chemical fertilizers is one of primary factors causing extensive plant invasions in China (Chen et al. 2013).

Besides soil fertility, soil biota has also been evidenced to affect invasibility. For example, certain microbes can alleviate the allelopathic effects of *A. adenophora* against other plants (Zhu et al. 2011; Li et al. 2015c). Presence of pathogen fungi in soil may reduce biomass of *M. micrantha* and change biomass allocation (Gao et al. 2013), thereby reduce the competitive capacity of this plant over native plants. Therefore, presence of certain soil biota can reduce invasibility of the ecosystem.

2.4 Invasiveness and Underlying Invasion Mechanisms of IAS in China

Invasion capacity and the underlying mechanisms have attracted much research attention in China. Based on the analysis of the biology of top ten invasive insect pests in China agroecosystem, Wan and Yang (2016) claimed there are some evident traits common to successful insect invaders, including wide host range, high reproductivity, strong adaptation to diverse stress (e.g., adverse temperature, insecticides and pathogens), release from natural enemies, and mutualism established with other organisms. Besides, there are some species-specific traits that have significantly enhanced the invasion of some IAS. For example, rice water weevil *Lissorhoptrus oryzophilus* Kuschel can reproduce parthenogenetically, which allows a single female adult to produce offspring without the need of finding mates in the newly introduced range (Chen et al. 2012). The invasive MEAM1 of *B. tabaci* has the capacity of interfering with indigenous whiteflies' reproduction through a mechanism "asymmetric mating interactions" (Liu et al. 2007b). MEAM1 has also developed an indirect vector-virus mutualism with geminiviruses (Jiu et al. 2007; Zhang et al. 2012; Luan et al. 2013). Both these interspecific interactions were believed to have driven the invasion of MEAM1 in China.

New insights have also been obtained for many invasive plant pathogens. For example, *Phytophthora sojae* (causing phytophthora root rot in soybean) was found to be able to secrete a number of effector proteins when interacting with host plants, such as Avr3b that is a secreted NADH and ADP-ribose pyrophosphorylase capable of modulating plant immunity (Dong et al. 2011b), and an isochorismatase effector that disrupts plant salicylate metabolism pathway to suppress host immunity (Liu et al. 2014b).

Below, we present more details on the invasive plants in China. We focused on such plants that (i) have serious threat to agriculture, ecosystems or human/livestock health, (ii) have a broad distribution range, or have a limited range but are rapidly spreading currently, and (iii) have attracted much research attention in the past decade regarding their invasiveness. They are reviewed in seven aspects, i.e., adaptation to physical stresses, allelopathic effects, phenotypic plasticity, enemy release, energy-use capacity/strategy, associations with microbial community, and clonal integration.

2.4.1 Adaptation to Physical Stresses

A number of invasive plants in China were showed to be highly tolerant to adverse environments. For example, *A. philoxeroides* can tolerate submergence and high levels of salt, heavy metals and nitrogen (Ding et al. 2014; Gao et al. 2014; Yang et al. 2014b; Fan et al. 2015). Due to this capacity, this plant can get established and grow very well in a diversity of habitats, including fields that are heavily disturbed by flooding, saline soils, heavy-metal polluted soil, and eutrophic water.

Some natural *A. adenophora* populations in China can tolerate low temperature (Li et al. 2008). In southwest China, for example, this plant has evolved to attain some ecotypes that are capable of tolerating freezing through physiological adaptation. This plant can also tolerate high temperature (Chen et al. 2016). This suggests that *A. adenophora* may be benefiting from these features during its spread and has the potential of reaching further northern and southern regions beyond the current distribution range.

Strong adaptation has also been found in other invasive plants. For example, the Canada goldenrod *Solidago canadensis* is highly tolerant to Pb stress, capable of excluding or reducing uptake of this metal. This capacity may enable it to outperform native plants [e.g., *Kummerowia striata* (Thunb.) Schindler] and colonize successfully in Pb contaminated soil (Yang et al. 2007b; Zhang et al. 2008). Seeds of buffalobur *Solanum rostratum* Dunal can adapt to high salt stress (Wei et al. 2009). *M. micrantha* can also tolerate saline to certain degree, with the potential of spreading in coastal saline regions (Hu et al. 2014).

2.4.2 Allelopathic Effects

Allelopathic effects against native plants have been frequently observed in *S. canadensis* (Sun et al. 2006; Yang et al. 2007a; Zhang et al. 2009b; Wang et al. 2016), *A. adenophora* (Yang et al. 2006, 2013; Zhao et al. 2009), and *M. micrantha* (Shao et al. 2005; Wu et al. 2009). The general conclusion is that, under the influence of allelopathic chemicals, seed germination, seedling growth and/or reproduction of native plants may be remarkably suppressed. Besides, other outcomes may also be generated. For examples, allelochemicals exuded from *S. canadensis* may suppress local soilborne pathogens (Zhang et al. 2011b), and inhibit colonization of arbuscular mycorrhizal fungi *Glomus* spp. to native plants barnyard grass *Echinochloa crusgalli* (L.) Beauv., *K. striata* and *A. conyzoides* (Zhang et al. 2007). Certain allelochemicals produced by *M. micrantha* can increase soil nutrients available to itself (Chen et al. 2009). All these kinds of effects would benefit invasive plants when competing with natives, exploiting resource, and resisting pathogens. For this reason, allelopathic effects should have played important even crucial roles in the plant invasions in China (Xie et al. 2010).

Interestingly, some invasive plants in China have evolved to be more allelopathic in the introduced range. For example, as compared with the *S. canadensis* plants from the USA (native range), those from China (introduced range) could produce more allelochemicals and exert greater allelopathic effects against native plant *K. striata* (Yuan et al. 2013).

2.4.3 Phenotypic Plasticity

Phenotypic plasticity is one of most important invasion mechanisms of plants (Funk 2008; Hulme 2008). In China, this aspect has been much learned in a few species, such as *A. philoxeroides* and *A. adenophora*.

A. philoxeroides in China were found to be highly plastic in response to water, nutrients, and shading. This feature could be observed not only among geographic populations, but also among the individuals within a population that grow in different habitats. It may exist in plant morphology, anatomy (Geng et al. 2006; Pan et al. 2006b; Zhang et al. 2015a, b; Du et al. 2016), and biomass allocation strategy (Geng et al. 2007a, b; Zhang et al. 2015b). Furthermore, some studies showed that *A. philoxeroides* is more plastic than native congeners (Geng et al. 2006; Pan et al. 2006a). Owing to these attributes, *A. philoxeroides* has successfully invaded a number of highly diverse habitats, e.g., aquatic versus terrestrial habitats, and nutrient-rich versus nutrient-poor habitats. Now in China it is widely accepted that phenotypic plasticity has played critical roles in the invasion of *A. philoxeroides* (Geng et al. 2007a, 2016). Recently, the underlying mechanisms for the phenotypic plasticity of this weed were investigated. Gao et al. (2010) found that the weed adopts an environmentally sensitive, flexible "epigenetic regulatory system", through which it undergoes genome-wide epigenetic reprogramming so as to adapt rapidly to changing environments.

A. adenophora in China also possesses high levels of plasticity (e.g., Feng et al. 2007; Zhao et al. 2012, 2013). It may take place in the responses of the plant to irradiance (Feng et al. 2007; Zheng et al. 2009), soil nutrients (Wang and Feng 2005; Wang et al. 2006), and elevation (Li and Feng 2009b; Zhang et al. 2009a). However, the plants of a higher plasticity do not have to be more invasive, suggesting that the roles of plasticity for *A. adenophora* need further clarification (Zheng et al. 2012).

Phenotypic plasticity has also been reported in *A. artemisiifolia* (Qin et al. 2012; Li et al. 2015b). As compared with the native plant Caesarweed *Urena lobata* L., *A. artemisiifolia* is more plastic in response to irradiance in the traits relating to biomass allocation, growth, and plant structure. Thus, this plant can live well in shaded conditions while maintain the ability of growing rapidly at elevated irradiance levels, which enables it to outcompete over other plants (Qin et al. 2012).

2.4.4 Natural Enemy Release and Corresponding Changes in Growth and Defence Characters

The evolution of increased competitive ability (EICA) hypothesis suggests that, as compared with the plants in native range, the invasive plants out of native range would evolve to have increased competitive ability while decreased defense/tolerance to herbivory following natural enemy release. This hypothesis has been tested in several invasive plants in China to analyze its associations with invasiveness, including *A. adenophora*, *C. odorata* and *A. philoxeroides*.

In *A. adenophora*, the EICA hypothesis has been tested to be true. As an invasion mechanism, this plant has evolved to increase N allocation to photosynthesis (growth) while reduce allocation to cell walls, which results in poorer structural defenses (Feng et al. 2009).

The success of *A. philoxeroides* invasion in China has been partially attributed to natural enemy release (Pan et al. 2013). At the same time, in some region this plant seems to have not evolved following EICA hypothesis, because they can be stimulated to acquire a compensative capacity when exposed previously to native herbivory insects and introduced natural enemies (Lu and Ding 2012). Therefore, *A. philoxeroides* is probably diverging in China with the potential of evolving a more delicate growth and defense strategy (Pan et al. 2013).

In *C. odorata*, which has grown for over 150 years in nonnative ranges with few natural enemies (McFadyen 1988), the EICA hypothesis has been tested to be true by overseas scientists. However, no evidence was detected for such an evolution in the Chinese populations (Li et al. 2012b). Actually, *C. odorata* plants in China have become more resistant to generalists and more tolerant to damage than the plants from native range (Liao et al. 2014). Moreover, they have evolved to have the ability of producing novel allelochemicals in response to native competitors and new enemy regimes, which in turn increases their competitive ability. This suggests that after being introduced to China, *C. odorata* has benefited from both EICA and "novel chemical weapons" in its invasions (Qin et al. 2013; Zheng et al. 2015).

2.4.5 Energy-Use Capacity and Strategy

Association of energy-use capacity and strategy with invasions has been investigated in several notorious invasive plants in China. A number of interesting findings come from *A. adenophora*, which was demonstrated to have a high resource (e.g., irradiances and nitrogen) capture and use efficiency (Feng et al. 2007, 2011; Feng 2008a, b; Wang et al. 2013). This plant is more efficient in photosynthetic N partitioning and tends to allocate more leaf N photosynthesis (growth), as compared with native plants, *Eupatorium chinense* L. and *Eupatorium heterophyllum* DC. Therefore, *A. adenophora* has a higher light-saturated photosynthetic rate, and a high efficiency in photosynthetic energy use, water use, respiration, *etc.* (Feng 2008b). Moreover, *A. adenophora* might have acquired a quicker return energy-use strategy after being introduced to nonnative ranges (Feng et al. 2011). Clearly, these advantages have helped *A. adenophora* become one of the most successful invasives in China.

High resource capture and use efficiency has also been observed in water hyacinth *E. crassipes* (Xie et al. 2004; Fan et al. 2013) and *Wedelia trilobata* (L.) Hitchc. (Song et al. 2010). As the concentration of available nutrients increases, *E. crassipes* can duly regulate biomass allocation to optimize resource acquisition (Xie et al. 2004), and at high nutrient levels it can increase photosynthetic rate as well as the photosynthetic nitrogen use efficiency (Fan et al. 2013).

2.4.6 Associations with Microbial Community

A number of invasive plant species in China are able to modify microbial community in the soil of invaded areas, which, in turn, generates positive feedbacks (benefits) to the invasive plants themselves. This has been referred as a novel invasion mechanism of plants. A typical example was reported from *A. adenophora*. After the microbial community is modified by *A. adenophora*, the soil will become more suitable for growth of this plant but may inhibit the growth of co-occurring native plants, eventually *A. adenophora* acquires a higher competitive capacity over natives (Yu et al. 2005; Niu et al. 2007; Sun et al. 2013; Xiao et al. 2014a; Yu et al. 2014). Similar results have also been reported from *S. canadensis* (Zhang et al. 2010b), *M. micrantha* (Li et al. 2006, 2007), *C. odorata* (Xiao et al. 2014b), and *A. artemisiifolia* L. (Li et al. 2014a). Some of the microbes playing crucial roles such an association have been identified, such as arbuscular mycorrhizal fungi (AMF), which are present in rhizosphere of *A. adenophora* (Yu et al. 2014) and *S. canadensis* (Jin et al. 2004; Zhang et al. 2010b; Yang et al. 2014a).

Besides soil microbes, endophytes may also contribute to invasion. Dai et al. (2016) showed that an endophytic *Bacillus* sp. strain could significantly promote *W. trilobata* growth but had no effects on the native congener *Wedelia chinensis*.

2.4.7 Clonal Integration

Under low nutrient conditions, *A. philoxeroides* can improve net photosynthetic rate, transpiration rate, stomatal conductance, and ramets' minimal and maximal chlorophyll fluorescence through clonal integration (Liu et al. 2008). Some studies suggest that clonal integration probably have played important roles in the expansion of *A. philoxeroides* from terrestrial to aquatic habitats (Wang et al. 2009), and in the expansion in heterogeneous and frequently disturbed environments (You et al. 2016; Zhang et al. 2016a).

The roles of different aspects described above may vary with the plant species analyzed. Based on the work so far reported by Chinese scientists, we summarize the relative significance of some of these aspects for the major invasive plants in China (Table 2.2).

2.5 Management of IAS in China Agroecosystems

2.5.1 Risk Analysis

Since the beginning of this century, at least 100 invasive species were assessed for the potential of being introduced, establishing, spreading and causing damage in China agroecosystems. Of them, more than 60 species were determined to have high risk and should be managed with priority in the future. Management strategy and techniques have been developed for each of these major IAS (Wan et al. 2009b).

Table 2.2 The relative significance of different invasion strategies for major invasive plants in China[a]

Species	Allelopathy	Plasticity	Energy-use strategy/capacity	Changing soil microbial community	Clonal integration
Crofton weed *A. adenophora*	+++	+++	+	+++	
Alligator weed *A. philoxeroides*	+	+++	+++		+++
Canada goldenrod *S. canadensis*	+++			++	
Siam weed *C. odorata*	++		+++	+	
Mile-a-minute *M. micrantha*	+++		+	++	
Common ragweed *A. artemisiifolia*		+		+	
Giant ragweed *A. trifida*	++				
Water hyacinth *E. crassipes*	+		+		+
Yellowtop *F. bidentis*	+				

[a]The number of plus sign (+) stand for the significance degree. The number of this sign was determined according to the number of articles published in China and overseas

2.5.2 Detection and Monitoring

PCR-based rapid detection techniques have been available for more than 70 invasive plant pathogens, insects and weeds which can not be reliably identified using conventional methods (Wan et al. 2011). Moreover, highly effective kits have been developed for a number of pathogens for the rapid detection in entry & exit ports and the fields, such as *P. sojae*, *Tilletia controversa* (an internationally quarantined pathogenic fungus causing dwarf bunt of wheat), *Erwinia amylovora* (a pathogenic bacterium causing fire blight), race 4 of *Fusarium oxysporum* f. sp. *cubense* Schlechtend (Smith) Snyder & Hansen (*Foc*, a pathogenic fungus causing Fusarium wilt of banana), burrowing nematode *Radopholus similis*, potato ring rot *Clavibacter michiganense* subsp. *sepedonicus*, *Xanthomonas axonopodis*, and *Clavibacter michiganensis* subsp. *michiganensis* (Smith) Davis et al. (a pathogenic causing bacterial canker of tomato), *etc*.

Rapid molecular detection techniques have also been available for the invasive insects that are too small to be identified or have to be identified at juvenile stages which may have few diagnostic characters, such as thrips, whiteflies, fruit flies and leafminers. Recently, DNA barcoding was developed for nearly 200 insects in China (e.g., Zhang et al. 2011; Qiao et al. 2012; Li et al. 2014b; Liu et al. 2014a).

Field-monitoring systems have been established nationwide for more than 80 invasive species. In these systems, multiple technical methods are adopted depending on the biology of targeted insects, which include (i) real-time data-collecting from the field; (ii) rapid molecular detecting; (iii) chemical pheromone monitoring using sex pheromones, plant volatile odors, and protein-based attractants; (iv) physical monitoring using yellow stick traps, spectroscopic technique, and ground penetrating radar; and (v) remote data transferring, analyzing and diagnosis (Wan and Yang 2016). Based on these techniques, a national monitoring network has been established, which greatly strengthens the China's capacity of early warning and monitoring of IAS.

2.5.3 Regional Prevention

In 2007, a national project commenced to construct a large scale of "prevention belts" that span the coastal provinces and the ones neighboring other countries. The objectives were to prevent IAS from being introduced into China, and to prevent the already introduced dangerous species (e.g., CPB, RIFA, codling moth *Cydia pomonella* (L.), and burrowing nematode *R. similis*) from spreading to other regions of the country. The belts have been working effectively in some regions. For example, to prevent CPB spread, a 250 km "controlling and intercepting belt" was established at the frontier of CPB occurring regions, which extends from eastern Mori Kazak autonomous county to western Barkol county. In this belt, the CPB population density has been suppressed to the utmost via the use of chemicals, clearing of CPB's wild hosts (e.g., black henbane *Hyoscyamus niger* L.), and eradicating of the CPB on non-potato crops whenever possible. Currently the CPB is retrained in a limited region of Xinjiang and has been successfully prevented from spreading eastwards (Guo et al. 2011).

2.5.4 Biological Control

Introducing and releasing natural enemies is an environment-friendly, sustainable approach for control of IAS. Since 1950s China has introduced more than 40 natural enemy species from overseas (Wan and Yang 2016). Of them, some have been practically used in fields or greenhouses, such as the flea beetle *Agasicles hygrophila* Selman and Vogt, ragweed borer *Epiblema strenuana* Walker, ragweed leaf beetle *Ophraella communa* LeSage, water hyacinth weevils *Neochetina eichhorniae* Warner and *N. bruchi*, and the parasitoid wasp *Encarsia formosa* Gahan (Table 2.3).

A. hygrophila is one of the most successful biological agents in China. Both larvae and adults feed on leaves of alligator weed *A. philoxeroides* and mature larvae also bore stems prior to pupation, from which growth of alligator weed is substantially suppressed. *A. hygrophila* can be mass reared in ponds and canals abundant

Table 2.3 Major natural enemies introduced from overseas to China for control of IAS in agroecosystems

Natural enemy	Introduction source and year	Targeted invasives
Flea Beetle *Agasicles hygrophila* Selman and Vogt (Coleoptera: Chrysomelidae)	The US, 1986	alligator weed *A. philoxeroides*
Ragweed borer *Epiblema strenuana* Walker (Lepidoptera: Tortricidae)	Australia, 1990	common ragweed *A. artemisiifolia*
Ragweed leaf beetle *Ophraella communa* LeSage (Coleoptera: Chrysomelidae)	Unclear; recorded the earliest in 2001 on mainland	*A. artemisiifolia*
Ragweed leaf beetle *Zygogramma suturalis* F. (Coleoptera: Chrysomelidae)	Canada, 1987; former Soviet Union, 1988; Australia, 1990	*A. artemisiifolia*
Water hyacinth weevils *Neochetina eichhorniae* Warner and *N. bruchi* Hustache (Coleoptera: Curculionidae)	The US and Argentina, 1995	Water hyacinth *E. crassipes*
Butterflies *Actinote anteas* (Doubleday and Hewitson) and *Actinote thalia pyrrha* (Fabricius) (Lepidoptera: Acraeidae)	Indonesia, 1996 and 2001	Mile-a-minute *M. micrantha*
Predatory mite *Amblyseius cucumeris* (Oudemans) (Acari: Phytoseiidae)	British, 1997	Western flower thrips *F. occidentalis*
Parasitoid *Encarsia formosa* Gahan (Hymenoptera: Aphelinidae)	British, 1978	Whiteflies *B. tabaci* and *T. vaporariorum*
Parasitoid *Encarsia sophia* (Girault and Dodd) (Hymenoptera: Aphelinidae)	The US, 2006, 2008	*B. tabaci* and *T.vaporariorum*
Parasitoid *Fopius arisanus* (Sonan) (Hymenoptera: Braconidae)	The US, 2005	Oriental fruit fly *B. dorsalis*
Parasitoid *Diachasmimorpha longicaudata* (Ashmead) (Hymenoptera: Braconidae)	Thailand, 2004	*B. dorsalis*
Parasitoid *Aphelinus mali* (Haldeman) (Hymenoptera: Aphelinidae)	Japan, 1942	Woolly apple aphid *E. lanigerum*

with alligator weed. From 1988 to 1997, alligator weed in some regions of China was successfully controlled by releasing this beetle. Currently *A. hygrophila* has been widespread in the released regions (Wang et al. 2008).

Work is continuing in China with respect to mass rearing and releasing of *A. hygrophila*. Jin et al. (2015) developed a more efficient mass-rearing method, where mature larvae are allowed to pupate outside of stems and thus pupal survival rate can be increased as compared with conventional rearing method. Meanwhile, the rearing steps can be simplified. Li et al. (2011) proposed a more rational releasing density of the beetle. They suggested taking the beetles' self-dispersing capacity and the type of habitats (marsh, paddy land, and dry land) into accounts when determining the release density, so as to guarantee the establishment of a sustainable post-release beetle population and meanwhile reduce the control's economic cost.

Epiblema strenuana and *Ophraella communa* are the most important control agents of common ragweed *A. artemisiifolia*. They damage different parts of the host: *E. strenuana* bores stems while *O. communa* feed on leaves. Based to this feeding habit, some have suggested using them together so as to increase the overall control efficiency. During 2007–2010, several release trials were performed in Hunan, Guangxi, Jiangxi, and Guangdong, with each site reporting good control results (Zhou et al. 2015). However, both insects have not been widely released yet.

N. eichhorniae and *N. bruchi*, both native to Argentina and nearby aquatic regions, are specific biological control agents of water hyacinth. In China, first release trial was performed in 1996 in Wenzhou, Zhejiang, at which 1000 *N. eichhorniae* adults were released to a river 1372 m^2 with the water surface absolutely occupied by hyacinths (Ding et al. 2001). Two years later hyacinths were reduced by 25%, and in the next year nearly all the hyacinths were removed by the weevils. In following years the weevils were introduced to other regions of China, such as Ningbo of Zhejiang (Chen et al. 2005) and Kunming of Yunnan (Zhao 2006). The key problem to be resolved is to increase weevils' overwintering survival so that they can produce a sufficient population in following years.

E. formosa is one of the most successful parasitoid wasps so far used in China. It has been commercialized for the control of whiteflies *B. tabaci* and *T. vaporariorum* in greenhouses (Wan et al. 2008). However, it may not work very well in northern China due to the stress of low temperature met at release times. To overcome this weakness, Zhang et al. (2010a) recommended many approaches to be employed jointly with parasitoid release, including (i) using whitefly-free vegetable seedlings, (ii) preventing the released parasitoids from escaping from greenhouses, and (iii) using yellow stick traps and bio-insecticides to reduce whitefly density prior to parasitoid release. Moreover, some were seeking more friendly chemical use methods to reduce negative impacts on *E. formosa* (e.g., Rao et al. 2012). More recently, it was suggested using *E. formosa* together with other natural enemies, such as *Orius sauteri* (Poppius), an anthocorid predator of *B. tabaci*, to achieve high control efficiency (Li et al. 2014c).

Besides the introduced biological control agents, a large number of native natural enemies have also been screened in China (Wan et al. 2008; Wan and Yang 2016). Some of them has the great potential of being used as biological control agents in the future, such as *O. sauteri* for the control of thrips *F. occidentalis* (Wang et al. 2014), and some parasitoid wasps for the control of oriental fruit fly *Bactrocera dorsalis* (Hendel) (Zheng et al. 2014). However, none of them have been practically used in the fields yet.

2.5.5 Replacement of Invasive Plants

Replacing invasive plants with selected native plants is a promising control method around the world (Ogden and Rejmánek 2005; Gosper and Vivian-Smith 2009). In China, such work focused on common ragweed *A. artemisiifolia* and crofton weed *A. adenophora*. A number of native plants proved to be effective for replacement of

A. artemisiifolia, such as Indigobush *Amorpha fruticosa* L., common sea-buckthorn *Hippophae rhamnoides* L., crownvetch *Coronilla varia* L., Kentucky bluegrass *Poa pratensis* L., and Jerusalem artichoke *Helianthus tuberosus* L.. As early as early 1990s, some of them were planted in Liaoning province, northeastern China, for replacing the *A. artemisiifolia* and *A. trifida* growing along highways (Guan et al. 1995).

For *A. adenophora*, more than 40 native plant species were tested for replacing efficacy, of which pasture plants were found to be most promising (e.g., Zhong et al. 2007). Some tree species also have high potential, such as Chinese red pine *Pinus massoniana* Lamb. and swamp mahogany *Eucalyptus robusta* Sm., which have been recommended for use in southern mountainous regions of Guizhou province, southwestern China, for replacement of *A. adenophora* (Ou et al. 2012).

Work is also on the way for seeking native plants against other invasive plants, including tropic ageratum *A. conyzoides* (Li et al. 2012a, 2013), praxelis *E. catarium* (Li et al. 2009a), *M. micrantha*, and the yellowtop *Flaveria bidentis* (L.) Kuntze (Ma et al. 2011).

2.5.6 Breeding and Use of Resistant Crop Cultivars

Using resistant cultivars is an economical, sustainable and environment-friendly way for control of invasive crop pathogens. Since 1990s, screening work has been extensively conducted in China for the germplasms resistant to *P. sojae*, *R. similis*, and *Foc* race 4. A number of germplasms resistant to *P. sojae* have been identified (See chapter 32). Banana germplasms highly resistant to *R. similis* and *Foc* race 4 have also been identified recently (Fu et al. 2015; Li et al. 2015a).

China are also seeking resistant stuff for the control of invasive insect pests. A Siberian crabapple *Malas baccata* (L.) line, *Jin 67*, was found to be resistant to woolly apple aphid *Eriosoma lanigerum* Hausm., and has the potential of being used as apple stock for control of this aphid (Deng et al. 1997). More recently, the apple variety *Zhaojin 108* was found to have certain resistance to *E. lanigerum* (Yin et al. 2010b). Several transgenic potato lines that express Cry3A protein targeting CPB have been generated (Zhou et al. 2012; Guo et al. 2016).

2.5.7 Integrated Use of Invasive Plants

In addition to the control methods described above, exploiting invasive plants as biological resources has also been a promising, sustainable managing approach of some invasive plants. Yin et al. (2010a) reported *A. adenophora* as an ideal biofuel stuff for the production of methane and hydrogen. *A. adenophora* can also be used for producing synthetic fodders, organic manure, pesticides, and particle boards (Zhang et al. 2006; Luo et al. 2009; Yin et al. 2009). Extracts from *S. canadensis* were showed to have antimicrobic activity (Deng et al. 2015) and the cytotoxic activity against human tumor cells (Liu et al. 2007c; Huang et al. 2012). Some suggested using *S. canadensis* to produce activated carbon (Luo et al. 2011). Water hyacinth might be a promising hyperaccumulator for purifying polluted water (Deng et al. 2009).

2.6 Conclusion and Future Perspectives

China agroecosystem is a kind of large, frequently disturbed, highly heterogeneous and resource-rich ecosystems. It has been intensively invaded by IAS since 1850s in particular from 1980s when the reforming and opening-up policy was initiated in this country, and suffered huge economic losses. To deal with this challenge, Chinese scientists conducted extensive investigations over IAS invasiveness, ecosystem invasibility, and the problems associated with managements. The obtained results have remarkably deepened our understanding of the ecological processes of biological invasions in China, and have enabled us to build the fundamentals essential for successful management.

As has been speculated more than once, China will be probably invaded by a great number of novel IAS in coming decades (Jenkins and Mooney 2006; Lin et al. 2007; Ding et al. 2008; Weber and Li 2008; Wan and Yang 2016). Under this background, China agroecosystem will remain to be a hot spot for new arrivers. Actually, in the past decade we have witnessed such a trend: from 2006 to 2015 at least 26 novel invasive species invaded China agroecosystems, at a high rate that has never been reached before (data provided by Xiaoqing Xian). Particularly, establishment of an increasing number of free-trade zones, the "One Belt One Road", and new international transport lines between China and overseas, would even cause more intensive introductions of IAS to China. It is the time to target the potential IAS in these settings. Their invasion risk is to be assessed urgently.

China agroecosystems are changing tremendously. The ongoing building of "new countrysides" and rapid urbanization in China will continue to change local farmland landscape and cropping systems, which will in turn change habitats of IAS and their associations with crop systems (Wan and Yang 2016). How and to what a range these changes will affect invasibility of agroecosystems has to be investigated.

Effective management systems have been developed for most serious IAS in China. However, there exist many problems that have to be resolved in time. For example, new control options have to be developed for the invasive insect pests that have developed high levels of resistance to insecticides, such as *B. tabaci*, *L. decemlineata*, *L. sativae*, and *F. occidentalis* (Wan and Yang 2016). To increase control efficacy of biological control agents in northern China, their tolerance to low temperatures is to be elevated, such as for the parasitoid wasps of *A. artemisiifolia* and *B. tabaci* (Kidane et al. 2015; Zhou et al. 2015). Much effort is still required to improve present chemical control methods so as to reduce their negative impacts on natural enemies.

To strengthen management over biological invasions, the central government, local governments, and the public are required to contribute more in the future. Ministry of Agriculture (MOA), as the core decision maker, coordinator and manager of top-down control programs of IAS in China, is expected to develop a much closer relationship with local governments. The public is still to be encouraged to increase their awareness over biological invasions and engage more in management programs (Wan and Yang 2016).

Acknowledgments We thank Xiaoqing Xian, Xiaowei Wang, Baoping Li, Yuanchao Wang, Baishi Hu, Li Gao and Haigen Xu for providing data essential for this chapter.

References

Ansong M, Pickering C (2013) Are weeds hitchhiking a ride on your car? A systematic review of seed dispersal on cars. PLoS One 8:e80275

Cao KH, Birchenall JA (2013) Agricultural productivity, structural change, and economic growth in post-reform China. J Dev Econ 104:165–180

Chen RX, Ye GY, Wang YJ, Gu BQ, Zhang CF, Ying XL, Shen J (2005) The biology and host range of water hyacinth weevil, *Neochetina eichhorniae*, an insect agent for the control of the weed. Acta Agric Zhejiangensis 17:74–78

Chen BM, Peng SL, Ni GY (2009) Effects of the invasive plant *Mikania micrantha* HBK on soil nitrogen availability through allelopathy in South China. Biol Invasions 11:1291–1299

Chen SJ, Lu F, Cheng JA, Jiang MX, Way MO (2012) Identification and biological role of the endosymbionts *Wolbachia* in rice water weevil (Coleoptera: Curculionidae). Environ Entomol 41:469–477

Chen GQ, He YH, Qiang S (2013) Increasing seriousness of plant invasions in croplands of eastern China in relation to changing farming practices: a case study. PLoS One 8:e74136

Chen SG, Yang J, Zhang MS, Strasser RJ, Qiang S (2016) Classification and characteristics of heat tolerance in *Ageratina adenophora* populations using fast chlorophyll a fluorescence rise O-J-I-P. Environ Exp Bot 122:126–140

Dai ZC, Fu W, Wan LY, Cai HH, Wang N, Qi SS, Du DL (2016) Different growth promoting effects of endophytic bacteria on invasive and native clonal plants. Front Plant Sci 7:706

Davis MA, Grime JP, Thompson K (2000) Fluctuating resources in plant communities: a general theory of invasibility. J Ecol 88:528–534

Deng JQ, Rui GS, Guan YT, Yu YQ, Zhang DM (1997) The selection of an apple stock line, Siberian crabapple *Jin 67*, immune from woolly apple aphid. Acta Phys Sin 20:217–222

Deng FT, Sun PS, Qing XY, Deng FS (2009) Pilot-scale study on purification of the polluted water in Dianchi Lake by planting *Eichhornia crassipes* and its reutilization. J Wuhan Univ Technol 12:84–86

Deng Y, Zhao YY, Padilla-Zakourc O, Yang GY (2015) Polyphenols, antioxidant and antimicrobial activities of leaf and bark extracts of *Solidago canadensis* L. Ind Crop Prod 74:803–809

Ding JQ, Chen ZQ, Fu WD, Wang R, Zhang GL, Lv XJ, Fang YJ, Xia WQ (2001) Control *Eichhornia crassipes*, an invasive aquatic weed in South China with *Neochetina eichhorniae*. Chin J Biol Control 17:97–100

Ding H, Xu HG, Liu ZL (2007) Impacts of invasion of *Eupatorium adenophorum* on vegetation diversity. J Ecol Rural Environ 23(2):29–32, 75

Ding JQ, Mack RN, Lu P, Ren MX, Huang HW (2008) China's booming economy is sparking and accelerating biological invasions. Bioscience 58:317–324

Ding WJ, Zhang HY, Zhang FJ, Wang LJ, Cui SB (2014) Morphology of the invasive amphiphyte *Alternanthera philoxeroides* under different water levels and nitrogen concentrations. Acta Biol Cracov Ser Bot 56:136–147

Dong SK, Li JP, Li XY, Liu SL, Zhao QH (2011a) Impacts of geo-physical factors and human disturbance on composition and diversity of roadside vegetation: a case study from Xishuangbanna National Nature Reserve of Southwest China. Afr J Biotechnol 10:16228–16235

Dong SM, Yin WX, Kong GH, Yang XY, Qutob D, Chen QH, Kale SD, Sui YY, Zhang ZG, Dou DL et al. (2011b) *Phytophthora sojae* avirulence effector Avr3b is a secreted NADH and ADP-ribose pyrophosphorylase that modulates plant immunity. PLoS Pathog 7:e1002353

Du LS, Yang BF, Guan WB, Li JM (2016) Phenotypic variation and water selection potential in the stem structure of invasive alligator weed. Acta Oecol 71:22–30

Eschtruth AK, Battles JJ (2009) Assessing the relative importance of disturbance, herbivory, diversity, and propagule pressure in exotic plant invasion. Ecol Monogr 79:265–280

Fan SF, Liu CH, Yu D, Xie D (2013) Differences in leaf nitrogen content, photosynthesis, and resource-use efficiency between *Eichhornia crassipes* and a native plant *Monochoria vaginalis* in response to altered sediment nutrient levels. Hydrobiologia 711:129–137

Fan SF, Yu HH, Liu CH, Yu D, Han YQ, Wang LG (2015) The effects of complete submergence on the morphological and biomass allocation response of the invasive plant *Alternanthera philoxeroides*. Hydrobiologia 746:159–169

Feng YL (2008a) Nitrogen allocation and partitioning in invasive and native *Eupatorium* species. Physiol Plant 132:350–358

Feng YL (2008b) Photosynthesis, nitrogen allocation and specific leaf area in invasive *Eupatorium adenophorum* and native *Eupatorium japonicum* grown at different irradiances. Physiol Plant 133:318–326

Feng YL, Wang JF, Sang WG (2007) Biomass allocation, morphology and photosynthesis of invasive and noninvasive exotic species grown at four irradiance levels. Acta Oecol 31:40–47

Feng YL, Lei YB, Wang RF, Callaway RM, Valiente-Banuet A, Inderjit LYP, Zheng YL (2009) Evolutionary tradeoffs for nitrogen allocation to photosynthesis versus cell walls in an invasive plant. Proc Natl Acad Sci U S A 106:1853–1856

Feng YL, Li YP, Wang RF, Callaway RM, Valiente-Banuet A, Inderjit (2011) A quicker return energy-use strategy by populations of a subtropical invader in the non-native range: a potential mechanism for the evolution of increased competitive ability. J Ecol 99:1116–1123

Fu MY, Fu SJ, Rui K, Chen MC (2015) Identification of banana germplasm resource resistant to *Radopholus similis*. J Anhui Agric Sci 43(20):166–167, 170

Funk JL (2008) Differences in plasticity between invasive and native plants. J Ecol 96:1162–1173

Gao LX, Geng YP, Li B, Chen JK, Yang J (2010) Genome-wide DNA methylation alterations of *Alternanthera philoxeroides* in natural and manipulated habitats: implications for epigenetic regulation of rapid responses to environmental fluctuation and phenotypic variation. Plant Cell Environ 33:1820–1827

Gao L, Zan QJ, Li MG, Guo Q, Hu L, Jiang L, Zhou S, Liu HJ (2013) Soil fungi of three native tree species inhibit biomass production and shift biomass allocation of invasive *Mikania micrantha* Kunth. Chin Sci Bull 58:758–765

Gao JM, Li JZ, Xiao Q, Luo CB, Zhang SQ, Chen HL, Yi KX (2014) Isolation and characterization of a cryptogein-like gene from drought and salt-treated *Alternanthera philoxeroides* roots. Biotechnol Lett 36:2351–2356

Geng YP, Pan XY, Xu CY, Zhang WJ, Li B, Chen JK (2006) Phenotypic plasticity of invasive *Alternanthera philoxeroides* in relation to different water availability, compared to its native congener. Acta Oecol 30:380–385

Geng YP, Pan XY, Xu CY, Zhang WJ, Li B, Chen JK (2007a) Plasticity and ontogenetic drift of biomass allocation in response to above- and below-ground resource availabilities in perennial herbs: a case study of *Alternanthera philoxeroides*. Ecol Res 22:255–260
Geng YP, Pan XY, Xu CY, Zhang WJ, Li B, Chen JK, Lu BR, Song ZP (2007b) Phenotypic plasticity rather than locally adapted ecotypes allows the invasive alligator weed to colonize a wide range of habitats. Biol Invasions 9:245–256
Geng YP, van Klinken RD, Sosa A, Li B, Chen JK, Xu CY (2016) The relative importance of genetic diversity and phenotypic plasticity in determining invasion success of a clonal weed in the USA and China. Front Plant Sci 7:213
Gosper CR, Vivian-Smith G (2009) Approaches to selecting native plant replacements for fleshy-fruited invasive species. Restor Ecol 17:196–204
Guan GQ, Han YG, Yin R, Huo Q, Miao Q, Wang WH, Wang SZ, Goa DG, Li SD (1995) Studies on displacing and controlling of the ragweeds with economic plant. J Shenyang Agric Univ 26:277–283
Guo WC, Tu EX, Xu YM, Liu J, Xu JJ, Wang PL, He J, Xia ZH, Fu WJ, Jing XY et al. (2011) Study and application on sustained and integrated control techniques of Colorado potato beetle. Xinjiang Agric Sci 48:197–203
Guo WC, Wang ZA, Luo XL, Jin X, Chang J, He J, Tu EX, Tian YC, Si HJ, Wu JH (2016) Development of selectable marker-free transgenic potato plants expressing *cry3A* against the Colorado potato beetle (*Leptinotarsa decemlineata* Say). Pest Manag Sci 72:497–504
Horvitz N, Wang R, Zhu M, Wan FH, Nathan R (2014) A simple modeling approach to elucidate the main transport processes and predict invasive spread: river-mediated invasion of *Ageratina adenophora* in China. Water Resour Res 50:9738–9747
Hu L, Li MG, Wei PP (2014) Salt tolerance of the invasive vine *Mikania micrantha* Kunth. Ecol Environ Sci 23(1):7–15
Huang QQ, Qian C, Wang Y, Jia X, Dai XF, Zhang H, He L, Peng SL, Wang GX (2010) Determinants of the geographical extent of invasive plants in China: effects of biogeographical origin, life cycle and time since introduction. Biodivers Conserv 19:1251–1259
Huang BK, Lei YL, Qin LP, Liu J (2012) Chemical composition and cytotoxic activities of the essential oil from the inflorescences of *Solidago canadensis* L., an invasive weed in Southeastern China. J Essent Oil Bear Plants 15:667–671
Huang QQ, Shen YD, Li XX, Li SL, Fan ZW (2016) Invasive *Eupatorium catarium* and *Ageratum conyzoides* benefit more than does a common native plant from nutrient addition in both competitive and non-competitive environments. Ecol Res 31:145–152
Hulme PE (2008) Phenotypic plasticity and plant invasions: is it all Jack? Funct Ecol 22:3–7
Jenkins PT, Mooney HA (2006) The United States, China, and invasive species: present status and future prospects. Biol Invasions 8:1589–1593
Jin L, Gu YJ, Xiao M, Chen JK, Li B (2004) The history of *Solidago canadensis* invasion and the development of its mycorrhizal associations in newly-reclaimed land. Funct Plant Biol 31:979–986
Jin QA, Jin T, Lin YY, Wen HB, Tang ZZ (2015) Optimization of the method for pupation of *Agasicles hygrophila* at the outside stems. J Anhui Agric Sci 43:111–112
Jiu M, Zhou XP, Tong L, Xu J, Yang X, Wan FH, Liu SS (2007) Vector-virus mutualism accelerates population increase of an invasive whitefly. PLoS One 2:e182
Kidane D, Yang NW, Wan FH (2015) Effect of cold storage on the biological fitness of *Encarsia sophia* (Hymenoptera: Aphelinidae), a parasitoid of *Bemisia tabaci* (Hemiptera: Aleyrodidae). Eur J Entomol 112:460–469
Lei YB, Wang WB, Feng YL, Zheng YL, Gong HD (2012) Synergistic interactions of CO_2 enrichment and nitrogen deposition promote growth and ecophysiological advantages of invading *Eupatorium adenophorum* in Southwest China. Planta 236:1205–1213
Levine JM, D'Antonio CM (1999) Elton revisited: a review of evidence linking diversity and invasibility. Oikos 87:15–26

Li YP, Feng YL (2009b) Differences in seed morphometric and germination traits of crofton weed (*Eupatorium adenophorum*) from different elevations. Weed Sci 57:26–30
Li WH, Zhang CB, Jiang HB, Xin GR, Yang ZY (2006) Changes in soil microbial community associated with invasion of the exotic weed, *Mikania micrantha* HBK. Plant Soil 281:309–324
Li WH, Zhang CB, Gao GJ, Zan QJ, Yang ZY (2007) Relationship between *Mikania micrantha* invasion and soil microbial biomass, respiration and functional diversity. Plant Soil 296:197–207
Li H, Qiang S, Qian YL (2008) Physiological response of different croftonweed (*Eupatorium adenophorum*) populations to low temperature. Weed Sci 56:196–202
Li GY, Mo KL, Zhang GH, Li QF (2009a) Competition between *Paspalum atuatum* and *Eupatorium catarium* Veldkamp. Weed Sci 4:10–14
Li YN, Fu JW, Guo JY, Zhu X, Huang SD, Wan FH (2011) Effects of release density on the population dynamics of the biocontrol agent, *Agasicles hygrophila* (Coleoptera: Chrysomelidae). Entomol J East China 20:275–280
Li QF, Li GY, Liu ZD (2012a) Field replacement control of *Ageratum conyzoides* with five pasture species. Chin Agric Sci Bull 28(35):13–18
Li YP, Feng YL, Barclay G (2012b) No evidence for evolutionarily decreased tolerance and increased fitness in invasive *Chromolaena odorata*: implications for invasiveness and biological control. Plant Ecol 213:1157–1166
Li GY, Hou XW, Zou YK, Liu ZD, Li QF (2013) Replacement control effects of three pasture species combinations on *Ageratum conyzoides*. Weed Sci 31(2):19–25
Li HN, Xiao B, Liu WX, Wan FH (2014a) Changes in soil biota resulting from growth of the invasive weed, *Ambrosia artemisiifolia* L. (Compositae), enhance its success and reduce growth of co-occurring plants. J Integr Agric 13:1962–1971
Li XF, Tian H, Zhang JL, Zhang GF, Chen MM, Wan FH (2014b) Identification of common whitefly species (Hemiptera: Aleyrodidae) in China based on the 5′- and 3′-end sequences of the COI gene. Acta Entomol Sin 57:466–476
Li S, Lao SB, Wang S, Guo XJ, Zhang F (2014c) Control effect of *Orius sauteri* collaborated with *Encarsia formosa* on *Bemisia tabaci* in the greenhouse. Nat Enemies Insects 36:978–982
Li WM, Dita M, Wu W, Hu GB, Xie JH, Ge XJ (2015a) Resistance sources to *Fusarium oxysporum* f. sp. *cubense* tropical race 4 in banana wild relatives. Plant Pathol 64:1061–1067
Li XM, She DY, Zhang DY, Liao WJ (2015b) Life history trait differentiation and local adaptation in invasive populations of *Ambrosia artemisiifolia* in China. Oecologia 177:669–677
Li YP, Feng YL, Chen YJ, Tian YH (2015c) Soil microbes alleviate allelopathy of invasive plants. Sci Bull 60:1083–1091
Liao ZY, Zheng YL, Lei YB, Feng YL (2014) Evolutionary increases in defense during a biological invasion. Oecologia 174:1205–1214
Lin W, Zhou GF, Cheng XY, Xu RM (2007) Fast economic development accelerates biological invasions in China. PLoS One 2:e1208
Liu J, Liang SC, Liu FH, Wang RQ, Dong M (2005) Invasive alien plant species in China: regional distribution patterns. Divers Distrib 11:341–347
Liu J, Cui BS, Dong SK, Wang J, Zhao SQ (2007a) The changes of community components and their horizontal patterns caused by highway construction in the Longitudinal Range-Gorge Region. Chin Sci Bull 52:213224
Liu SS, de Barro PJ, Xu J, Luan JB, Zang LS, Ruan YM, Wan FH (2007b) Asymmetric mating interactions drive widespread invasion and displacement in a whitefly. Science 318:1769–1772
Liu XY, Zhu HK, Wu SH, Cheng YL, Liu FY, Wu P (2007c) Antitumor activity of diterpenoids from *Solidago canadensis* L. J Zhejiang Univ (Science Edition) 34:661–664
Liu J, He WM, Zhang SM, Liu FH, Dong M, Wang RQ (2008) Effects of clonal integration on photosynthesis of the invasive clonal plant *Alternanthera philoxeroides*. Photosynthetica 46:299–302

Liu N, Li YC, Zhang RZ (2012) Invasion of Colorado potato beetle, *Leptinotarsa decemlineata*, in China: dispersal, occurrence, and economic impact. Entomol Exp Appl 143:207–217

Liu SS, Zhang GF, Wan FH (2014a) DNA barcoding and phylogenetic analysis of common species of the genus *Bactrocera* (Diptera: Tephritidae) based on mtDNA COI gene sequences. Acta Entomol Sin 57:343–355

Liu TL, Song TQ, Zhang X, Yuan HB, Su LM, Li WL, Xu J, Liu SH, Chen LL, Chen TZ et al. (2014b) Unconventionally secreted effectors of two filamentous pathogens target plant salicylate biosynthesis. Nat Commun 5:4686

Lowe S, Browne M, Boudjelas S, De Poorter M (2004) 100 of the world's worst invasive alien species. A selection from the Global Invasive Species Database. Published by the Invasive Species Specialist Group (ISSG) a specialist group of the Species Survival Commission (SSC) of the World Conservation Union (IUCN). http://www.issg.org/worst100_species.html. Retrieved 3 Aug 2016

Lu XM, Ding JQ (2012) History of exposure to herbivores increases the compensatory ability of an invasive plant. Biol Invasions 14:649–658

Lu ZJ, Ma KP (2005) Scale dependent relationships between native plant diversity and the invasion of croftonweed (*Eupatorium adenophorum*) in southwest China. Weed Sci 53:600–604

Lu ZJ, Ma KP (2006) Spread of the exotic croftonweed (*Eupatorium adenophorum*) across southwest China along roads and streams. Weed Sci 54:1068–1072

Luan JB, Yao DM, Zhang T, Walling LL, Yang M, Wang YJ, Liu SS (2013) Suppression of terpenoid synthesis in plants by a virus promotes its mutualism with vectors. Ecol Lett 16:390–398

Luo Y, Liu Z, Gao L, Wang HH, Li CH, Liu GD (2009) Quality evaluation of the *Eupatorium adenophorum* Sprengel as organic manure. Chin Agric Sci Bull 25(7):179–182

Luo LS, Sun LH, Yu Y, Zhao XX, Zhou MH (2011) Preparation and characterization of activated carbon from *Solidago canadensis* by means of microwave irradiation activation. Chin J Environ Eng 5:1161–1165

Ma J, Huangfu CH, Yi J, Yang DL (2011) Effects of four replacement plants on nutrient and enzymatic activities of soil invaded by *Flaveria bidentis*. Ecol Environ Sci 20:805–812

McFadyen R (1988) Phytophagous insects recorded from *C. odorata*. Chromolaena odorata Newslett 2:5–23

Niu HB, Liu WX, Wan FH, Li B (2007) An invasive aster (*Ageratina adenophora*) invades and dominates forest understories in China: altered soil microbial communities facilitate the invader and inhibit natives. Plant Soil 294(1–2):73–85

Ogden JAE, Rejmánek M (2005) Recovery of native plant communities after the control of a dominant invasive plant species, *Foeniculum vulgare*: implications for management. Biol Conserv 125:427–439

Ou GT, Zhao YX, Jiang Y, Han Y, Xu DQ, Cai WD (2012) Biological substitution control of *Eupatorium adenophorum* Spreng. in the southern mountainous regions of Guizhou Province. For Pest Dis 31(2):23–26

Pan XY, Geng YP, Zhang WJ, Li B, Chen JK (2006a) The influence of abiotic stress and phenotypic plasticity on the distribution of invasive *Alternanthera philoxeroides* along a riparian zone. Acta Oecol 30:333–341

Pan XY, Liang HZ, Sosa A, Geng YP, Li B, Chen JK (2006b) Patterns of morphological variation of alligator weed (*Alternanthera philoxeroides*): from native to invasive region. Biodivers Sci 14(3):232–240

Pan XY, Geng YP, Sosa A, Zhang WJ, Li B, Chen JK (2007) Invasive *Alternanthera philoxeroides*: biology, ecology and management. Acta Phy Sin 45:884–900

Pan XY, Jia X, Fu DJ, Li B (2013) Geographical diversification of growth-defense strategies in an invasive plant. J Syst Cs Evol 51:308–317

Qiang S, Chen GQ, Li BP, Meng L (2010) Invasive alien species in Chinese agricultural ecosystems and their management. Biodivers Sci 18:647–659

Qiao WN, Wan FH, Zhang AB, Min L, Zhang GF (2012) Application of DNA barcoding technology for species identification of common thrips (Insecta: Thysanoptera) in China. Acta Entomol Sin 55:344–356

Qin Z, Mao DJ, Quan GM, Zhang JE, Xie JF, DiTommaso A (2012) Physiological and morphological responses of invasive *Ambrosia artemisiifolia* (common ragweed) to different irradiances. Botany-Botanique 90:1284–1294

Qin RM, Zheng YL, Valiente-Banuet A, Callaway RM, Barclay GF, Pereyra CS, Feng YL (2013) The evolution of increased competitive ability, innate competitive advantages, and novel biochemical weapons act in concert for a tropical invader. New Phytol 197:979–988

Quan GM, Mao DJ, Zhang JE, Xie JF, Xu HQ, An M (2015) Response of invasive *Chromolaena odorata* and two coexisting weeds to contrasting irradiance and nitrogen. Photosynthetica 53:419–429

Rao Q, Xu YH, Zhang F, Luo C, Zhang HY, Devine GJ, Gorman K (2012) Parasitism of *Encarsia formosa* against *Bemisia tabaci* as affected by systemic and foliar applications of imidacloprid. Chin J Biol Control 28:467–472

Shao H, Peng SL, Wei XY, Zhang DQ, Zhang C (2005) Potential allelochemicals from an invasive weed *Mikania micrantha* HBK. J Chem Ecol 31:1657–1668

Song LY, Li CH, Peng SL (2010) Elevated CO_2 increases energy-use efficiency of invasive *Wedelia trilobata* over its indigenous congener. Biol Invasions 12:1221–1230

Sun BY, Tan JZ, Wan ZG, Wan ZG, Gu FG, Zhu MD (2006) Allelopathic effects of extracts from *Solidago canadensis* L. against seed germination and seedling growth of some plants. J Environ Sci 18:304–309

Sun X, Gao C, Guo LD (2013) Changes in soil microbial community and enzyme activity along an exotic plant *Eupatorium adenophorum* invasion in a Chinese secondary forest. Chin Sci Bull 58:4101–4108

The Ministry of Agriculture of China. The alien species to be managed with priority in China (1st group). Announced 10 Sep 2013. http://www.moa.gov.cn/sydw/stzyzz/stzybh/201608/t20160831_5259805.htm. Retrieved 5 Feb 2017

Wan FH, Yang NW (2016) Invasion and management of agricultural alien insects in China. Annu Rev Entomol 61:77–98

Wan FH, Li BP, Guo JY (eds) (2008) Biological invasions: biological control theory and practice. Science Press, Beijing

Wan FH, Guo JY, Zhang F (eds) (2009a) Research on biological invasions in China. Science Press, Beijing

Wan FH, Peng DL, Wang R (eds) (2009b) Biological invasions: risk analysis and early prevention. Science Press, Beijing

Wan FH, Feng J, Xu J (eds) (2011) Biological invasions: detection, surveillance and monitoring. Science Press, Beijing

Wang ML, Feng YL (2005) Effects of soil nitrogen levels on morphology, biomass allocation and photosynthesis in *Ageratina adenophora* and *Chromoleana odorata*. Acta Phy Sin 29:697–705

Wang ML, Feng YL, Li X (2006) Effects of soil phosphorus level on morphological and photosynthetic characteristics of *Ageratina adenophora* and *Chromolaena odorata*. Chin J Appl Ecol 17:602–606

Wang H, Lai XL, Ma RY, Hu W (2008) Bionomics of *Agasicles hygrophila*, as the natural enemy for invasive weed, *Alternanthera philoxeroides*. Chin Bull Entomol 45:480–482

Wang N, Yu FH, Li PX, He WM, Liu J, Yu GL, Song YB, Dong M (2009) Clonal integration supports the expansion from terrestrial to aquatic environments of the amphibious stoloniferous herb *Alternanthera philoxeroides*. Plant Biol 11:483–489

Wang HF, Lopez-Pujol J, Meyerson LA, Qiu JX, Wang XK, Ouyang ZY (2011a) Biological invasions in rapidly urbanizing areas: a case study of Beijing, China. Biodivers Conserv 20:2483–2509

Wang R, Wang JF, Qiu ZJ, Meng B, Wan FH, Wang YZ (2011b) Multiple mechanisms underlie rapid expansion of an invasive alien plant. New Phytol 191:828–839
Wang WB, Wang RF, Lei YB, Liu C, Han LH, Shi XD (2013) High resource capture and use efficiency and prolonged growth season contribute to invasiveness of *Eupatorium adenophorum*. Plant Ecol 214:857–868
Wang R, Wang XL, Wang S, Zhang F (2014) Evaluation of the potential biocontrol capacity of *Orius sauteri* (Hemiptera, Anthocoridae) on *Frankliniella occidentalis* (Thysanoptera, Thripidae). J Environ Entomol 36:983–989
Wang CY, Xiao HG, Zhao LL, Liu J, Wang L, Zhang F, Shi YC, Du DL (2016) The allelopathic effects of invasive plant *Solidago canadensis* on seed germination and growth of *Lactuca sativa* enhanced by different types of acid deposition. Ecotoxicology 25:555–562
Weber E, Li B (2008) Plant invasions in China: what is to be expected in the wake of economic development? Bioscience 58:437–444
Wei SH, Zhang CX, Li XJ, Cui HL, Huang HJ, Sui BF, Meng QH, Zhang HJ (2009) Factors affecting buffalobur (*Solanum rostratum*) seed germination and seedling emergence. Weed Sci 57:521–525
Wu XW, Luo J, Chen JK, Li B (2006) Spatial patterns of invasive alien plants in China and its relationship with environmental and anthropological factors. J Plant Ecol 30:576–584
Wu TM, Ding H, Liu ZL, Xu HG, Bian XM (2007) Effects of alien invasive plant *Eupatorium adenophorum* on soil nutrients. J Ecol Rural Environ 23(2):94–96
Wu AP, Yu H, Gao SQ, Huang ZY, He WM, Miao SL, Dong M (2009) Differential belowground allelopathic effects of leaf and root of *Mikania micrantha*. Trees Struct Funct 23:11–17
Xiao B, Zhou W, Liu WX, Jiang ZL, Wan FH (2014a) Feedback of *Ageratina adenophora* soil microbe on *A. adenophora* and native plants. J Agric Sci Technol 16(4):151–158
Xiao HF, Feng YL, Schaefer DA, Yang XD (2014b) Soil fungi rather than bacteria were modified by invasive plants, and that benefited invasive plant growth. Plant Soil 378:253–264
Xie YH, Wen MZ, Yu D, Li YK (2004) Growth and resource allocation of water hyacinth as affected by gradually increasing nutrient concentrations. Aquat Bot 79:257–266
Xie LJ, Zeng RS, Bi HH, Song YY, Wang RL, Su YJ, Chen M, Chen S, Liu YH (2010) Allelochemical mediated invasion of exotic plants in China. Allelopath J 25:31–50
Xu HG, Qiang S (2011) China's invasive alien species. Science Press, Beijing
Xu HG, Qiang S, Han ZM, Guo JY, Huang ZG, Sun HY, He SP, Ding H, Wu HR, Wan FH (2004a) The distribution and introduction pathway of alien invasive species in China. Biodivers Sci 12:626–638
Xu KY, Ye WH, Cao HL, Deng X, Yang QH, Zhang Y (2004b) The role of diversity and functional traits of species in community invisibility. Bot Bull Acad Sin 45:149–157
Xu HG, Ding H, Li MY, Qiang S, Guo JY, Han ZM, Huang ZG, Sun HY, He SP, Wu HR, Wan FH (2006) The distribution and economic losses of alien species invasion to China. Biol Invasions 8:1495–1500
Xu YJ, Huang J, Zhou A, Zeng L (2012a) Prevalence of *Solenopsis invicta* (Hymenoptera: Formicidae) venom allergic reactions in mainland China. Fla Entomol 95:961–965
Xu HG, Qiang S, Genovesi P, Ding H, Wu J, Meng L, Han ZM, Miao JL, Hu BS, Guo JY et al. (2012b) An inventory of invasive alien species in China. NeoBiota 15:1–26
Yang GQ, Wan FH, Liu WX, Zhang XW (2006) Physiological effects of allelochemicals from leachates of *Ageratina adenophora* (Spreng.) on rice seedlings. Allelopath J 18:237–245
Yang RY, Mei LX, Tang JJ, Chen X (2007a) Allelopathic effects of invasive *Solidago canadensis* L. on germination and growth of native Chinese plant species. Allelopath J 19:241–247
Yang RY, Tang HJ, Yang YS, Chen X (2007b) Invasive and non-invasive plants differ in response to soil heavy metal lead contamination. Bot Stud 48:453–458
Yang GQ, Qiu WR, Jin YN, Wan FH (2013) Potential allelochemicals from root exudates of invasive *Ageratina adenophora*. Allelopath J 32:233–241
Yang RY, Zhou G, Zan ST, Guo FY, Su NN, Li J (2014a) *Arbuscular mycorrhizal* fungi facilitate the invasion of *Solidago canadensis* L. in southeastern China. Acta Oecol 61:71–77

Yang SX, Liang SC, Yi LB, Xu BB, Cao JB, Guo YF, Zhou Y (2014b) Heavy metal accumulation and phytostabilization potential of dominant plant species growing on manganese mine tailings. Front Environ Sci Eng 8:394–404

Yin F, Huang M, Xu R, Liu SQ, Li JC, Chen YB, Zhang WD (2009) Analysis on *Eupatorium* hazards and development in its comprehensive utilization. J Catal 24(4):63–67

Yin F, Hu J, Zhang WD, Li JC, Xu R, Chen YB, Liu SQ (2010a) The study on co-fermentation of hydrogen and methane production by *Eupatorium adenophorum* Spreng. Chin Sci Bull 55:3469–3476

Yin XW, Li XY, Zhang L, Chen B (2010b) Resistance determination of the apple varieties against the woolly apple aphid, *Eriosoma lanigerum*. J Yunnan Agric Univ 25:29–33

You WH, Han CM, Fang LX, Du DL (2016) Propagule pressure, habitat conditions and clonal integration influence the establishment and growth of an invasive clonal plant. *Alternanthera philoxeroides*. Front Plant Sci 7:568

Yu XJ, Yu D, Lu ZJ, Ma KP (2005) A new mechanism of invader success: exotic plant inhibits natural vegetation restoration by changing soil microbe community. Chin Sci Bull 50:1105–1112

Yu WQ, Wan FH, He XH, Liu WZ, Liu WX, Zhang LL (2014) Soil microbes enhance competition ability of the exotic *Ageratina adenophora* Sprengel against native plant species. J Biosaf 23(3):156–164

Yuan YG, Wang B, Zhang SS, Tang JJ, Tu C, Hu SJ, Yong JWH, Chen X (2013) Enhanced allelopathy and competitive ability of invasive plant *Solidago canadensis* in its introduced range. J Plant Ecol 6:253–263

Zhang PH, Luo WF, Yang YL (2006) Inhibition effect of leaf juice extracts of *Eupatorium adenophorum* on *Phytophthora infestans*. Southwest Chin J Agric Sci 19:246–250

Zhang Q, Yao LJ, Yang RY, Yang XY, Tang JJ, Chen X (2007) Potential allelopathic effects of an invasive species *Solidago canadensis* on the mycorrhizae of native plant species. Allelopath J 20:71–77

Zhang Q, Yang RY, Tang JJ, Chen X (2008) Competitive interaction between the invasive *Solidago canadensis* and native *Kummerowia striata* in lead contaminated soil. Bot Stud 49:385–391

Zhang CL, Li YP, Feng YL, Zhen YL, Lei YB (2009a) Roles of phenotypic plasticity and local adaptation in *Eupatorium adenophorum* invasions in different altitude habitats. Acta Ecol Sin 29:1940–1946

Zhang SS, Jin YL, Tang JJ, Chen X (2009b) The invasive plant *Solidago canadensis* L. suppresses local soil pathogens through allelopathy. Appl Soil Ecol 41:215–222

Zhang JM, Zhang F, Wang B (2010a) Controlling of *Trialeurodes vaporariorum* on tomato in protected areas by collaboratively using *E. formosa* and other methods. Vegetable 7:34–35

Zhang Q, Yang RY, Tang JJ, Yang HS, Hu SJ, Chen X (2010b) Positive feedback between Mycorrhizal fungi and plants influences plant invasion success and resistance to invasion. PLoS One 5:e12380

Zhang GF, Meng XQ, Wan FH (2011a) Advances in identification techniques of the western flower thrips, *Frankliniella occidentalis* (Pargande). J Biosafety 20(1):81–88

Zhang SS, Zhu WJ, Wang B, Tang JJ, Chen X (2011b) Secondary metabolites from the invasive *Solidago canadensis* L. accumulation in soil and contribution to inhibition of soil pathogen *Pythium ultimum*. Appl Soil Ecol 48:280–286

Zhang T, Luan JB, Qi JF, Huang CJ, Li M, Zhou XP, Liu SS (2012) Begomovirus-whitefly mutualism is achieved through repression of plant defences by a virus pathogenicity factor. Mol Ecol 21:1294–1304

Zhang HJ, Wang RQ, Wang X, Du N, Ge XL, Du YD, Liu J (2015a) Recurrent water level fluctuation alleviates the effects of submergence stress on the invasive riparian plant *Alternanthera philoxeroides*. PLoS One 10:e0129549

Zhang ZY, Zhang ZJ, Pan XY (2015b) Phenotypic plasticity of *Alternanthera philoxeroides* in response to shading: introduced vs. native populations. Biodivers Sci 23:18–22

Zhang LL, Chen XY, Wen DZ (2016a) Interactive effects of rising CO_2 and elevated nitrogen and phosphorus on nitrogen allocation in invasive weeds *Mikania micrantha* and *Chromolaena odorata*. Biol Invasions 18:1391–1407

Zhang HJ, Liu FH, Wang RQ, Liu J (2016b) Roles of clonal integration in both heterogeneous and homogeneous habitats. Front Plant Sci 7:551

Zhao WJ (2006) Studies on controlling *Eichhornia crassipes* with introducing natural enemies of *Neochetina eichhorniae*. Southwest Chin J Agric Sci 19:900–905

Zhao X, Zheng GW, Niu XM, Li WQ, Wang FS, Li SH (2009) Terpenes from *Eupatorium adenophorum* and their allelopathic effects on *Arabidopsis* seeds germination. J Agric Food Chem 57:478–482

Zhao YJ, Yang XJ, Xi XQ, Gao XM, Sun SC (2012) Phenotypic plasticity in the invasion of crofton weed (*Eupatorium adenophorum*) in China. Weed Sci 60:431–439

Zhao XJ, Liu WY, Zhou M (2013) Lack of local adaptation of invasive crofton weed (*Ageratina adenophora*) in different climatic areas of Yunnan Province, China. J Plant Ecol 6:316–322

Zheng YL, Feng YL, Liu WX, Lao ZY (2009) Growth, biomass allocation, morphology, and photosynthesis of invasive *Eupatorium adenophorum* and its native congeners grown at four irradiances. Plant Ecol 203:263–271

Zheng YL, Feng YL, Lei YB, Liao ZY (2012) Comparisons of plastic responses to irradiance and physiological traits by invasive *Eupatorium adenophorum* and its native congeners. J Plant Physiol 169:884–891

Zheng Y, Li DS, Zhang YP, Zhao YL (2014) Interspecific competition among three parasitoids of *Bactrocera dorsalis*. Chn J Biol Control 30:713–717

Zheng YL, Feng YL, Zhang LK, Callaway RM, Valiente-Banuet A, Luo DQ, Liao ZY, Lei YB, Barclay GF, Pereyra CS (2015) Integrating novel chemical weapons and evolutionarily increased competitive ability in success of a tropical invader. New Phytol 205:1350–1359

Zhong S, Kui JX, Zhou ZW (2007) Planting displacement techniques for *Eupatorium adenophorum* in pasture. Plant Prot 33(3):16–19

Zhou ZS, Guo JY, Li BP, Meng L, Fu JW, Chen HS, Ma MY, Shi MZ, Li M, Guo W et al. (2011) Distribution and regional control strategies of common ragweed *Ambrosia artemisiifolia* L. and alligator weed *Alternanthera philoxeroides* (Mart.) in China 20:263–266

Zhou ZX, Pang JH, Guo WC, Zhong NQ, Tian YC, Xia GX, Wu JH (2012) Evaluation of the resistance of transgenic potato plants expressing various levels of Cry3A against the Colorado potato beetle (*Leptinotarsa decemlineata* Say) in the laboratory and field. Pest Manag Sci 68:1595–1604

Zhou ZS, Guo JY, Wan FH (2015) Review on management of *Ambrosia artemisiifolia* using natural enemy insects. Chn J Biol Control 31:657–665

Zhu XZ, Zhang JT, Ma KP (2011) Soil biota reduce allelopathic effects of the invasive *Eupatorium adenophorum*. PLoS One 6:e25393

Chapter 3
Biological Invasions in Forest Ecosystem in China

Min Lu and Jianghua Sun

Abstract With the increasing volume of international trade in forest products around the world, hidden organisms in cargos were brought into the destination countries. These alien organisms, including insects, plants, and microorganisms, were introduced to new ecosystems, spread and threaten local biodiversity (also known as biological invasions). Nowadays, vulnerable forests in China are being threatened by large numbers of invasive species. In this chapter, we introduce current situation of invasive species in forest ecosystems in China, propose novel theoretical researches on biological invasions in forest ecosystems and several invasive species, as well as discuss main strategies and policies for management of forest pest invasions.

Keywords Biological invasions • Invasive species • Forest • Invasive pest • Forestry invasive alien species

3.1 Introduction

China has a total forest land cover about 208 million hectares with the forest stock volume reaching 15,137 million cubic meters. However, the forest resources in this country are largely scarce due to the big population. The forest area per capita was estimated to be only 1/4 of the world's average, while the forest stock volume per capita be only 1/7 (State Forestry Administration of the People's Republic of China 2014). Increase in international trade is a major factor causing threats of biological invasion to China forests (Sun et al. 2002). Each kind of invasive species hiding in soil, seeds, bonsais and other nursery stocks, logs, pallets, and many wood products may be unintentionally introduced.

M. Lu • J. Sun (✉)
State Key Laboratory of Integrated Management of Pest Insects and Rodents, Institute of Zoology, Chinese Academy of Sciences, 1 Beichen West Road, Chaoyang District, Beijing 100101, China
e-mail: sunjh@ioz.ac.cn

F. Wan et al. (eds.), *Biological Invasions and Its Management in China*,
Invading Nature - Springer Series in Invasion Ecology 11,
DOI 10.1007/978-94-024-0948-2_3

Exotic forest pests can be particularly devastating since the native host species have not evolved under selection pressure to acquire resistance to invasive species, and the invade forests generally lack natural enemies against exotic pests. Strikingly, half of the most harmful forest pests in China are exotic. It is estimated that exotic pests cause a loss of about 120 billion RMB each year (Cui et al. 2009). Besides scientists, the government and the general public are also becoming more concerned about the threats to forests posed by alien invasive species (Wingfield et al. 2015).

3.2 Several Forest Invasive Species in China

In last decades, numerous invasive species had been detected in China, and some have been paid much attention since their first reports. The major invasive species are listed in Table 3.1 (Zhao et al. 2015). Of them, a few are occurring very seriously such as the red turpentine beetle (RTB), *Dendroctonus valens* LeConte (Coleoptera: Scolytidae), the pinewood nematode and the invasive plant *Eupatorium adenophorum* Spreng (Compositae). In this section, we introduce briefly four invasive species, mainly focusing on their invasive histories, biology, threats and management strategies.

3.2.1 Fall Webworm Hyphantria cunea *(Drury)*

The fall webworm *Hyphantria cunea* (Drury), native to North America, has spread to many countries through human activities. In China, it was first recorded in Liaoning province in the 1970s, and so far has spread to eastern coastal provinces and part of Shaanxi province (Zhang et al. 2013). It has become a serious invasive pest to agricultural crops, forests and city ornamental trees. Fall webworm is a typical polyphagous defoliator, and its host plants include shrubs, crops, vegetables, orchard trees, conifers and broadleaved trees with a preference for broadleaved trees. A total of 175 species of plants are recorded as its hosts in China (Zhang et al. 2013).

This moth in China have caused serious damages to local forests, including some valuable ornamental trees. The web formed by its larvae during feeding is a problem, especially on city ornamental trees. When its population outbreaks, the feeding larvae defoliate all leaves on a tree, which damages its aesthetic value in cities. Moreover, larvae may crawl around and sometime get into buildings becoming a nuisance to humans. The cost of controlling the moth is largely increasing.

Detection, monitoring and management of this invasive pest have been extensively performed in China. The simplest detection method involves surveying the

Table 3.1 China's major forest pest invasion species

Species	Classification	Native area	Time for first record	Invasion way	Place first detected
Eriosoma lanigerum	Pest	USA	1914	Imported seedlings	Shandong Province
Eichhornia crassipes	Plant	South America	1930s	Introduced as livestock feed	unknown
Eupatorium odoratum	Plant	Central America	1934	May be introduced as spices plant	Yunnan Province
Solidago canadensis	Plant	North America	1935	Introduced as an ornamental	East China area
Eupatorium adenophorum	Plant	Mexico	1940s	Natural transmission invasion from the Burmese border	Yunnan Province
Matsucoccus matsumurae	Pest	Asia	1942	Introduced naturally by wind	Liaoning Province
Cydia pomonella	Pest	Asia Minor	1953	Passengers	Xinjiang Province
Spartina anglica	Plant	The south coast of England	1963	Introduced as the bank revetment plants	East China area
Botryosphaeria laricina	Fungus	Janpan	1970s	Unknown	Northeast China area
Poplar mosaic virus	Virus	Canada	1972	Poplar	unknown
Trialeurodes vaporariorum	Pest	North America	1975	Unknown	Beijing City
Lecanosticta acicola	Fungus	USA	1978	Unknown	Fujian Province
Hyphantria cunea	Pest	North America	1979	Introdeced with wood packaging materials;introduced naturally by wind from North Korea	Liaoning Province
Heterobostrychus aequalis	Pest	Southeast Asia	1980s	Introduced with wood packaging materials	Guangdong Province and Hainan Province
Bursaphelenchus xylophilus	Nematode	North America	1982	Introduced with wood packaging materials	Jiangsu Province
Hemiberlesia pitysophila	Pest	Japan	1982	Christmas Tree from Japan	Guangdong Province
Mikania micrantha	Plant	South America	1984	Introduced naturally by wind from southeastern Asia	Guangdong Province

(continued)

Table 3.1 (continued)

Species	Classification	Native area	Time for first record	Invasion way	Place first detected
Opogona sacchari	Pest	Africa	1987	Brazilwood	Guangdong Province
Oracella acuta	Pest	USA	1988	*Pinus elliottii*	Guangdong Province
Chilades pandava	Pest	India, Thailand, myanmar, Malaysia and Sri Lanka	1980s	Unknown	Guangdong Province
Liriomyza sativae	Pest	South America	1993	Host	Hainan Province
Synanthedon tipuliformix	Pest	Europe	1995	*Ribes nigrum*	Xinjiang Province
Dendroctonus valens	Pest	North America	1998	Pinewood	Shanxi Province
Rhynchophorus ferrugineus	Pest	Philippines	1998	Imported seedlings	Hainan Province
Octodonta nipae	Pest	Indonesia and Papua New Guinea	2001	Imported seedlings	Hainan Province
Brontispa longissima	Pest	Malaysia	2002	Imported seedlings	Hainan Province
Rhabdoscelus lineaticollis	Pest	Philippines	2002	Unknown	Guangdong Province
Frankliniella occidentalis	Pest	North America	2003	Unknown	Beijing City
Solenopsis invicta	Pest	Paraguay and the panama canal	2004	The means of transport or seedling with soil	Guangdong Province
Quadrastichus erythrinae	Pest	Africa, Mauritius and reunion	2005	Woods	Guangdong Province
Obolodiplosis robiniae	Pest	USA	2006	Unknown	Hebei Province and Liaoning Province
Carpomyia vesuviana	Pest	India	2007	Passengers	Xinjiang Province
Corythucha ciliata	Pest	North America	2007	Unknown	Hubei Province
Leptocybe invasa	Pest	The Middle East and countries along the Mediterranean coast	2007	May fly by itself or imported from Vietnam by wind	Guangxi Province
Phenacoccus solenopsis	Pest	North America	2008	Imported seedlings	Guangdong Province
Sirex noctilio	Pest	Europe	2013	May be introduced with the wood or wood packaging materials	Heilongjiang Province
Opisina arenosella	Pest	India, Sri Lanka and Burma	2013	Imported seedlings	Hainan Province

webs. Population monitoring is mainly conducted using black-light trapping or pheromone trapping. An effective and cost-efficient management plan has been developed over the past 20 years, since fall webworm's introduction to China. The measures employed in management plans include: strict quarantine procedures specifically targeting late-instar larvae and pupae to prevent human activity-induced long distance spread; encouraging the planting of mixed forest stands to restructure the current plantation estate; manual removal of webs during the 3rd-4th instar larval stage; and spraying NPV virus during larval stages and releasing the parasitoid *Chouioia cunea* during the pupal stage. *C. cunea* has proven to be very effective and has been applied in several epidemic areas.

3.2.2 Sirex Woodwasp **Sirex noctilio *F.***

The Sirex woodwasp (*Sirex noctilio*) F. is a species of horntail, native to Europe, Asia, and northern Africa (Bernard et al. 2007). In China, it was firstly detected in Heilongjiang province in 2013 (Li et al. 2015). So far, this pest is mainly distributed in Heilongjiang province, Jilin province and eastern Inner Mongolia Autonomous Region. The Sirex woodwasp only attacks conifers, especially pines. The major hosts include *Pinus sylvestris*, *P. pinaster* and *P. nigra*, all of which are native pines. In comparison, in the Southern Hemisphere and North America, the wasp attacks both exotic and domestic pine species, generally in plantations.

The Sirex woodwasp and white rot fungus *Amylostereum areolatum* have a mutualistic symbiotic relationship (Li et al. 2015). The wasp creates the optimal conditions for the infestation through the fungus by drilling into the underlying wood layers and weakening the host tree. Infestation damage can be divided into four categories or phases, depending on whether it is caused by the imago, fungus, larvae or secondary parasites.

(a) The first reaction of the host tree is traced to the adult wasp and occurs after 10–14 days. A toxin secreted by the wasp impairs metabolism in the shoots and needles, causing loss of water balance. The attacked pines tend to develop flagging. Tip dieback begins with needles which become chlorotic and change from green to yellowish-red, finally turn completely brown over a 3–6-month period. The wasp bores holes in the tree. As with many other wood pests, fine resin drops are present in wasp boreholes of the central trunk. Unstressed trees may be attacked uniformly along the main stem, while trees with low osmotic phloem pressure are preferentially attacked, with denser clusters of boreholes be generated (Madden 1974).
(b) During this process, fungal spores germinate in the boreholes. The fungus breaks down the lignin, causing white rot. It moves towards the vertically aligned xylem. The vertical profile shows reddish and white streaks which run in the direction of growth.

(c) At the third stage, larvae begin to bore into the wood. By doing this, they eat in their paths, which at first proceed towards the trunk center, before turning and running back to the bark. The paths are not visible in cross-section, because they are heavily blocked with wood flour; they may also be unobserved during wood processing.
(d) Stressing of host trees and visible larval boreholes appear at the fourth stage. The infestation is reinforced further by insects or fungi, which in turn may cause more symptoms. Imago, fungus, and larvae together can cause tree death in a period ranging from 2 weeks to 8 months (Taylor 1981; Berryman 2013).

Several biological control agents have been employed to reduce populations of Sirex woodwasp. *Deladenus siricidicola* (Bedding) has been shown to infect up to 70% of the wasps. The introductions of parasitoid wasps, such as *Megarhyssa nortoni* (Cresson), *Rhyssa persuasoria* (L.) and *Ibalialeu cospoides* (Hochenwarth) have been successful at hyperparasitism, but it is unable to reduce local wasp populations to <40% (Haugen and Underdown 1990; Cláudia et al. 2012). Although some success has been detected in slowing the population growth of the wasp, overall these measures can not stop the spread of the wasp (Bernard et al. 2007). Other control methods have also been attempted with bromomethane (CH_3Br), through heat, or by removing the bark (Haugen and Underdown 1990).

3.2.3 Blue Gum Chalcid **Leptocybe invasa** *Fisher & La Salle*

The blue gum chalcid *Leptocybe invasa* Fisher& La Salle (Hymenoptera, Eulophidae) is native to Queensland and Australia. It is currently spreading through Africa, Asia and the Pacific, Europe, Latin America and the Caribbean, Near East and North America (Mendel et al. 2004). In China, this species was first recorded in Guangxi Zhuang Autonomous Region in 2000, and has been introduced into adjacent regions such as Guangdong, Hainan, and Fujian provinces (Wu et al. 2009; Gu et al. 2012). Transportation of plant materials infected with larvae is the dominant spread path of this pest. It is therefore very important to avoid the movement of infested material from infested to uninfested areas, and screening of seedlings for infestation is necessary before transporting plants among such regions.

A great number of plants may be attached by *L. invasa*. Suitable host species include *Eucalyptus saligna*, *E. botryoides*, *E. bridgesiana*, and various clones and hybrids (Wylie and Speight 2012). *Eucalyptus* seedlings and trees of less than 3 years old appear to be most vulnerable to *L. invasa* damage. The adult female wasp lays her eggs on the midrib, petioles and stem of young trees, as well as on newly produced coppices and seedlings, resulting in the formation of bump-shaped galls. These lead to stunted growth of the trees. Five stages of gall development have been identified prior to wasp emergence (TPCP 2005):

i) One to 2 weeks after egg laying, cork tissue appears at the egg insertion point and gall development begins within the plant tissue; ii) Development of a typical bump shape until the gall reaches its maximum size; iii) Fading of green color on the gall surface, changing to a glossy pink color; iv) Loss of gall glossiness with a color change occurring to light or dark red; and v) Emergence holes of wasps visible.

No effective measures have been established. Insecticides have limited effect in preventing infestations on nursery saplings. There is potential for implementation of biocontrol through introduction of suitable parasitoid agents, and/or selection of resistant tree genotypes. The parasitoid *Selitrichodes neseri* (Kelly & La Salle) is currently being reared in quarantine in South Africa for potential release against *L. invasa* (Kelly et al. 2012).

3.2.4 Loblolly Pine Mealybug **Oracella acuta** *(Lobdell)*

The loblolly pine mealybug, *Oracella acuta* (Hemiptera: Pseudococcidae), was first reported on pine twigs collected in Mississippi, USA (Lobdell 1930) and is believed to be native in the Mississippi River basin. It was introduced into Guangdong province, China, in 1988 through an accidental importation of infested scions of slash pine, *P. elliottii*, from Georgia, USA (Xu et al. 1992; Sun et al. 1996). Currently, around 264,320 ha of forests across four provinces are infested in China (You et al. 2013). Although *O. acuta* can feed on a wide range of pine species, damage in China is most severe on introduced *P. elliottii*, *P. taeda*, *P. caribaea*, and native *P. massoniana*. Both adults and nymphs are found to completely cover needles and shoots on pine branches; they feed at the base of old needles and cause copious resin flow, at which point the needles turn brown and abscise, resulting in 70–80% of needle loss (Xu et al. 1992). Feeding directly affects the growth of pine trees by the extraction of nutrients from shoots and twigs as well as cones. Infestation can cause growth reduction or malformation, loss of plant vigor, stunting, defoliation, reduced seed production, and timber yield, and even death of the tree. Sooty mold induced by honeydew secreted from *O. acuta* also hampers photosynthesis in needles and indirectly affects tree growth. Although there is no direct evidence of correlation between the mortality of various pine species in China with the presence of *O. acuta*, growth and production is believed to decline significantly in heavily infested forests (You et al. 2013).

Upon the discovery of *O. acuta* in southern China in 1990, efforts were rapidly initiated to confirm its identity and then to eradicate it, and finally to limit its spread. Initial eradication efforts involved foliar sprays and trunk injections of 40% omethoate EC and 80% dichlorvos EC at its first arrival site (Xu et al. 1992). When the insecticide applications proved ineffective against early infestations of *O. acuta*, most of the infested branches and limbs were physically removed, and eventually the entire infested trees were cut and destroyed (Sun et al. 1996). Because of the failure to achieve early eradication, testing of insecticides was conducted at the *P. elliottii* nursery and demonstrated that the combination of 40% omethoate EC and 80% dichlorvos EC could achieve a mortality of 85–90%, with a relatively low cost

(Xu et al. 1992). Work funded by the National Ministry of Forestry led to the development of an integrated *O. acuta* control program mainly based on chemical and biological approaches in 1993 (Sun et al. 1996). Main chemicals used for *O. acuta* control include diesel oil, rotenone, methidathion, methomyl and omethoate (You et al. 2013). Three parasitoids (*Acerophagus coccois* (Smith), *Zarhopalus debarri* (Sun) and *Allotropa oracellae* (Masner)) and two pathogens (*Cladosporium dadosporioides* and *Verticillium lecanii*) have also successfully been tested as potential microbial biological control agents of *O. acuta* (You et al. 2013). The discovery of natural enemies in local habitats provides a significant foundation for the potential augmentation and implementation of biological control into a sustainable management strategy to better suppress populations of this exotic pest in China, although further research is required to determine the ecological mechanisms behind the interaction between natural enemies and *O. acuta*.

3.3 Basic Theoretical Research on Biological Invasions in Forest Ecosystem in China

What is the reason for the forest pest becoming invasive in its introduced ranges, but seldom causing severe damages in its native ranges? Because of our specialized training, entomologists, pathologists, ecologists, and weed scientists are among the few who recognize the magnitude of the threat that exotic pests pose to forest ecosystems. Since the early twenty-first century, many efforts have been made to illuminate invasion mechanism on the forest exotic pests in China, and two potential invasion mechanisms including multi-species synergy and microevolution of alien species were proposed.

3.3.1 Multi-species Synergy in Biological Invasions

Alien species would meet various obstacles during invasions into new environments. In this process, some organisms living together could help each other in adapting to a new environment. For example, pine wilt disease (PWD) is one of the most serious threats to pine forests worldwide. Its causative agent, the pinewood nematode (PWN), engages in a symbiotic partnership with its insect vector, the Monochamus beetle, as well as associated bacteria and ophiostomatoid fungi. Chemical and molecular signals positively influencing the survival, reproduction, and spread of PWN in the symbiosis system was described (Zhao et al. 2014). Longhorn beetles of the genus *Monochamus* (Insecta: Coleoptera: Cerambycidae) transmit PWN to susceptible, live pine trees as they engage in maturation feeding that occurs during the summer. PWN enters through the feeding wound, migrates mainly via the resin canals of the tree, and feeds on the epithelial cells that line the resin ducts, ultimately disrupting the water transport system of the tree (Futai 2013). Especially at later stages of infection in weak pine trees, PWN feeds on various wood infesting fungi, such as blue stain fungi (Noritoshi and Kazuyoshi 1997, Zhao et al. 2014). During the propagative phase of the life cycle, PWN reproduces rapidly, developing from the egg, proceeding through four larval stages (J_I–J_{IV}), and finally becoming a reproductive adult. Female Monochamus beetles lay eggs on weak pine trees in autumn (Futai 2013), and the resulting beetle larvae develop within pine trees. Under unfavorable conditions such as low food or high population density, PWN enters the dispersal phase of its life cycle, in which it molts from J_{II} into third stage dispersal juveniles (J_{III}). These juveniles aggregate around the pupal chambers of the beetle larvae during winter. Once the beetle larvae molt to the late pupa or young adult stages during the following spring, the J_{III} juveniles then develop into fourth stage dispersal juveniles (J_{IV}), which enter the tracheal system of the beetle for dispersal (Zhao et al. 2007, 2014). One adult *Monochamus* beetle can carry thousands of nematodes in its tracheal system with an average of 1628 for each beetle (Futai 2013).

Upon entering a healthy pine tree, J_{IV} nematode larvae recover and become propagative J_n larvae, which are attracted to terpenes produced by the pine tree. PWN interacts with beneficial microbiota that may be introduced to the pine tree by the nematode or its vector beetle. Blue stain fungus, *Sporothrix* sp. 1, induces the xylem tissue of the pine tree to produce diacetone alcohol (DAA), which may increase PWN propagation and beetle larvae growth. Blue stain fungi also provide food to PWN at later stages of infestation after pine tree death. As conditions deteriorate inside the pine tree, PWN enters the dispersal phase of its life cycle. J_{III} nematode larvae form in response to unknown signals and aggregate around the pupal chamber of the vector beetle in response to terpene signals produced by the beetle larvae. Then, J_{III} molt to J_{IV} in response to C_{16} and C_{18} chemicals released from the surface of the emerging Monochamus beetle adult (Fig. 3.1).

3.3.2 Microevolution in Invasive Species

In the process of invasiveness, some invasive species could change their genetic traits. These changes may happen at very fine scales, but it will make differences in improving invasive species' fitness. For example, *Leptographium procerum*, a fungus vectored by an invasive bark beetle, *D. valens*, contributed to the invasive success of the beetle–fungal complex in China (Lu et al. 2009, 2010, 2011). The fungus was introduced by the beetle from the United States to China, where several novel genotypes were identified using microsatellite markers. These novel genotypes were more pathogenic to Chinese host seedlings than other genotypes and they also induced the release of higher amounts of 3-carene, the primary host attractant for the beetle vector, from inoculated seedlings. Thus, more beetles are attracted by 3-carene released from pine. Based on the evolution of a novel genotype during the two or three decades since its introductions, the invasion success of the

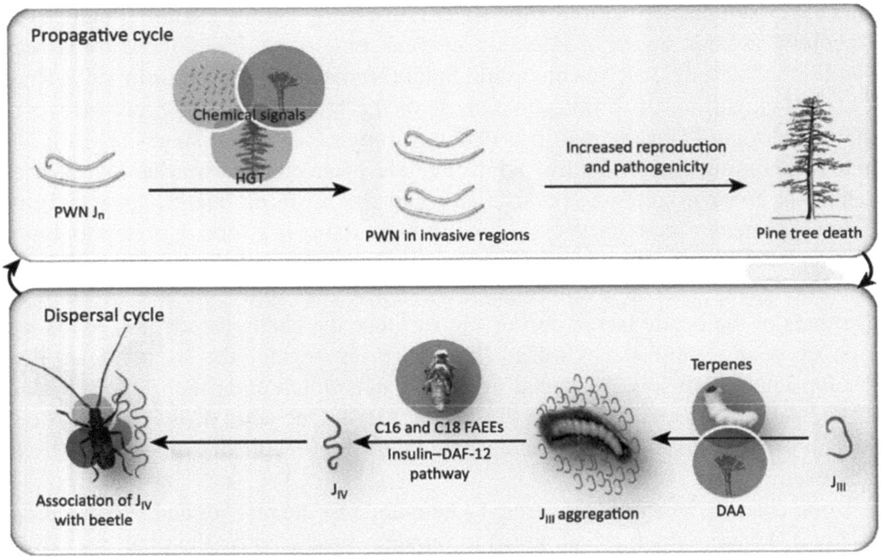

Fig. 3.1 Multispecies interactions in PWD (Modified from Zhao et al. 2014 Trends in Parasitology 30, 299–308)

beetle–fungal complex in its introduced region could be explained. Another example is the pinewood nematode. The PWN has caused serious damage to the conifer forests in many countries. According to the latest researches, some key genes evolved in pathogenicity have changed, and these changes may explain the increased threating brought by this pest (unpublished data) (Fig. 3.2).

3.4 Strategies and Policies for Management of Forest Pest Invasions

Given the tremendous potential impact of forest biology invasions in China, the main method for tackling this problem is prevention and early detection, backed by integrated management. Several aspects should be paid great attention:

The Invasion Risk Precontrol Management Risk precontrol management is the first barrier to prevent pest invasion, and it is the basis for the prevention and control work. According to the origin country of the main forest pest analysis, current work should focus on prevention of invasive species from North American continent countries where latitude is similar to China's geographic location. Areas with similar latitudes may have similar climate, which would provide suitable ecological niches for alien species. Based on scientific and accurate risk assessment, a pest outbreak data sharing platform including North America, South America, Europe,

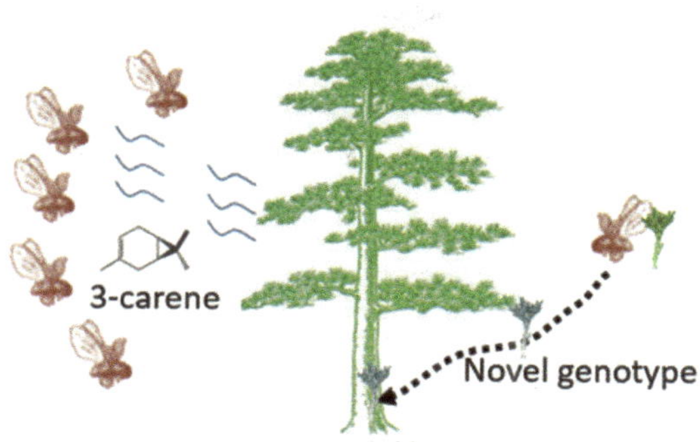

Fig. 3.2 Beetle–fungus–host tree relationships mediated by induction of 3-carene by novel *L. procerum* genotypes (Modified from Lu et al. 2011 Ecology 92, 2013–2019)

Africa and Oceania should be established. For species which is of high adaption and fitness and high degree of risk should be included in the "Entry Plant Quarantine Pest List" which was redacted by General Administration of Quality Supervision, Inspection and Quarantine of the People's Republic of China (AQSIQ). According to such a list, local government of each province could focus on the alien species listed, which would implement directional management measures in quarantine.

Strengthen the Management of Entry Quarantine Inspection Human carriage such as wood import and wooden packing is the main way of forest pest invasions, which is important to prevent pest invasion. The introduction of seedling and breeding material are the most important way of forest pest invasion. During the process of seedling introduction into a new environment, some other organisms associated with these seedlings might also be introduced, which may become invasive species. Main forestry regulatory department shall pay great attention to strengthen the approval and supervision for introduced plants from abroad and its breeding material. All biological materials including seedlings or bleedings should be sterilized. To speed up the quarantine development, accurate and practical testing as well as the elimination of processing including quarantine technology and equipment should be improved.

Intensive Basic Research on Exotic Pests For example, sufficient biological information on pests concerned can be used to guide decision-making, monitoring and detection techniques. The goal is to build an efficient and broad early warning and detection system. A comprehensive database on forest exotics is being compiled, in collaboration with Chinese Academy of Sciences (http://www.chinaias.cn/wjPart/index.aspx). This database presently contains information on 400 species.

Legislation and Regulatory System for Invasive Species As there were so many entry ports, complete quarantine was almost impossible. Thus, legislation and control procedures regarding quarantine, inspection, and intentional introduction of plants or animals need to be strengthened, with more attention to risk assessment before approval and follow-up monitoring after introductions. Efforts are underway to strengthen on-site quarantine inspection and domestic quarantine for established exotic pests, in order to prevent their further spread. However, new legislation is also required to better prevent exotic pests, including restricting the use of alien species in protected areas and promoting the use of native species in reforestation programs. In addition, it is also necessary to establish a mandatory requirement for scientific trials and risk assessment prior to large-scale planting of exotic trees.

Monitoring and Detection System Monitoring timely and in place can effectively prevent forest pest colonization, which is a final barrier to prevent negative effects caused by biological invasions. To do the work, one step is to strengthen the inspection in entry port and the monitoring of the surrounding forest. More than 1000 forest-pest monitoring and detection centers have been established at county level and a further 8000 monitoring stations have been set up across the country. These centers and stations can monitor both native and exotic forest pests all year round for early detection. In 2012, the National Inspection and Quarantine Bureau released the specified port of adjusted 45 invasive plant seedlings. According to the distribution of ports, carrying out the monitoring work would be efficient. Another policy is to establish the traceability of plant entry management platform. Entry-exit inspection and quarantine information can be shared with the competent Forestry Department. According to this information, traceability to conduct inspection, monitoring and prevention work would be easier. In addition, to introduce alien species without identification, quarantine and implementation of diversifying the traceability management shall be in accordance with the relevant provisions of the state.

International Cooperation and Vertical Control Network It is very important to strengthen the coordination and information sharing between the relevant government agencies for forestry, agriculture, trade, environmental protection and research institutions in dealing with forest invasive species. It is also important to strengthen regional and international cooperation. Much work had been conducted in managing biological invasions. For example, a global conference, the International Congress on Biological Invasions, had been conducted since 2013 each year. Cause of forest pest invasion factor, route of transmission, prevention and control work is complicated. Thus, a multilevel and multidisciplinary collaboration vertical platform to prevent pest invasion hazards should be established including agriculture, forestry, inspection, transportation, civil aviation, postal and other relevant departments, which can effectively prevent and control harmful biological invasions. Agriculture, forestry, and quality inspection departments, according to the division of responsibilities, should establish epidemic information sharing mechanisms to strict entry plant quarantine examination and approval and the prison work. Transport, post, civil aviation departments shall strictly examine the relevant certificate of plant quarantine, for plants and their products without inspection and quarantine shall be rejected shipment or reject mail.

3.5 Conclusion and Perspectives

Tremendous efforts have been devoted to the monitor and management of forest invasive species, and obtained great achievement. Fundamental researches, such as biology of invasive species, should be conducted continually. Regulators cannot make reasonable judgments about pest risk without basic information about the biology of the pests. We must also provide good data that can be used to weigh the value of specific mitigation measures against the costs. It is essential to provide decision makers with loss estimates, so they can compare the costs and benefits of regulatory and control options (Zhang and Kang 2003). Today it is difficult to obtain funding to provide scientific information that demonstrates the importance of exotic pests. But as we accumulate data that document their ecological and economic impact, this task will become easier. Once exotic pests become established beyond the reach of eradication efforts, biological control is desired as part of an integrated pest management strategy (Zhao et al. 2005). It is encouraging to see the development of collaborative projects, such as the China-US *D. valens* international project.

References

Bernard S, Hurley BP, Wingfield MJ (2007) A comparison of control results for the alien invasive woodwasp, *Sirex noctilio*, in the southern hemisphere. Agric For Entomol 9:159–171

Berryman AA (2013). Dynamics of forest insect populations: patterns, causes, implications. Springer Science & Business Media

Cláudia V, Margarida E, Paulo V, Manuel M (2012) Pine wilt disease: a threat to European forestry. Eur J Plant Pathol 133:89–99

Cui Y, Zhao Y, Hu X (2009) Exotic forest pest invasion in China and control strategies. Forest Pest Dis 28(3):40–42. (in Chinese)

Futai K (2013) Pine wood nematode, *Bursaphelenchus xylophilus*. Annu Rev Phytopathol 51:61–83

Gu P, Jiang X, Li D, Yao G, Yao D (2012) Research situation and prospect of *Leptocybe invasa* Fisher & La Salle. Jiangxi For Sci Technol 1:32–34

Haugen D, Underdown M (1990) *Sirex noctilio* control program in response to the 1987 Green Triangle outbreak. Aust For 53:33–40

Kelly J, Salle JL, Harney M, Der GD-S, Hurley B (2012) *Selitrichodes neserin* sp., a new parasitoid of the eucalyptus gall wasp *Leptocybe invasa* Fisher & La Salle (Hymenoptera: Eulophidae: Tetrastichinae). Zootaxa 3333:50–57

Li D, Shi J, Luo Y (2015) Mutualism between the Eurasian woodwasp, *Sirex noctilio* (Hymenoptera: Siricidae) and its fungal symbiont *Amylostereum areolatum* (Russulales: Amylostereaceae). Acta Entomol Sin 58:1019–1029

Lobdell GH (1930) Twelve new mealybugs from Mississippi. (Homoptera: Coccoidea.). Ann Entomol Soc Am 23:209–236

Lu M, Zhou XD, De Beer ZW, Wingfield MJ, Sun J (2009) Ophiostomatoid fungi associated with the invasive pine-infesting bark beetle, *Dendroctonus valens*, in China. Fungal Divers 38:133–145

Lu M, Wingfield MJ, Gillette NE, Mori SR, Sun J (2010) Complex interactions among host pines and fungi vectored by an invasive bark beetle. New Phytol 187:859–866

Lu M, Wingfield MJ, Gillette N, Sun J (2011) Do novel genotypes drive the success of an invasive bark beetle–fungus complex? Implications for potential reinvasion. Ecology 92:2013–2019

Madden J (1974) Oviposition behaviour of the woodwasp, *Sirex noctilio* F. Aust J Zool 22:341–351

Mendel Z, Protasov A, Fisher N, La Salle J (2004) Taxonomy and biology of *Leptocybe invasa* (Hymenoptera: Eulophidae), an invasive gall inducer on Eucalyptus. Aust J Entomol 43:101–113

Noritoshi M, Kazuyoshi F (1997) Effect of fungal interactions on the numbers of the pinewood nematode, *Bursaphelenchus xylophilus* (Nematoda: Aphelenchoididae), carried by the Japanese pine sawyer, *Monochamus alternatus* (Coleoptera: Cerambycidae). Fundam Appl Nematol 20:611–618

State Forestry Administration of the People's Republic of China (2014) Forest resources in China

Sun J, DeBarr GL, Liu X, Berisford CW, Clarke SR (1996) An unwelcome guest in China: a pine-feeding mealybug. J For 94:27–32

Sun J, Yuan D, Ouyang H (2002) Exotic forest pest invasion: a threat to forest ecosystems in China. For Pest Dis 6:32–35

Taylor K (1981) The Sirex woodwasp: ecology and control of an introduced forest insect, The ecological of pests some Australian cases histories. CSIRO, Melbourne, pp 231–248

TPCP (2005) Tree Protection Co-operative Programme (TPCP). In Pest alert. Blue gum chalcid. Tree Protection News 10:13. Updated 2007

Wingfield M, Brockerhoff E, Wingfield B, Slippers B (2015) Planted forest health: the need for a global strategy. Science 349:832–836

Wu D, Jiang Y, Li X, Luo M, Zhou J, Chang G, Yang Z (2009) *Leptocybe invasa*, a new invasive forest pest making galls on twigs and leaves of eucalyptus trees in China (Hymenoptera: Eulophidae). Sci Silvae Sin 45:161–165

Wylie FR, Speight MR (2012) Insect pests in tropical forestry. CABI, Cambridge, MA

Xu J, Ding K, Situ R (1992) A preliminary study on biology of *Oracella acuta*. Guangdong Forest Sci Technol:22–24

You S, Liu J, Huang D, Ke F, Vasseur L, Gurr G, Clarke SR, Sun J, Wu K, You M (2013) A review of the mealybug *Oracella acuta*: invasion and management in China and potential incursions into other countries. For Ecol Manag 305:96–102

Zhang R, Kang L (2003) On the strategy of prevention and legislative management for invasion of alien species. Bull Chin Acad Sci 6:413–415

Zhang J, Dong Q, Zhao H, Peng H, Zhang W, Li J, Zhang Y, Ren B (2013) A summary of distribution, harm and control measures of *Hyphantria cunea* in mainland China. J Jilin For Sci Technol 42:27–47

Zhao Y, Li X, Zhang Z (2005) The damage of major adventitious invasive species in our country and countermeasure for control. For Invent Plan 30:113–117

Zhao L, Wei W, Kang L, Sun J (2007) Chemotaxis of the pinewood nematode, *Bursaphelenchus xylophilus*, to volatiles associated with host pine, *Pinus massoniana*, and its vector *Monochamus alternatus*. J Chem Ecol 33:1207–1216

Zhao L, Mota M, Vieira P, Butcher RA, Sun J (2014) Interspecific communication between pinewood nematode, its insect vector, and associated microbes. Trends Parasitol 30:299–308

Zhao Y, Wu J, Luo Y, Tao J (2015) The identification of risk sources of forestry invasive alien species in China and their controlling countermeasures. Plant Quarantine 29:42–47

Chapter 4
Biological Invasions in Aquatic Ecosystems in China

Aibin Zhan, Ping Ni, Wei Xiong, Yiyong Chen, Yaping Lin, Xuena Huang, Yuzhan Yang, and Yangchun Gao

Abstract Aquatic ecosystems have been largely influenced by non-native invasive species in China. Despite the fact that negative ecological and economic effects caused by aquatic invasive species are increasing in number and area affected, the problem of biological invasions in aquatic ecosystems is highly overlooked in China – introduction vectors are not (well) regulated or managed, and lessons from reported disasters are slowly learned. In this chapter, we discuss the general information and background of biological invasions in aquatic ecosystems in China. We divide this chapter into four sections. Section 4.2, non-native species in aquatic ecosystems: we summarize the general information and distribution patterns of non-native species in aquatic ecosystems in China. Section 4.3, major vectors: we focus on three major vectors including shipping, aquaculture, and aquarium and ornamental trades, and their roles for introductions and spread of non-native species in China. Section 4.4, known and possible impacts of invasive species: we review representative disasters caused by aquatic invasive species and discuss possible risks caused by non-native species introduced into new environments. Section 4.5, management of biological invasions in aquatic ecosystems: we generally summarize the available management solutions, point out existing problems, and propose possible management solutions for future management. Finally, we discuss common research gaps among different vectors.

Keywords Aquatic ecosystem • Biological invasions • Invasion vector • Invasion management • Invasive species • Non-native species

---

A. Zhan (✉) • P. Ni • W. Xiong • Y. Chen • Y. Lin • X. Huang • Y. Yang • Y. Gao
Research Center for Eco-Environmental Sciences (RCEES), Chinese Academy of Sciences (CAS), 18 Shuangqing Road, Haidian District, Beijing 100085, China
e-mail: azhan@rcees.ac.cn; zhanaibin@hotmail.com

F. Wan et al. (eds.), *Biological Invasions and Its Management in China*, Invading Nature - Springer Series in Invasion Ecology 11,
DOI 10.1007/978-94-024-0948-2_4

4.1 Introduction

Increased human activities have largely facilitated introductions and spread of non-indigenous species (NIS) at a global scale in the past three decades (Levings et al. 2002; Ruiz and Carlton 2003; Carlton and Ruiz 2005). Aquatic ecosystems are among the most invaded habitats due to various introduction/spread vectors mainly operated by human activities, such as shipping, aquaculture, and aquarium or ornamental trades (Sylvester et al. 2011; Rhyne et al. 2012; Lin et al. 2015). For example, only for aquaculture-mediated introductions, more than 5600 records have been collected by Food and Agriculture Organization of the United Nations (FAO, http://www.fao.org/fishery/dias/en). Owing to language barriers, difficulties in data collection and many other reasons, the number of introduction events can be largely underestimated. For shipping-mediated introductions, every hour in every day, more than 7000 marine and coastal species are estimated to travel unnoticed with shipping vessels across the world's oceans (World Wildlife Fund 2009). Based on the survey conducted by World Wildlife Fund (WWF) in 2008, marine invasive species have been reported for at least 84% of the world's 232 marine eco-regions (World Wildlife Fund 2009).

Due to frequent introductions and population outbreaks of introduced invasive species, many countries/areas are suffering from huge ecological and economic damages in aquatic ecosystems (e.g. Lodge et al. 2006; Xu et al. 2006; Hulme 2009). In addition, the cost is extremely high for prevention, early detection and rapid response, and control/management of aquatic invasive species. In 2010 alone, USA committed $78.5 million in investments to prevent introductions of Asian carps to the Great Lakes (U.S. Fish & Wildlife Service 2012).

Among many countries suffering from biological invasions in aquatic ecosystems, China is among the most affected ones. China has a large number of water bodies, including more than 2800 inland lakes (total coverage area > 78,000 km^2), 53.6 million km^2 wetlands, 45,000 river basins with coverage area > 50 km^2, and >3 million km^2 territorial sea waters (>32,000 km coastal lines). As China's economy develops fast in the past four decades, these water bodies have been highly influenced by human activity-mediated introductions of an extremely large number of non-native species (see Sect. 5.2 for more detail). For example, a total of 179 species (virus and bacteria excluded) were introduced internationally into China for aquaculture or due to aquaculture-related activities (Lin et al. 2015). As China spans five climatic zones (50° latitude), covers approximately 5200 km from east to west, and supports extremely divergent habitats, non-native may find suitable habitats to successfully establish and subsequently become invasive, although initial entry points may not provide suitable environmental conditions for introduced non-native species. Here we define non-native species as those that historically have never occurred in a particular region and differentiate them from invasive species which we define as high impact species.

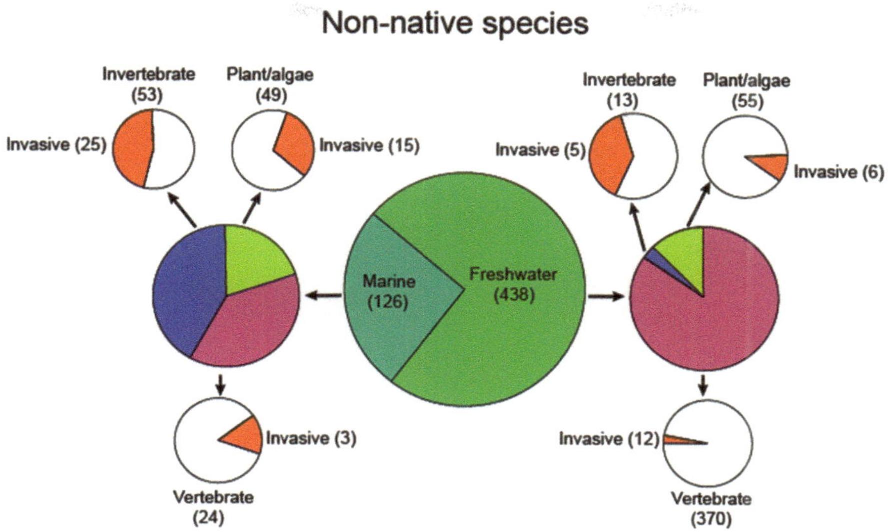

Fig. 4.1 Taxonomic group composition of non-native species internationally introduced into China

4.2 Non-native Species and Their Distributions in Aquatic Ecosystems

4.2.1 Aquatic Non-native Species

Following a comprehensive review of available literature, we identified a total of 564 non-native species in aquatic ecosystems in China, 438 and 126 of which were recovered from freshwater and marine habitats, respectively (Fig. 4.1). Because clear historical records are not available in many parts of China, the number of non-native species is likely much greater than that summarized here. In addition, compared to freshwater habitats, the relatively poorer records in marine and coastal ecosystems are likely responsible for the smaller number of marine non-native species. Indeed, the lack of historical records is globally common in many oceans and coastal regions, where we cannot even make a simple list of which species are non-native (e.g. Carlton 2009; Geller et al. 2010). Although domestic transfers of non-native species have caused significant ecological and economic loss in China (see Lin et al. 2015), both the poor historical records and large numbers of water bodies in China make it extremely difficult to generate a clear list (except for aquaculture-mediated introductions). We therefore focus on non-native species which were introduced internationally in this section. For domestic transfers, relatively clear records are available for aquaculture-mediated introductions. We discuss such a type of introduction in Sect. 5.3.2.

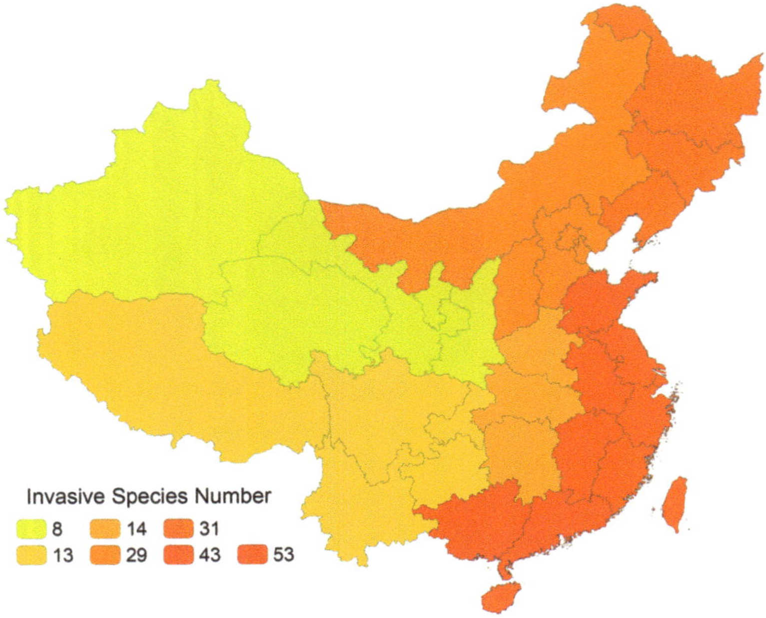

Fig. 4.2 Geographic distribution of invasive species at the provincial scale

In freshwater ecosystems, the most abundant group of non-native species was vertebrates (370 species), accounting for 84.5% of all freshwater non-native species (Fig. 4.1). In contrast, only 13 (3.0%) freshwater non-native invertebrate species were recovered from available literature (Fig. 4.1). In marine and coastal ecosystems, the number of non-native plants/algae, invertebrates and vertebrates were 49, 53 and 24, accounting for 38.9%, 42.1% and 19.0% of all non-native species, respectively.

Among the 564 non-native species in China, 69 species (12.2%) have caused negative effects in recipient water bodies, 25 and 44 of which were recovered from freshwater and marine ecosystems, respectively. The percentages of invasive species in each group largely varied. For both marine and freshwater ecosystems, the largest ratios were detected in invertebrates, with 5 species (38.5%) in freshwater ecosystems and 25 species (47.2%) in marine ecosystems. However, the ratios of invasive species were relatively low for some groups, for example, only 3.2% for freshwater vertebrates (Fig. 4.1).

4.2.2 Distribution of Aquatic Non-native Species

In general, the geographic distribution of invasive species varies largely at the provincial scale (Fig. 4.2). The distribution of non-native species is very similar to that of invasive species (figure not shown). Coastal provinces and provinces in South

China received more invasive species than northern and inland provinces (Fig. 4.2). For example, a total of 53 invasive species were detected in coastal provinces in South China (Fig. 4.2). Marine organisms, including 15 red tide-causing algae and 20 fouling species, dominate the invasive species in these provinces. These invasive species, which were likely introduced by shipping-mediated vectors such as ballast water and hull fouling, were detected in bays and coastal waters with a high risk of red tide blooming, such as the Daya Bay and Dapeng Bay in South China. Compared to South China, the number of invasive species in Middle, West and Northwest China were comparatively smaller (Fig. 4.2). The smaller number of invasive species in these provinces is not due to the smaller number (area) of water bodies. For example, the Qinghai-Tibetan Plateau has the largest number of inland lakes but smallest number of invasive species (Fig. 4.2). Compared to coastal and southern provinces, water bodies in inland provinces receive less human activity-mediated disturbance (e.g. Lin et al. 2007), leading to a smaller number of non-native species introduced into water bodies located in these areas.

4.3 Vectors for Introductions and Spread

As natural dispersal such as free-swimming larvae/juveniles advected by water currents and active swimming for large animals (e.g. fish) occurs only in a limited number of species and may only affect local and regional scales, here we focus on human activity-mediated vectors that commonly facilitate introductions and spread of numerous non-native species in China. The human activity-mediated vectors in China mainly include aquaculture (e.g. Lin et al. 2015), aquarium and ornamental trades (e.g. Cheung and Dudgeon 2006), those associated with shipping such as ballast water and hull fouling (e.g. Zhang et al. 2014a, b), and several others such as intentional introductions for ecosystem restoration and wetland conservation (e.g. Feng et al. 2014).

In general, vectors for introductions and spread of non-native largely vary between freshwater and marine ecosystems (Fig. 4.3). In freshwater ecosystems, aquarium and ornamental trades are responsible for the largest proportion of introduced non-native species (306 species, 69.9% of all freshwater non-native species; Fig. 4.3), followed by aquaculture-mediated introductions (116 species, 26.5%; Fig. 4.3). Only 10 species (2.3%) were introduced for particular purposes such as wetland restoration (Fig. 4.3). For marine ecosystems, 66 species were introduced for aquaculture, accounting for 52.4% of all introduced marine non-native species. A total of 55 species (43.0%) have clear records that were introduced *via* shipping-mediated vectors, mainly thorough ballast water and hull fouling (Fig. 4.3). However, the percentage may be largely underestimated, owing to many huge ports newly constructed in China and incomplete survey of vast areas (see below subsection for more detail). Only two species (1.6%) were introduced *via* aquarium and ornamental trades, and two species, and two species (1.6%) were intentionally introduced for ecosystem restoration and wetland conservation (Fig. 4.3).

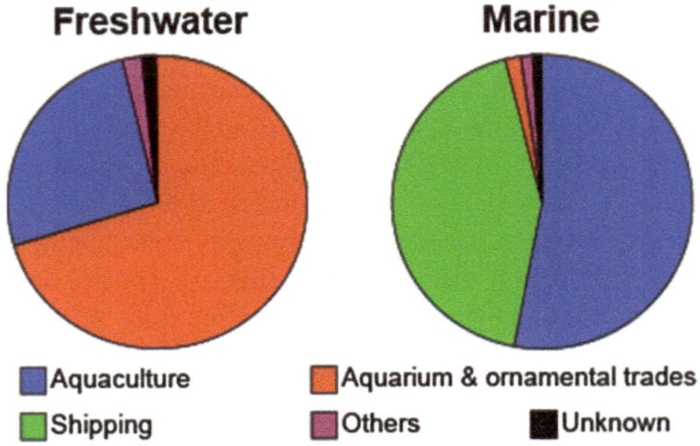

Fig. 4.3 Composition of major vectors responsible for introductions of non-native species in both marine and freshwater ecosystems in China

4.3.1 Shipping-Based Vectors

Owing to fast population growth and improved efficiency in shipping, global trade has increased dramatically since 1950s (Hulme 2009). The increased global trade, as well as rapid development of novel shipping routes, creates great opportunities for the introduction and spread of aquatic non-native species globally. As one of the most important nodes of many newly developed shipping routes and networks, China has become a major destination and transfer terminal for shipping. When ranked based on port throughput in 2013, eight out of the ten world's largest ports are located in China (Table 4.1). Thus, a large number of non-native species must be introduced into China through shipping-mediated vectors, such as ballast water, hull fouling and fellow travelers (i.e. non-native species that inadvertently accompany the shipment of objectives).

As expected, among 44 invasive species in marine and coastal ecosystems, the major introduction vector is shipping: 34 species (77.3%) were introduced by ballast water (13 species) and hull fouling (21 species; Table 4.2, Fig. 4.3). Bryozoans (13 species) dominate the marine invasive species introduced *via* hull fouling, while all detected red tide-causing algae algae (13 species) were introduced by ballast water (Table 4.2, Fig. 4.3). The introduced species *via* shipping, especially red tide-causing algae, have caused significant ecological and economic damages to local environments (see Sect. 5.4 for more detail).

Compared to intentional introductions such as those mediated by aquaculture and ornamental trades (see below sections for more detail), non-native species introduced *via* shipping are usually detected after they cause environmental problems (i.e. become invasive). Such detection lag is more obvious in China, mainly owing to the lack of effective surveillance/monitoring systems and management policies along long coastal lines. Among 34 invasive species introduced *via* shipping, the invasion dates of 23 species (67.6%) remain unknown (Table 4.3). Consequently, the invasion histories of these species remain largely unknown.

Table 4.1 Top ten ports ranked based on the total cargo in 2012

Rank	Port	Country	Total cargo (million metric tons)	
			2011	2012
1	Ningbo-Zhoushan	China	691	744
2	Shanghai	China	727.6	736
3	Singapore	Singapore	531.2	538
4	Tianjin	China	451	476
5	Rotterdam	Netherland	434.6	441.5
6	Guangzhou	China	429	434
7	Qingdao	China	375	402
8	Dalian	China	338	373
9	Tangshan	China	308	364.6
10	Yingkou	China	261	301.1

4.3.2 Aquaculture

Aquaculture develops extremely fast in both farming area and production during the past two decades in China. The total production in 2012 was 53.95 million tons in quantity and US$ 69.1 million in value, 13 and 22 times more than those in 1984, respectively (data from FAO at http://www.fao.org). The total farming area for marine and freshwater aquaculture in 2012 reached 21.8 million hm^2 and 59.1 million hm^2, six and 19 times more than those in 1980, respectively (data from China Fisheries Yearbook 2013). By 2012, China has dominated the reported aquaculture production, accounting for almost 60% of the global production (see Lin et al. 2015 and references therein).

Coupled with several factors such as improved breeding and farming technologies, the introduction and use of non-native species contributes significantly to rapid development of aquaculture in China. More than 25% of total production in China is derived from non-native species (see Lin et al. 2015 and references therein). Two well-known examples are the Pacific white shrimp *Penaeus vannamei* and red swamp crayfish *Procambarus clarkia*. The total production of the former species reached >74% (marine) and 44% (freshwater) of the total production of the category 'shrimp, prawn, crayfish and lobster' (Lin et al. 2015). Although the read swamp crayfish has caused significant ecological and economic damages to various aquatic habitats in China, its farming area and production keep increasing, reaching more than 0.5 million ton and accounting more than 30% of the total production of the category 'shrimp, prawn, crayfish and lobster' (Lin et al. 2015).

Compared to other vectors, there are relatively clear records for aquaculture-mediated introductions. A total of 179 species (virus and bacteria excluded) were internationally introduced into China for aquaculture (i.e. intentional introductions) or due to aquaculture-related activities (such as fellow travelers; Lin et al. 2015). More than half of these intentionally introduced species has been successfully farmed in China (see Table 4.3 in Lin et al. 2015).

The introduction and use of non-native species for aquaculture in China can be traced back as early as 1920s, but the number of species introduced before 1970s is

Table 4.2 A list of freshwater invasive species in China

Species	Native range	Year of introduction/Vector	Distribution in China	Negative effects in local environment	Major reference
Plant-6					
Pistia stratiotes	South America	1593/Aquaculture as fodder plant	Nationwide	Effect water quality, outcompete native species	Liu et al. (2006)
Eichharnia crassipes	South America	1901/Aquaculture as fodder plant	Nationwide	Effect water quality, outcompete native species	Xu et al. (2004b)
Cabomba caroliniana	America	1993/Aquarium & ornamental trade	Southeastern regions	Block waterways	Xu et al. (2004a)
Alternanthera philoxeroides	South America	1930s/Aquaculture as fodder plant	Nationwide	Effect water quality, outcompete native species	Xu et al. (2004b)
Spartina alterniflora	USA	1979/ Wetland restoration	Coast from Tianjin to Guangdong	Significant alteration of both marsh composition and structure	Xu et al. (2004b)
Spartina anglica	England	1963/Wetland restoration	Natiionwide	Significant alteration of both marsh composition and structure	Xu et al. (2004b)
Invertebrate-5					
Pomacea canaliculata	South America	1977/Aquaculture	Southern China	Pest of rice, intermediate hosts for parasites, and exclude native species	Li et al. (2007)
Biomphalaria straminea	South America	1973/Aquarium & ornamental trade-related	South China	Intermediate hosts for parasites	Attwood et al. (2015)
Physa acuta	North America or Europe	Unknown/Unknown	Beijing and Taiwan	Intermediate hosts for parasites	Guo et al. (2009)

Achatina fulica	Africa	1930/Aquaculture	South China	Pest in agriculture and horticulture, vector for disease-causing pathogens	Guo et al. (2009)
Procambarus clarkii	USA, Central & South America	1929/Aquaculture	Nationwide	Destroy dams and agriculture crops	Li et al. (2007)
Vertebrate-12					
Rana catesbeiana	North America	1959/Aquaculture	Nationwide	Predation on native species	Li et al. (2007)
Trachemys scripta	America	1987/Aquarium & ornamental trades	South China	Predation on native species	Cheung and Dudgeon (2006)
Pygocentrus nattereri	South America	1990/Aquarium & ornamental trades	Nationwide	Predation on almost all native species,attack human beings	Li et al. (2007)
Cyprinus carpio	Russian	1958/Aquaculture	Northwest China	Genetic introgression into local gene pools	Li et al. (2007)
Cyprinus carpio var. mirror	Ukraine	1958/Aquaculture	Nationwide	Genetic introgression into local gene pools	Li et al. (2007)
Cyprinus carpio var. specularis	German	1982/Aquaculture	Nationwide	Genetic introgression into local gene pools	Li et al. (2007)
Gambusia affinis	Central & South America	1927/Pest control	Nationwide	Attack native species, potential hosts of helminth parasites	Li et al. (2007)
Channa striatus	Southeast Asia	1992/Aquaculture	South China	Predation on fish,amphibians and crustaceans	Li et al. (2007)

(continued)

Table 4.2 (continued)

Species	Native range	Year of introduction/Vector	Distribution in China	Negative effects in local environment	Major reference
Micropterus salmoides	USA & Canada	1983/Aquaculture	Central & South China	Compete with native fish for food and space	Li et al. (2007)
Oreochromis niloticus	Africa	1978/Aquaculture	Nationwide	Exclude native species	Li et al. (2007)
Pterygoplichthys pardalis	South America	1980/Aquaculture	Nationwide	Alteration of bank structure and erosion and competition with native species	Aquatic NIS database (2015)
Oncorhynchus mykiss	West coast of North America	1959/Aquaculture	Nationwide	Competition with native species	Li et al. (2007)

The native ranges, year of introduction, current distribution ranges, major negative effects on local environment, and major references for each species are shown

Table 4.3 A list of marine invasive species in China

Species	Native range	Year of introduction/Vector	Major distribution in China	Negative effects in local environment	Major reference
Plant/algea-15					
Chaetoceros concavicornis	North America	1996/Ballast water	Zhejiang & Guangdong Provs.	Cause red tide	Xu and Qiang (2011)
Cyclindrotheca closterium	North America	1996/Ballast water	Zhejiang Prov.	Cause red tide	Xu et al. (2004a)
Melosiar cancellate	North America	1996/Ballast water	Zhejiang & Guangdong Provs.	Cause red tide	Xu et al. (2004b)
Nitzschia delicatissima	North America	1996/Ballast water	Yellow & BoHai Seas	Cause red tide	Xu et al. (2004a)
Pinnularia viridis	North America	1996/Ballast water	Zhejiang Prov.	Cause red tide	Xu et al. (2004a)
Alexandrum catenclla	North America	Unknown/Ballast water	Nationwide	Cause red tide	Xu et al. (2004b)
Peridinium perardiforme	North America	1996/Ballast water	Nationwide	Cause red tide	Xu et al. (2004b)
Prorocentrum balticum	North America	1996/Ballast water	East China Sea	Cause red tide	Xu et al. (2004b)
Prorocentrum minimum	North America	1977/Ballast water	South, East & Bohai Seas	Cause red tide	Xu et al. (2004b)
Prorocentrum sigmoides	North America	Unknown/Ballast water	Unclear	Cause red tide	Xu et al. (2004b)
Scrippsiella trochoidea	North America	2000/Ballast water	Southeast coast	Cause red tide	Xu et al. (2004b)

(continued)

Table 4.3 (continued)

Species	Native range	Year of introduction/Vector	Major distribution in China	Negative effects in local environment	Major reference
Noctiluca scintillans	Unclear	Unknown/Ballast water	Nationwide	Cause red tide	Qiao et al. (2010)
Heterosigma akashiwo	Japan	1980/Ballast water	South China Sea & Hongkong	Cause red tide	Aquatic NIS database (2015)
Desmarestia ligulata	Japan	1997/Aquaculture-related	Liaoning Prov.	Release sulphuric acid after death	Shao and Li (2000)
Undaria pinnatifida	Japan & Korea	1940s/Aquaculture	Liaoning & Shandong Provs.	Reduce native seaweed diversity	Liang and Wang (2001)
Invertebrate-25					
Hydroides elegans	Unclear	Unknown/Hull fouling	Unknown	Fouling on cultured mollusks	Zhang et al. (2013)
Bugula californica	Unclear	Unknown/Hull fouling	Nationwide	Fouling on underwater facilities	Aquatic NIS database (2015)
Bugula neritina	Unclear	Unknown/Hull fouling	South China	Fouling on cultured mollusks	Xu et al. (2004a)
Bugula stolonifera	Unclear	Unknown/Hull fouling	South China	Fouling on cultured mollusks	Xu et al. (2004a)
Celleporella hyalina	Unclear	Unknown/Hull fouling	South China	Fouling on cultured mollusks	Xu et al. (2004a)
Crisia eburneo-denticulata	Unclear	Unknown/Hull fouling	South China	Fouling on underwater facilities	Xu et al. (2004a)
Tubulipora flabellaris	Unclear	Unknown/Hull fouling	Nationwide	Fouling on cultured mollusks	Xu et al. (2004a)
Schizoporella unicornis	Unclear	Unknown/Hull fouling	Nationwide	Fouling on cultured mollusks	Xu et al. (2004a)
Lichenopora radiata	Unclear	Unknown/Hull fouling	South China	Fouling on cultured mollusks	Xu et al. (2004a)

Membranipora savarii	Unclear	Unknown/Hull fouling	South China	Fouling on cultured mollusks	Xu et al. (2004a)
Tricellaria occidentalis	Unclear	Unknown/Hull fouling	South China	Fouling on cultured mollusks	Xu et al. (2004a)
Watersipora subtorquata	Unclear	Unknown/Hull fouling	South China	Fouling on cultured mollusks	Xu et al. (2004a)
Microporella orientalis	Unclear	Unknown/Hull fouling	South China	Fouling on cultured mollusks	Xu et al. (2004a)
Zoobotryon verticellatum	Unclear	Unknown/Hull fouling	South China	Fouling on underwater facilities	Xu et al. (2004a)
Strongylocentrotus intermedius	Japan	1989/Aquaculture	Liaoning & Shandong Provs.	Destroy seagrass beds and aquaculture facilities, compete with native species for space and food	Chang et al. (2000)
Balanus amphitrite	Pacific coast of North America	1978/Hull fouling	Nationwide	Fouling on ships and harbor infrastructures	Xu et al. (2004a)
Balanus eburneus	North America	Unknown/Hull fouling	Nationwide	Fouling on ships and harbor infrastructures	Xu et al. (2004a)
Balanus improvisus	Atlantic & Arctic Oceans, Baltic, North, Mediterranean & Black Seas	Unknown/Hull fouling	Nationwide	Fouling on ships and harbor infrastructures	Xu et al. (2004a)
Sphaeroma walkeri	North Indian Ocean	Unknown/Hull fouling	Unclear	Affect the growth of mangroves	Xu et al. (2004)
Haliotis discus discus	Japan	1996/Aquaculture	Shandong & Liaoning Provs.	Genetic introgression into local gene pools	Zhang et al. (2004)

(continued)

Table 4.3 (continued)

Species	Native range	Year of introduction/Vector	Major distribution in China	Negative effects in local environment	Major reference
Mytilus galloprovincialis	Mediterranean, Black, Adriatic Seas	Unknown/Unknown	Nationwide	Outcompete native mollusks	Li et al. (2007)
Crassostrea gigas	Japan, Australia	1979/Aquaculture	Nationwide	Fouling species with high abundance on vessels and aquaculture facilities	Guo (2009)
Mytilopsis sallei	Caribbean, Central America	Unknown/Hull fouling	South China	Fouling on almost everything, outcompete native species	Huang et al. (2006)
Styela canopus	Indo-West Pacific	Unknown/Hull fouling	Nationwide	Fouling on vessels and aquaculture species/ equipment	Aquatic NIS database (2015)
Molgula manhattensis	North-east America	Unknown/Hull fouling	Nationwide	Fouling on vessels and aquaculture species/ equipment	Aquatic NIS database (2015)
Vertebrate-3					
Anguilla anguilla	Europe	1991/Aquaculture	Southeast China	Genetic introgression into local gene pools	Li et al. (2007)
Cynoscion nebulosus	North America	1997/Aquaculture	Shandong, Zhejiang, Fujian, Beijing & Liaoning Provs.	Predation native species	Li et al. (2007)
Morone saxatilis	USA & Canada	1997/Aquaculture	Yellow & East China Seas	Predation on and competition with native species	Li et al. (2007)

The native ranges, year of introduction, current distribution ranges, major negative effects on local environment, and major references for each species are shown. *NA* not available

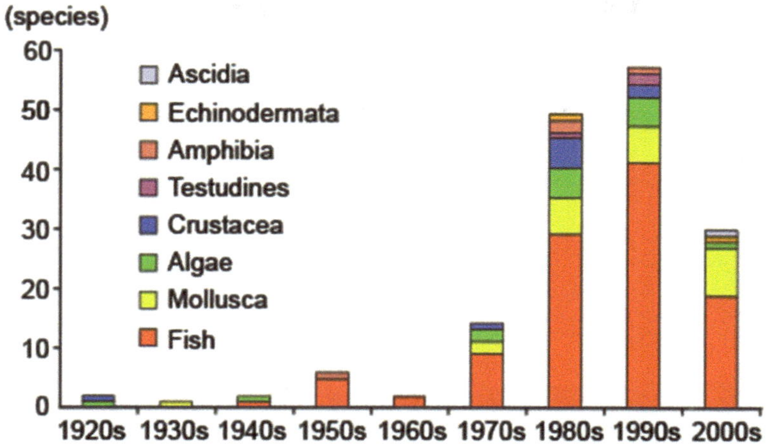

Fig. 4.4 Number of non-native species introduced into China due to aquaculture or related activities (Figure was adapted and modified from Lin et al. 2015)

small (13 species; Fig. 4.4). More than 150 species were introduced in the past 40 years, with an average 3.8 species per year (Fig. 4.4). Among the 179 species, fish (111 species) is the most abundant, followed by molluscs (27 species), algae (16 species), and crustaceans (14 species). The introduced fish and molluscs cover a wide range of taxonomic groups – ten orders for fish and eight infraorders/families/superfamilies for molluscs (Fig. 4.5). So far, 17 species (9.5%) have caused negative effects on local environments (Tables 4.2 and 4.3; see next section for more detail on negative effects).

The international introductions of non-native species have gained significant attention; however, the domestic transfers of species to local water bodies where transferred species do not naturally occur have been highly overlooked. Indeed, there exists a large scale of domestic transfers in geographically large countries such as China, where valued aquaculture species are usually restricted to limited geographical regions (Lin et al. 2015). Due to the lack of policies and easy transportation, the number of individuals per introduction, geographical scales involved, and the frequency of transfers are much larger than those of international introductions (See Lin et al. 2015 and references therein). All provinces in China have records of introductions of non-native species for aquaculture.

Based on all available records, 73 species have been domestically transferred for aquaculture within China (Fig. 4.5). Among these 73 species, fish is the most abundant and also diverse group, accounting for 83.6% of all transferred species and covering 12 orders (Fig. 4.5). Despite that 12 species have clear records of negative effects/history of biological invasions in native and/or introduced ranges, they have been frequently introduced into new water bodies (see Table 4.2 in Lin et al. 2015 for more detail). Similarly to non-native species introduced internationally, the domestically transferred species have caused significant ecological and economic damages to local water bodies (see Lin et al. 2015 and references therein).

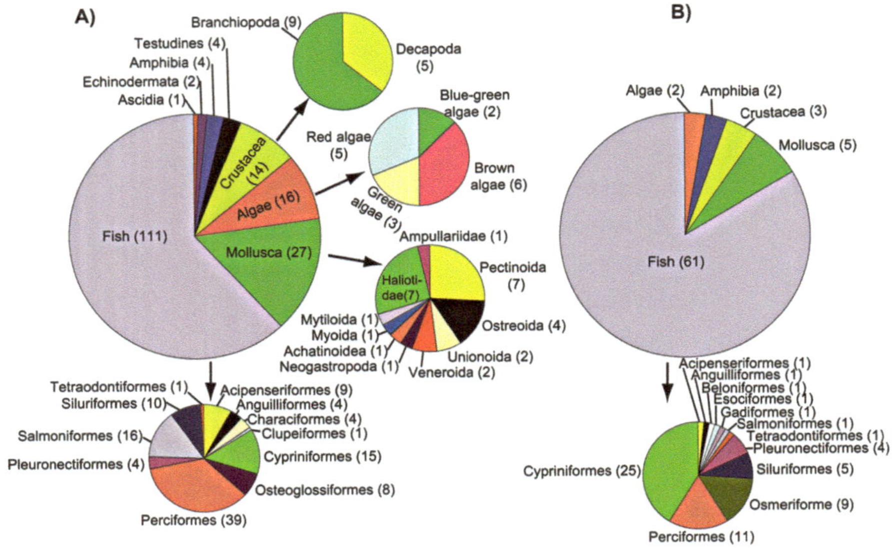

Fig. 4.5 Taxonomic composition of non-native species internationally introduced into China (**a**) and domestically transferred within China (**b**) due to aquaculture or related activities. The number of species in each taxon is shown in brackets after taxon name (Figure was adapted and modified from Lin et al. 2015)

4.3.3 *Aquarium and Ornamental Trades*

Intentional release from aquarium and ornamental trades is an important vector for the spread of aquatic non-native species (McDowall 2004). The aquarium and ornamental trades has a long history of transporting and introducing aquatic non-native species into regions where they are not native. Thus far, species from the aquarium and ornamental trades is responsible for one-third of the world's aquatic non-native species (Padilla and Williams 2004). Studies showed that the most popular species sold in the aquarium trade had a higher probability to establish in local ecosystems (e.g. Duggan 2010). Despite that the aquarium and ornamental trades have been considered as one of the most important vectors for introductions of non-native invasive species, such a vector has received relatively little attention from both scientists and policy makers (e.g. Padilla and Williams 2004; Thomas et al. 2009; Strecker et al. 2011).

In China, aquarium and ornamental trades has become the biggest vectors responsible for introductions of freshwater non-native species (Fig. 4.3, Table 4.4). The traded pets cover a large number of specie and diverse taxonomic groups (Table 4.4). Among 306 species introduced through aquarium and ornamental trades, turtle is the biggest group (115 species, 37.6%), followed by fish (112 species, 36.6%), frogs (47 species, 15.4%) and aquarium plants (33 species, 10.8%). Compared to several countries such as U.S.A. where statistical data is available for the number of hobbyists, market size and revenue derived from aquarium and ornamental trades (e.g.

Table 4.4 Aquatic non-native species introduced into China *via* aquarium and ornamental trades.

Habitat	Phylum/division	Class	Family	Species
Freshwater	Chordata	Osteichthyes	13	72
		Chondrichthyes	1	2
		Reptilia	10	115
		Amphibia	17	47
		Actinopterygii	10	32
		Mammalia	1	1
	Mollusca	Gastropoda	1	2
	Angiospermae	Dicotyledons	5	10
	Gymnospermae	Monocotyledons	15	23
		Coniferopsida	1	2
Subtotal		10	74	306
Marine	Chordata	Osteichthyes	1	1
		Actinopterygii	1	1
Subtotal		2	2	2
Total		12	76	308

Padilla and Williams 2004), such statistical data is largely unavailable in China. However, the huge markets can be reflected by a survey on traded turtles conducted in 2000–2003 (Cheung and Dugeon 2006). After a 35-month survey of turtles traded in only two cities (Guangzhou and Shenzhen) and one special administrative region (Hong Kong), a total of 950,251 individuals of 157 turtle species were recorded, covering around 60% of the global chelonian fauna (Cheung and Dugeon 2006). Considering the large number of cities in China (> 600 cities), the number of species and individuals traded in aquarium and ornamental markets must be extremely large. Owing to the unavailability of data and "black markets (see below)", the number of non-native species that we report here may be only a small part of the traded ones.

Among 306 species introduced through aquarium and ornamental trades, the introduction dates of 284 species (92.8%) were unknown, mainly owing to illegal introductions and trades. Despite that China has relatively strict laws and regulations for introductions of live species and their inspection and quarantine, almost all of these traded species were not registered in and approved to be legally traded by authorities/agencies such as General Administration of Quality Supervision, Inspection and Quarantine of the People's Republic of China (AQSIQ) and Ministry of Agriculture of the People's Republic of China (MOA). Although great efforts have been made to stop illegal introductions and trades, such "money-driven activities" are increasing. For example, based on the statistics from AQSIQ, the number of illegally imported species intercepted by AQSIQ was 3972 in 2011, 4331 in 2012 and 4838 in 2013 (data from AQSIQ official website at http://www.aqsiq.gov.cn/).

The non-native species introduced *via* aquarium and ornamental trades may become invasive after they escape or are released into local water bodies. They may escape from tanks and breeding facilities, the drainage of water containing organisms from tanks, and public aquariums (Padilla and Williams 2004). Also, they are released into natural water bodies as unwanted pets for several reasons, such as too large to be fed, aggressiveness, and high reproductive rates (Padilla and Williams 2004; Gertzen et al. 2008). Moreover, the ritualistic release of species during religious practices is also one of the most important ways to have non-native species colonize local water bodies (Padilla and Williams 2004; Gertzen et al. 2008). One of the well-known examples is the wide spread of red-eared turtle *Trachemys scripta* in China. This turtle species is widely used as the first choice for ritualistic release of Buddhist activities. Based on an incomplete survey, there exist a total of 68 aquaculture farms in 17 provinces, 140 huge trading markets in 34 provinces, and 51 release ponds in 18 provinces in China (Liu et al. 2011). The number of red-eared turtles traded in China is more than 30 million (Liu et al. 2011 and references therein). This turtle species has successfully colonized a large number of freshwater bodies nationwide, and caused extinct of many native aquatic species (Liu et al. 2011 and references therein).

4.3.4 Other Vectors

In addition to the three major vectors mentioned above, non-native species were introduced into China due to several more purposes. One representative example is the introduction of the smooth cordgrass *Spartina alterniflora* for wetland/coastal conservation and restoration. *S. alterniflora*, native to the Atlantic and Gulf coasts of North America, was introduced into China in 1979 (An et al. 2007). Some biological characteristics, such as fast growth, high fecundity, well-developed belowground structures and high salt tolerance make *S. alterniflora* a suitable species for ecological restoration (Hinkle and Mitsch 2005). However, following its initial introduction, *S. alterniflora* spreads extremely quickly. Now, it has successfully established along coasts from Tianjin along Bohai Sea to Guangxi along South China Sea, and covers more than 112,000 ha by 2000 (An et al. 2007). The high invasiveness of *S. alterniflora* highly threats many related species/ecosystems such as mangrove forests. For example, fast growth and wide spread of *S. alterniflora* have thoroughly destroyed all mangrove forests in Ningde, Fujian Province (Du et al. 2006), and mangrove tidal flats in Qi'ao Mangrove Nature Reserve (Zhuhai, Guangdong Province) were covered entirely by *S. alterniflora* (Chen et al. 2009).

4.4 Influence of Aquatic Invasive Species

Detrimental effects of introduced aquatic species fall into two broad categories: ecological and socio-economical influence. However, these two categories are not independent. Ecological changes caused by introduced invasive species can cause socio-economic damages, and socio-economic damages are usually companied by ecological changes. Socio-economic damages can be extremely large, however, data for such damages is largely unavailable, scare and/or extremely difficult to collect in China. We therefore focus on ecological influence of introduced invasive species in this section.

Under the circumstance that increasing human activities have extensively altered aquatic ecosystems, disentangling the respective roles of biological invasions and other anthropogenic stressors in causing major ecological impacts is a key challenge to invasion ecology (see Ricciardi and MacIsaac 2011). In addition, established non-native species may become invasive after a lag phase of many years in new environments where populations remained small and localized (O'Dowd et al. 2003; Jeschke and Strayer 2005). Here we discuss well-known significant negative ecological effects caused by aquatic invasive species so far in China (Tables 4.2 and 4.3).

Generally, the well-known negative ecological effects caused by aquatic invasive species in China mainly include: (1) habitat damage; (2) community change (3) pests of agriculture and aquaculture; (4) biodiversity loss; (5) genetic pollution/extinction.

4.4.1 *Habitat Damage*

The well-known habitat damages caused by invasive species are those associated with red tide blooms. China has more than 150 red tide-causing species, covering a wide range of taxonomic groups such as dinoflagellates, diatoms, cyanobacteria, chrysophyta (Zhou et al. 2001). Many of these red tide-causing species remain cryptogenic, with only 13 species having clear records as non-native species (Table 4.3). Due to frequent outbreak, extremely long lasting time, large geographical area and significant ecological effects, China has become one major country seriously affected by red tide blooms. Since 1998, super-large red tide blooms (covering area > 1000 km^2) occur along Chinese coasts every year. For example, the invasive sea sparkle *Noctiluca scintillans* causes red tides along coasts of four major seas (i.e. Bohai, Yellow, East and South China Seas) almost every year (see Annual Bulletin of Marine Environmental Status of China at the State Oceanic Administration, People's Republic of China's official website at http://www.soa.gov.cn). One of the well-known red tides caused by this species occurred along the coast of Liaoning Province from the Yalu River estuary to the coast near Changhai County in 2011. This red tide covered more than 4000 km^2, and caused significant damage to marine ecosystems (see Annual Bulletin of Marine Environmental Status of China 2011 at http://www.soa.gov.cn), such as rapid changes of pH, luminous permeability and dissolved oxygen. In addition, after red tide blooms, algae cell decomposition consumed a large amount of water dissolved oxygen. All these changes have caused significant consequences on biodiversity and food webs (see Zhang 2007 and references therein).

4.4.2 *Community Change*

Some invasive species affect communities owing to their strong competitive abilities (Tables 4.2 and 4.3). One well-known example is significant community change caused by an introduced invasive small bivalve *Mytilopsis sallei* in China. *M. sallei* is considered a Central American origin, and was introduced into mainland China most likely *via* hull fouling at the beginning of the 1990s. Soon after its introduction, *M. sallei* overgrows and out-competes native species, and becomes dominant members of benthic communities (Wang et al. 1999; Cai et al. 2014). Field surveys showed that invasions of *M. sallei* have caused significant changes of benthic communities and associated biomass of benthic species (Cai et al. 2014). Interestingly, phytoplankton communities have been significant changed due to filtering feeding of huge populations (Lin and Yang 2006).

4.4.3 Pests of Agriculture or Aquaculture

Several introduced species, such as *Pomacea canaliculata* and *Strongylocentrotus intermedius*, has become pests for agriculture/aquaculture (see Lin et al. 2015 for a list of invasive species for aquaculture). For pest species for agriculture, one notorious example in China is the apple snail *Pomacea canaliculata*, which is a native of South America. Since its first introduction into Taiwan in 1977 and then into Guangdong Province for aquaculture in 1981, *P. canaliculata* has rapidly spread in South China and become a huge threat to rice farms (Hu et al. 2014). *P. canaliculata* has an extremely high fecundity, rapid growth rate and sexual maturity: an adult female can produce 2400–8700 eggs with successful hatchery rate as high as >90%, and after 4 months, new generation can reach sexual maturity and reproduce (Xian et al. 2008 and references therein). Consequently, *P. canaliculata* can form large populations in a short period of time, and subsequently can thoroughly destroy rice farms very quick by feeding on rice seedlings and leaves. For example, in Guangdong Province in 1991, more than 667,000 ha rice farms were thoroughly destroyed by *P. canaliculata* (Xian et al. 2008and references therein). Although great efforts have been put to control this pest species in China, it still threatens rice farms, especially in South China. In addition, *P. canaliculata* impacts public health by acting as a vector of disease-causing species such as *Angiostrongylus cantonensis* (Lv et al. 2009).

4.4.4 Biodiversity Loss

Despite that it remains difficult to disentangle respective roles and effects of each stressor, such as human activities, pollution and introductions of non-native species, on biodiversity loss in aquatic ecosystems (see Ricciardi and MacIsaac 2011 and references therein), biological invasions appear to be a principal contributor to biodiversity loss in some aquatic ecosystems in China. One well-known event is the negative effects caused by introductions of fish from the Yangtze River drainage into major lakes in Yunnan Province, a hotspot of biodiversity for freshwater fish in China (Chen et al. 1998). After introductions of the four major Chinese carps (black carp *Mylopharyngodon piceus*, grass carp *Ctenopharyngodon idellus*, silver carp *Hypophthalmichthys molitrix*, bighead carp *Aristichthys nobilis*), the number of native fish in Lake Dianchi, the biggest lake in Yunan Province, decreased extremely fast: from 23 to 15 in 1978, to eight in 1982, to two in 1997, and only one in 2007–2008, and all endemic species become extinct now (Chen et al. 1998; Yuan et al. 2010). Although many factors, such as overfishing, reclamation of land from lakes, pollution, and many others may contribute to the loss of endemic biodiversity, there exists a fact that both the number and production of non-native species increase (Chen et al. 1998; Yuan et al. 2010), suggesting that introductions of non-native species is a principal contributor to biodiversity loss.

4.4.5 Genetic Pollution/Extinction

Hybridization between native and non-native species has many biological and evolutionary consequences, one of which is rapid adaption to local environments (e.g. Kovach et al. 2015). Hybridization between native and non-native species is considered as one of the most important mechanisms for non-native species to survive Allee or stochastic effects (i.e. non-native species can be eliminated in the absence of hybridization mainly owing to low population density). Through hybridization, genetic pollution such as loss of genetic variation, breakdown of population structure, and even species extinction may occur as rapid introgression of non-native genetic materials into native gene pools (e.g. Huxel 1999; Bartley et al. 2001; Hails and Morley 2005; Yang et al. 2011). Using mathematical simulations, Huxel (1999) found that displacement of native gene pools could occur very rapidly: less than five generations. The disappearance of native gene pools leads to another extinction, i.e. genetic extinction. Genetic extinction, which is different from 'traditional extinction' (i.e. permanent disappearance of a species), is largely 'invisible' and cannot be verified without detailed genetic analyses (Lin et al. 2015). Indeed, genetic extinction occurred in China through both intra-specific and inter-specific hybridizations between native and non-native species. For genetic extinction through intra-specific hybridization, the Pacific abalone (*Haliotis discus hannai*) represents an example. Japanese stocks of Pacific abalone were introduced into North China to successfully resolve the high-mortality problem (>90%) caused by diseases in 1994 through intra-specific hybridization between Chinese and Japanese stocks (Zhang et al. 2004). However, after 20 years hybridization practice and aquaculture in China, almost all individuals caught from the wild are hybrids now (Wang 2011). For genetic pollution/extinction through inter-specific hybridization, carps in lakes in Yunan Province represent good cases. For example, almost all carp individuals (genus *Cyprinus*) in Lake Xingyun were derived from inter-specific hybrids (Yang et al. 2011).

4.5 Management of Biological Invasions in Aquatic Ecosystems

As non-native invasive species remain major environmental and ecological problems in China and such problems have become more severe with increased human activities as mentioned above, management of biological invasions represents one of the most important actions towards mitigating the problems. However, invasion biology generally suffers from an obvious "knowing-doing" gap (Esler et al. 2010), that is, scientific knowledge and recent research progress are often poorly translated into effective management strategies (Darling 2015). Such a gap is more obvious in aquatic ecosystems in China, as sufficient management solutions are significantly lacking when compared to invasive species in terrestrial ecosystems, especially those pests to agriculture and forest ecosystems (Wan et al. 2012). Although significantly negative effects have been well-known and/or reported in aquatic ecosystems, extremely slow responses, poor coordination between government and academia, as well as increased illegal activities (such as aquaculture, trade, etc.) make invasions to aquatic ecosystems more problematic and severe.

Management strategies can be categorized as two types: species-based and vector-based; however, vector-based management can be more effective than species-based management, especially aiming at management of a large numbers of species (e.g. Williams et al. 2013). Here we discuss the management solutions on aquatic invasive species in China based on vectors, although some management solutions and technologies are common among multiple vectors. Owing to the poor and/or unavailable management towards many vectors, we both summarize research progress and, more importantly, propose possible solutions/managements for each vector. Finally, we discuss some common research gaps among different vectors.

4.5.1 Management of Shipping-Based Vectors: Extremely Little Effort Versus Huge Problems

Despite the fact that eight out of the ten world largest ports are located in China and huge volumes of ballast water are discharged into Chinese coastal regions (Table 4.1), both scientific studies and sufficient management are largely lacking. A search based on multiple combinations of key words associated with shipping-based biological invasions in China resulted in only one article at the Web of Science (search done on March 31, 2015). In this paper, Zhang and her colleagues aimed to establish a testing system for ballast water treatment suitable for China based on three Chinese species (Zhang et al. 2014a, b). Articles and reports from Chinese libraries only focused on species identification/survey in very limited geographical regions, ports and/or vessels. In contrast, such little research effort has recovered a large number of non-native invasive species. For example, more than 240 phytoplankton species were detected in surveys on ports in Fujian, Rizhao and Shanhai, 62 and 21 of which were categorized as red tide-causing species and/or non-native species, respectively (Li et al. 2006; Yang et al. 2009; Zhou et al. 2012). These introduced species have caused significant damages to local water bodies: a total of 792 red tide blooming events were reported between 2000–2009, with an average of 79.2 events/year (Ge et al. 2013).

For the effectiveness of management, the laws and regulations in China clearly ban the illegal discharge of ballast water into any regions of Chinese coasts. However, these laws and regulations, mainly including Marine Environment Protection Law and Regulations on Administration of Prevention and Control of Pollution to the Marine Environment by Vessels, were made based on the prevention of chemical (especially petroleum) pollution and infectious diseases to human. For management of biological organisms, they provide directions for pathogenic microorganisms; however, they largely neglect other species such as algae and zooplankton species, although some species have been reported as invasive species causing problems in both Chinese and international coastal zones.

Compared to several countries such as U.S.A., Australia, Canada and New Zealand where shipping-based vectors are well managed, China should take immediate actions to improve the following issues: (1) conduct comprehensive studies on both spread and/or post-establishment dynamics, develop management technologies and solutions, and make science-based laws and regulations; (2) clarify the administrative and management roles of each agency; (3) enhance the mandatory report and inspection/quarantine for any incoming vessels into Chinese water bodies; (4) clearly make ballast water management (such as ballast water exchange) and hull husbandry mandatory for any incoming vessels into Chinese water bodies; (5) punish more strictly on any violation of laws and regulations, and clarify the responsibility for direct and indirect costs for any future problems.

4.5.2 Management of Aquaculture-Based Vectors: Clear Policies on Overlooked High-Risk Activities

The management of aquaculture-based vectors largely depends on the availability of policies and the implement of such policies. So far, both national and regional laws/regulations have clearly directed how to claim, evaluate and then introduce non-native species from international origins for aquaculture. Several representative national laws/regulations include Law of People's Republic of China on the Entry and Exit Animal and Plant Quarantine, Marine Environment Protection Law of the People's Republic of China, Fisheries Law of the People's Republic of China and Administrative Management Regulations on Stocks and Seedlings for Aquaculture. However, there are many crucial points need to be added and/or fixed during introduction practices, such as the checklist is not updated on time, risk assessment for species to be introduced and post-introduction surveillance programs are not mandatory, the trial culture period for invasion risk assessment is not defined properly, and responsibilities are not clearly defined if negative effects occur after introductions.

Several aquaculture-associated practices with high risks have been highly overlooked. These overlooked high-risk activities include large-scale domestic transfers of non-native species for aquaculture, artificial inter- and intra-specific hybridizations, and mass release of non-native species for ranching. In China, these activities are common and caused significant ecological and economic problems (see Lin et al. 2015). However, no policy is available to regulate and manage these high-risk activities. Consequently, the policy must be clearly made based on the principle that all activities are considered potentially harmful and therefore forbidden for practice unless risk is acceptably low by risk assessment. The basic principles and associated solutions have been recommended by Lin et al. (2015).

4.5.3 Management of Aquarium and Ornamental Trades: Effort Needed for Prevention of Illegal Introductions, Restrict Management and Public Education

Despite that the Law of People's Republic of China on the Entry and Exit Animal and Plant Quarantine clearly bans any illegal introductions of non-native species, the number of illegally introduced species sharply increased: from 3972 in 2011 to 4838 in 2013. Consequently, prevention of illegal introductions represents the first priority and also the most effective means of reducing the future costs of invasive species introduced through aquarium and ornamental trades in China. Although there remain extreme difficulties in the enforcement of regulations/laws in a large number of international ports in such a big country, extra effort should be made to improve current laws/regulations and their enforceability. In addition, extensive public education should be performed to clearly show the environmental and legal consequences of introductions of invasive species.

After invasive species are introduced into China and sold in pet markets, management becomes almost impossible. More seriously, many highly invasive species can be purchased over the internet. The management of internet trade is more difficult than that for "traditional shops". Indeed, it remains difficult to manage traded invasive species even in relatively smaller geographical scale. For example, the experience in California showed that prevention through regulation of the aquarium hobby and trade was difficult (Padilla and Williams 2004). Consequently, restrict guidelines should be implemented to regulate industry, buyers, sellers, and the public. For the enforcement of restrict regulations, for example, all breeding stocks must be certified as "invasive-free". In addition, the outflows from breeding facilities and large-scale aquariums must be strictly sterilized to prevent the accidental release of species. To prevent release from retailers and hobbyists, mono-sex populations (i.e. male or female only) should be allowed to be sold in markets. In addition, public education has approved as an effective measure. For example, educating both retailers and hobbyists about invasive species and steps they can take to reduce the risk had an immediate impact (Padilla and Williams 2004).

4.5.4 Common Research Gaps Among Vectors: Risk Assessment and Early Detection

Current risk assessment on invasive species in China is largely focused on terrestrial organisms. Owing to apparent differences in biological characteristics and environmental conditions between terrestrial and aquatic habitats, the risk assessment systems established for terrestrial organisms should modify before applying to aquatic habitats. Recently, several studies have tried to establish common risk assessment systems on aquatic invasive species in China (e.g. Ma et al. 2009; Wang et al. 2014). Several key points, such as universality across taxa and the elimination of biased results derived from experts' decisions, should be highly improved in these established systems.

Indeed, for deliberate introductions, the International Council for the Exploration of the Sea (ICES) has published a Code for of Practice on the Introductions and Transfers of Marine Organisms. This Code recommends effective procedures and practices to diminish potential risks caused by introductions of non-native species. The most up-to-date version of this Code (published in 2005) provides all concerns expressed in former versions and follows the precautionary approach adopted from FAO principles with the goal of reducing the spread of invasive species (ICES 2005). The ICES Code has become a well-recognized instrument, and its essential components have been widely applied to the evaluation of species introductions. Apparently, several major components should be well adopted and implemented into risk assessment programs based on the Code (Lin et al. 2015).

Recent studies have illustrated that prevention of invasive species can be more cost-effective than post-establishment management to control the establishment and spread of newly introduced non-native (Lodge et al. 2006; Finnoff et al. 2007). Consequently, early detection has been recommended as one of the first priorities for management plans for invasive species (e.g. Lodge et al. 2006; Vander Zanden et al. 2010). So far, there is no systematic and comprehensive early detection system for aquatic invasive species in China.

The biggest technical challenge for early detection is the identification of potential invasive species present at extremely low abundance in communities (Darling and Mahon 2011; Zhan et al. 2013). Such detection challenge becomes more obvious in some particular ecosystems such as aquatic habitats where organisms may be small, geographically constrained, and hidden beneath the water surface (Zhan et al. 2013; Xiong et al. 2016). Traditional methods, which rely on capture using different sampling instruments and then identification *via* assessment of morphological and/or anatomical characteristics, have been empirically proven as low-resolution means for detection of newly introduced non-native species in aquatic environments (e.g. $< 2\%$ using bulk samples, Harvey et al. 2009). Consequently, cost-effective, accurate, high-throughput and common methods across an array of taxa are desired to overcome the detection challenge of traditional methods (Darling and Mahon 2011; Zhan et al. 2013).

Recent technological advances have driven rapid development of many DNA-based methods designed to facilitate early detection of invasive species in aquatic environments (see review by Darling and Mahon 2011). The use of DNA-based methods have highly improved the detection efficiency and accuracy for rare species in aquatic communities (e.g. Zhan et al. 2013). Particularly, rapid advance in high-throughput sequencing technology has radically changed our ability to identify biodiversity in communities (e.g. Zhan et al. 2013, 2014a). Many advantages, such as extremely deep sequencing and low cost per sequence, have intensified researchers' interest to use high-throughput sequencing-based methods for early detection of invasive species (Zhan et al. 2013). Consequently, comprehensive early detection programs can be established based on genetic methods such as high-throughput sequencing to powerfully screen for all invasive species simultaneously, as long as technical issues are well addressed (e.g. Zhan et al. 2014a, b, Zhan and MacIsaac 2015).

4.6 Conclusions and Future Perspectives

China has become one of the most influenced countries by aquatic invasive species. As the disturbance to aquatic communities derived from frequent human activities is increasing, it is therefore expected that the introduction of non-native species becomes common and the influence of invasive species is expected to become more severe. Despite that the study of aquatic invasions has become a dynamic area in invasion biology in China, many fundamental questions remain to be answered and effective technologies and management strategies are largely required. Although huge efforts should be put into numerous research areas to address the problems caused by aquatic invasive species, several issues should be solved as priorities based on current situation in China.

Practically, the lack of risk assessment and sound management for non-native species is particularly obvious in China. Despite that significantly negative effects have been well recognized and widely reported in both literature and public media, lessons from known disasters are slowly learned and problems caused by aquatic invasive species are largely overlooked. Consequently, there is an extremely urgent need to construct, develop and implement risk assessment and sound management strategies for aquatic invasive species. In addition, policies should be implemented and put in place to make all essential points and issues mentioned above well functional.

Theoretically, the understanding of causes and consequences of biological invasions is extremely poor in China. For many highly invasive species, causative factors for successful invasions remain poorly understood. Future studies should focus on the contributions of each factor, including introduction vectors, biological/genetic characteristics, and environmental/ecological changes, to invasion success. Blackburn and his colleagues proposed a model for biological invasions (Blackburn et al. 2011). This model provides a good platform to unify a series of comprehensive studies to understand how these factors, either alone or by interactions/networks, underlie the invasion success.

Acknowledgments This work was partially supported by National Natural Science Foundation of China (31272665) and 100-Talent Program of the Chinese Academy of Sciences to AZ.

References

An SQ, Gu BH, Zhou CF, Wang ZS, Deng ZF, Zhi YB, Li HL, Chen L, Yu DH, Liu YH (2007) *Spartina* invasion in China: implications for invasive species management and future research. Weed Res 47:183–191

Aquatic non-native invasive species database (2015) http://bioinvasion.fio.org.cn/home/showonemodel.asp?cs_id=481

Attwood SW, Huo GN, Qiu JW (2015) Update on the distribution and phylogenetics of Biomphalaria (Gastropoda: Planorbidae) populations in Guangdong Province, China. Acta Trop 141:258–270

Bartley DM, Rana K, Immink AJ (2001) The use of inter-specific hybrids in aquaculture and fisheries. Rev Fish Biol Fish 10:325–337

Blackburn TM, Pyšek P, Bacher S, Carlton JT, Duncan RP, Jarosik V, Wison JRU, Richardson DM (2011) A proposed unified framework for biological invasions. Trends Ecol Evol 26:333–339

Cai L, Hwang J, Dahms H, Fu S, Zhuo Y, Guo T (2014) Effect of the invasive bivalve *Mytilopsis sallei* on the macrofaunal fouling communities and the environment of Yuandang Lagoon, Xiamen, China. Hydrobiologia 741:101–111

Carlton JT (2009) Deep invasion ecology and the assembly of communities in historical time. In: Rilov G, Crooks JA (eds) Biological invasions in marine ecosystems: ecological, management, and geographic perspectives. Springer-Verlag, Berlin, pp 13–56

Carlton JT, Ruiz GM (2005) Vector science and integrated vector management in bioinvasion ecology: conceptual frameworks. In: Mooney HA, Mack RN, McNeely JA, Neville LE, Johan Schei P, Waage JK (eds) *Invasive alien species: a new synthesis*. Island Press, Covelo, pp 36–58

Chang Y, Wang Z, Song J, Su Y, Wang J (2000) Cross breeding between four species of sea urchin, feasibility and early development of F1 offspring. J Fish China 24:211–216. (in Chinese with English abstract)

Chen Y, Yang J, Li Z (1998) The diversity and present status of fishes in Yunnan Province. Chin Biodivers 6:272–277. (in Chinese with English abstract)

Chen L, Wang W, Zhang Y, Lin G (2009) Recent progresses in mangrove conservation, restoration and research in China. J Plant Ecol 2:45–54

Cheung S, Dudgeon D (2006) Quantifying the Asian turtle crisis: market surveys in southern China, 2000–2003. Aquat Conserv Mar Freshwat Ecosyst 16:751–770

Darling JA (2015) Genetic studies of aquatic biological invasions: closing the gap between research and management. Biol Invasions 17:951–971

Darling JA, Mahon AR (2011) From molecules to management: adopting DNA-based methods for monitoring biological invasions in aquatic environments. Environ Res 111:978–988

Du W, Ma L, Liu J, Zhang Q, Zheng X (2006) Controlling technique of *Spartina alterniflora* Loisel in mangrove-spartina ecotones. Chin J Eco-Agric 14:154–156. (in Chinese with English abstract)

Duggan IC (2010) The freshwater aquarium trade as a vector for incidental invertebrate fauna. Biol Invasions 12:3757–3770

Esler KJ, Prozesky H, Sharma GP, McGeoch M (2010) How wide is the "knowing-doing" gap in invasion biology? Biol Invasions 12:4065–4075

Feng J, Guo J, Huang Q, Jiang J, Huang G, Yang Z, Lin G (2014) Changes in the community structure and diet of benthic macrofauna in invasive *Spartina alterniflora* wetlands following restoration with native mangroves. Wetlands 34:673–683

Finnoff D, Shogren JF, Leung B, Lodge DM (2007) Take a risk: preferring prevention over control of biological invaders. Ecol Econ 62:216–222

Ge B, Huang Y, Ran J, Zhang M, Lou S (2013) The spread of non-native species introduced via ballast water and possible regulatory strategies in China. Plant Quarantine 27:86–89

Geller JB, Darling JA, Carlton JT (2010) Genetic perspectives on marine biological invasions. Annu Rev Mar Sci 2:367–393

Gertzen E, Familiar O, Leung B (2008) Quantifying invasion pathways: fish introductions from the aquarium trade. Can J Fish Aquat Sci 65:1265–1273

Guo X (2009) Use and exchange of genetic resources in molluscan aquaculture. Rev Aquac 1:251–259

Guo Y, Wang C, Luo J, He H (2009) *Physa acuta* found in Beijing, China. Chin J Zool 44:127–128

Hails RS, Morley K (2005) Genes invading new populations: a risk assessment perspective. Trends Ecol Evol 20:245–252

Harvey CT, Qureshi SA, MacIsaac HJ (2009) Detection of a colonizing, aquatic, non-indigenous species. Divers Distrib 15:429–437

Hinkle RL, Mitsch WJ (2005) Salt marsh vegetation recovery at salt hay farm wetland restoration sites on Delaware Bay. Ecol Eng 25:240–251

Hu Y, Mu X, Luo D, Meng X, Yang Y, Gu D, Luo J, Zhang J (2014) Genetic variability of the invasive snail *Pomacea canaliculata* in South China based on mitochondrial 16S rDNA sequences. Biochem Syst Ecol 57:203–209

Huang Z, Hong R, Zhang L (2006) Study of species in Xiamen Bay. J Xiamen Univ (Nat Sci) 45:10–15

Hulme PE (2009) Trade, transport and trouble: managing invasive species pathways in an era of globalization. J Appl Ecol 46:10–18

Huxel GR (1999) Rapid displacement of native species by invasive species: effects of hybridization. Biol Conserv 89:143–152

ICES (2005) ICES code of practice on the introductions and transfers of marine organisms 2005. ICES, Copenhagen, 30p

Jeschke JM, Strayer DL (2005) Invasion success of vertebrates in Europe and North America. Proc Natl Acad Sci U S A 102:7198–7202

Kovach RP, Muhlfeld CC, Boyer MC, Lowe WH, Allendorf FW, Luikart G (2015) Dispersal and selection mediate hybridization between a native and invasive species. Proc Roy Soc B Biol Sci 282:20142454

Levings C, Kieser D, Jamieson G, Dudas S (2002) Marine and estuarine alien species in the Strait of Georgia, British Columbia. In: Claudi R (ed) *Alien species in Canada*. Natural Resources Canada, Ottawa, pp 111–131

Li W, Sun J, Wang D, Jia J, Lei Z, Wang X (2006) The characteristics of phytoplankton assemblage in Rizhao Harbor, Rizhao anchorage ground and entry ships' ballast water, China. Mar Sci 30:52–57. (in Chinese with English abstract)

Li J, Dong Z, Li Y, Wang C (2007) Alien aquatic plants and animals in China. Shanghai Scientific and Technical Publishers, Shanghai. (in Chinese)

Liang Y, Wang B (2001) Alien marine species and their impacts in China. Biodivers Sci 9:458–465. (in Chinese with English abstract)

Lin G, Yang Q (2006) Impacts of alien species *Mytilopsis sallei* on phytoplankton at Maluan Bay in Xiamen, Fujian, China. J Trop Oceanogr 25:63–67. (In Chinese)

Lin W, Zhou G, Cheng X, Xu R (2007) Fast economic development accelerates biological invasions in China. PLoS One 2:e1208

Lin Y, Gao Z, Zhan A (2015) Introduction and use of non-native species for aquaculture in China: status, risks and management solutions. Rev Aquac 7:28–58

Liu J, Dong M, Miao S, Li Z, Song M, Wang R (2006) Invasive alien plants in China: role of clonality and geographical origin. Biol Invasions 8:1461–1470

Liu D, Shi H, Liu X, Wang J, Gong S, Wang J, Shen L (2011) The distribution of *Trachemys scripta* in China. Bull Biol 46:18–21. (in Chinese)

Lodge DM, Williams S, MacIsaac HJ, Hayes KR, Leung B, Reichard S, Mack RN, Moyle PB, Smith M, Andow DA, Carlton JT, McMichael A (2006) Biological invasions: recommendations for U.S. policy and management. Ecol Appl 16:2035–2054

Lv S, Zhang Y, Liu H, Hu L, Yang K, Steinmann P, Wang L, Chen Z, Utzinger J, Zhou X (2009) Invasive snails and an emerging infectious disease: results from the first national survey on *Angiostrongylus cantonensis* in China. PLoS Negl Trop Dis 3:e368

Ma Y, Xiong H, Lin Y, Tang S (2009) Construction of invasive risk assessment system for alien marine species. J Fish China 33:617–623. (in Chinese with English abstract)

McDowall RM (2004) Shoot first, and then ask questions: a look at aquarium imports and invasiveness in New Zealand. N Z J Mar Fresh 38:503–510

O'Dowd DJ, Green PT, Lake PS (2003) Invasional "meltdown" on an oceanic island. Ecol Lett 6:812–817

Padilla DK, Williams SL (2004) Beyond ballast water: aquarium and ornamental trades as sources of invasive species in aquatic ecosystems. Front Ecol Environ 2:131–138

Qiao Y, Song W, Li W (2010) Discussion on biological invasion effect on aquatic organism: biodiversity and management countermeasure. J Agro Environ Sci S1:321–323. (In Chinese with English Abstract)

Ricciard A, MacIsaac HJ (2011) Impacts of biological invasions on freshwater ecosystems. In: Richardson DM (ed) *Fifty years of invasion ecology: the legacy of Charles Elton*. Wiley-Blackwell, Chichester, pp 211–224

Rhyne AL, Tlusty MF, Schofield PJ, Kaufman L, Morris JA Jr, Bruckner AW (2012) Revealing the appetite of the marine aquarium fish trade: the volume and biodiversity of fish imported into the United States. PLoS One 7:e35808

Ruiz GM, Carlton JT (2003) Invasion vectors: a conceptual framework for management. In: Ruiz GM, Carlton JT (eds) Invasive species: vectors and management strategies. Island Press, Washington, DC, pp 459–504

Shao K, Li X (2000) Seasonal variation of benthic seaweed community in the intertide in Dalian. J Dalian Fish Univ 15:29–34. (in Chinese with English abstract)

Strecker AL, Campbell PM, Olden JD (2011) The aquarium trade as an invasion pathway in the Pacific Northwest. Fisheries 36:74–85

Sylvester F, Kalaci O, Leung B, Lacoursière-Roussel A, Murray CC, Choi FM, Bravo MA, Therriault TW, MacIsaac HJ (2011) Hull fouling as an invasion vector: can simple models explain a complex problem? J Appl Ecol 48:415–423

Thomas VG, Vasarhelyi C, Niimi AJ (2009) Legislation and the capacity for rapid-response management of nonindigenous species of fish in contiguous waters of Canada and the USA. Aquat Conserv Mar Freshwat Ecosyst 19:354–364

U.S. Fish & Wildlife Service (2012) The Cost of Invasive Species.

Vander Zanden MJ, Hansen GJA, Higgins SN, Kornis MS (2010) A pound of prevention, plus a pound of cure: early detection and eradication of invasive species in the Laurentian Great Lakes. J Great Lakes Res 36:199–205

Wan F, Liu Q, Xie M (2012) Biological invasions: color illustrations of invasive alien plants in China. Science Press, Beijing. (in Chinese)

Wang J, Huang Z, Zheng C, Lin N (1999) Population dynamics and structure of alien species Mytilopsis sallei in Fujian, China. J Oceanogr Taiwan Strait 18:372–377. (in Chinese with English abstract)

Wang M (2011) The impacts of hybrids farming on genetic structure of native populations of Pacific abalone and genetic analysis of interspecific introgressive hybridization. Dissertation from the Ocean University of China, Qingdao, China (in Chinese with English abstract).

Wang Y, Shi H, Liu F, Zheng Z, Liao J (2014) An invasive risk assessment model of alien marine species. Chin J Nat 36:133–138. (in Chinese with English abstract)

Williams SL, Davison IC, Pasari JR, Ashton GV, Carlton JT, Crafton RE, Fontana RE, Grosholz ED, Miller AW, Ruiz GM, Zabin CJ (2013) Managing multiple vectors for marine invasions in an increasingly connected world. Bioscience 63:952–966

World Wildlife Fund (2009) Silent Invasion – The spread of marine invasive species via ships' ballast water. World Wildlife Fund, Gland, pp 1–19

Xian Z, Qin R, Leng F, Lv K, Zhou W (2008) Reasons for damages imposed by *Pomacea canaliculata* (Lamarck) in paddy field and its control strategy in China. Guangxi Agric Sci 40:1007–1009

Xiong W, Li H, Zhan A (2016) Early detection of invasive species in marine ecosystems using high-throughput sequencing: technical challenges and possible solutions. Mar Biol 163:139

Xu H, Qiang S (2011) China's invasive alien species. Science Press, Beijing. (in Chinese)

Xu H, Qiang S, Han Z, Guo J, Huang Z, Sun H, He S, Ding H, Wu H, Wan F (2004) The distribution and introduction pathway of alien invasive species in China. Biodivers Sci 12:626–638. (in Chinese with English abstract)

Xu H, Ding H, Li M, Qiang S, Guo J, Han Z, Huang Z, Sun H, He S, Wu H, Wan F (2006) The distribution and economic losses of alien species invasion to China. Biol Invasions 8:1495–1500

Yang Q, Cai L, Gao Y, Su Z, Chen C, Li B, Xu C, Lin G, Zheng H, Xiang P (2009) Study on distributional characteristics of organisms in ballast water of exotic ships entering major ports of Fujian and its potential invasion threat. T Oceanol Limnol 4:25–38. (in Chinese with English abstract)

Yang B, Chen X, Yang J (2011) Non-native carp of the genus Cyprinus in Lake Xingyun, China, as revealed by morphology and mitochondrial DNA analysis. Biol Invasions 13:105–114

Yuan G, Ru H, Liu X (2010) Fish diversity and fishery resources in lakes of Yunnan Plateau during 2007-2008. J Lake Sci 22:837–841. (in Chinese with English abstract)

Zhan A, MacIsaac HJ (2015) Rare biosphere exploration using high-throughput sequencing: research progress and perspective. Conserv Genet 16:513–522

Zhan A, Hulák M, Sylvester F, Huang X, Adebayo AA, Abbott CL, Adamowicz SJ, Heath DD, Cristescu ME, MacIsaac HJ (2013) High sensitivity of 454 pyrosequencing for detection of rare species in aquatic communities. Methods Ecol Evol 4:558–565

Zhan A, Bailey SA, Heath DD, MacIsaac HJ (2014a) Performance comparison of genetic markers for high-throughput sequencing-based biodiversity assessment in complex communities. Mol Ecol Resour 14:1049–1059

Zhan A, He S, Brown EA, Chain FJJ, Therriault TW, Abbott CL, Heath DD, Cristescu ME, MacIsaac HJ (2014b) Reproducibility of pyrosequencing data for biodiversity assessment in complex communities. Methods Ecol Evol 5:881–590

Zhang Y (2007) The effects of large-scale harmful algal bloom in the East China Sea on microzooplantkon community structure. Ph.D. thesis, Qingdao, China

Zhang G, Que H, Liu X, Xu H (2004) Abalone mariculture in China. J Shellfish Res 23:947–950

Zhang Q, Jiang S, Ju R, Pan X (2013) Diversity of invasive species in Shanghai. Biodivers Sci 21:732–737. (in Chinese with English abstract)

Zhang N, Zhang Y, Bai M, Zhang Z, Chen C, Meng X (2014a) Risk assessment of marine environments from ballast water discharges with laboratory-scale hydroxyl radicals treatment in Tianjin Harbor, China. J Environ Manag 145:122–128

Zhang Y, Wang Z, Cai L, Cai X, Sun W, Ma L (2014b) Tests to evaluate the ecological impact of treated ballast water on three Chinese marine species. Chin J Oceanol Limnol 32:1105–1117

Zhou M, Zhu M, Zhang J (2001) Status of harmful algal blooms and related research activities in China. Chin Bull Life Sci 13:53–59. (in Chinese with English abstract)

Zhou S, Wu J, Meng Q, Ma W, Qin Y, Gu X, Huang Y, Cao G, Chen J (2012) The investigation and risk analysis of the red tide species in ballast waters of foreign ships harboring in Shanghai port. Plant Quarantine 26:40–43. (In Chinese with English abstract)

Chapter 5
Biological Invasions in Desert Green-Islands and Grasslands

Amanulla Eminniyaz, Juan Qiu, Carol C. Baskin, Jerry M. Baskin, and Dunyan Tan

Abstract The desert ecosystem is an important component of the terrestrial ecosystems in China, as it contains a large number of rare species and important wild animal and plant genetic resources. Meanwhile, oases and grasslands, two major geographical landscapes in deserts areas, play an important role in the local economy and are sites where many invasive species have become established. To evaluate the current situation of alien invasive species in Chinese desert areas and to provide support for controlling alien invasive species, we have collected and analyzed the relevant references. There are 165 alien invasive species in Chinese desert areas, 88, 57 and 20 of which are plants, animals and microorganisms, respectively. The occurrence of invasive species differs in each province of the Chinese desert area; the number of invasive species is higher in Xinjiang, Gansu and Inner Mongolia than in Qinghai and Ningxia. Most of the alien invasive species in the desert region come from North America and Europe. The amount of time required for establishment of a new invasive species in the desert area is decreasing, but the number of new invasive species is increasing. In addition, alien invasive species in the Chinese desert area have caused serious negative impacts on local economy, agriculture, human health, social stability and minority ethnic group culture. Although researchers working in the Chinese desert area have achieved some remarkable goals in preventing establishment and spread of alien species and have thereby reduced their negative effects on agricultural production, many unsolved problems still exist.

A. Eminniyaz • J. Qiu • D. Tan (✉)
Xinjiang Key Laboratory of Grassland Resources and Ecology and Ministry of Education Key Laboratory for Western Arid Region Grassland Resources and Ecology, College of Grassland and Environment Sciences, Xinjiang Agricultural University, Ürümqi 830052, China
e-mail: tandunyan@163.com

C.C. Baskin
Department of Biology, University of Kentucky, Lexington, KY 40506, USA

Department of Plant and Soil Sciences, University of Kentucky, Lexington, KY 40546, USA

J.M. Baskin
Department of Biology, University of Kentucky, Lexington, KY 40506, USA

F. Wan et al. (eds.), *Biological Invasions and Its Management in China*,
Invading Nature - Springer Series in Invasion Ecology 11,
DOI 10.1007/978-94-024-0948-2_5

Therefore, we suggest establishment of regulatory policies, improved coordination between departments, and increased technical research and application.

Keywords Biological invasions • Chinese desert areas • Detrimental economic effects • Alien species • Invasive species • Prevention and control programs

5.1 Introduction

5.1.1 Geographical Scope and Division of Chinese Desert Areas

The purpose of this review is to gain an understanding of invasive species in the desert areas of China, but first some background information about the desert will be presented to show the context in which rapid invasion of alien species is occurring. Chinese desert areas occur in northwest China between longitudes 74° to 109°E and latitudes 33° to 48°N (Natural Division Working Committee of China Academy of Sciences 1959; The Central Weather Bureau 1978; Chen 1987; Yin 1997). They include the western part of Inner Mongolia Autonomous Region, western part of Ningxia Hui Autonomous Region, Gansu Province, Qinghai Province and most parts of Xinjiang Uyghur Autonomous Region (Fig. 5.1). The total area of the northwestern desert is about 1.09 × 10^6 km^2, accounting for 11.4% of the land area, and it includes basins (Tarim, Junggar, Qaidam, Hami, Yili, Barkol and

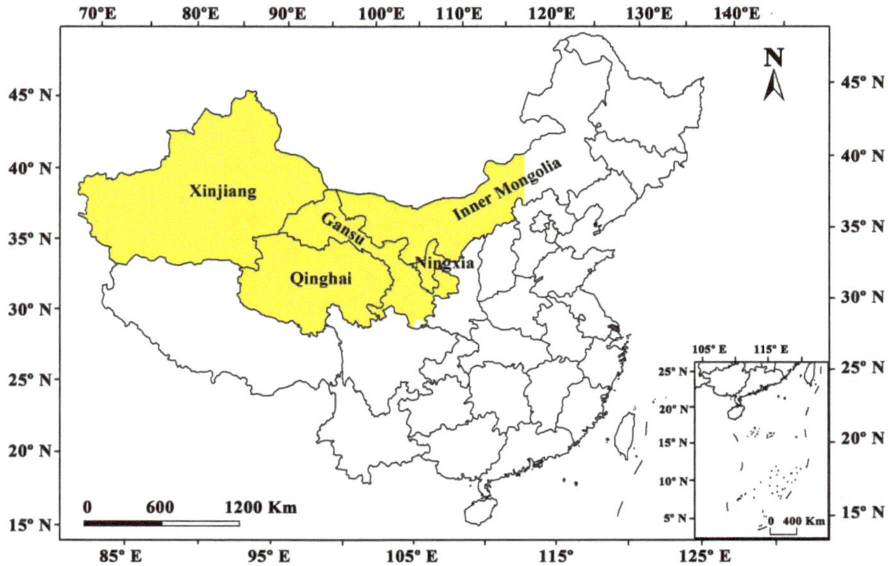

Fig. 5.1 Distribution maps of desert area in China

Turpan), sandy deserts (the Taklimakan Desert, the Gurbantunggut Desert, Badan Jilin Desert, Tenggeli Desert, Kumtag Desert and Ulan Buh Desert) and gravel deserts (Hashum Gobi, Beishan Gobi, Nuomin Gobi and Jiangjun Gobi) (Wu 1980; Liu 1982). By the year 2012, the total population in deserts of China was about 82.48 million, and the minority populations (25.40 million) accounted for 30.81% of the total (National Bureau of Statistics 2012).

Chinese desert areas are characterized by low precipitation, extreme temperature fluctuations, strong sunshine, and winter sand storms (Li et al. 2008). The soil is mainly gravelly alluvium on black stones, sandy aeolian deposits (sand dunes) and weathered residual material on bare rock, but there also are loess, river alluvium and lacustrine sediments. These soil matrixes form various types of soil, depending on climate. In the temperate arid climate zone of northwest China, grey brown desert soil, grey brown (desert type) aeolian sandy soil and sierozems soils are formed. In the warm temperate extreme arid zone of northwest China, brown desert soil and brown desert aeolian sandy soils are formed. In addition, saline soils are found in association with lakes, rivers and marshes (Yin 1997).

According to soil type, temperature conditions, amount of precipitation, vegetation cover, and life form of dominant plant species, Chinese desert areas can be divided into different types. According to soil type, Chinese desert areas can be divided into sandy desert (desert), gravel desert (Gobi), loamy desert (loess), clay desert (salt desert) and rock desert (stony butte) (Li et al. 2008). According to temperature conditions, they can be divided into temperate desert (northern Xinjiang and Alashan), warm temperate desert (southern and eastern Xinjiang) and plateau (highland) desert (Qaidam) (Natural Division Working Committee of China Academy of Sciences 1959; The Central Weather Bureau 1978). According to the amount of precipitation and vegetation cover, they can be divided into semi-desert (annual precipitation between 200 to 100 mm), common (true) desert (annual precipitation between 100 to 50 mm) and extremely arid desert (annual precipitation below 50 mm) (Chen1987). In the semi-desert, because the amount of precipitation is relatively high, vegetation growth is good reaching a coverage of 30–40%. Vegetation is mainly composed of strongly drought-tolerant, semi-lignified woody species, but it also includes a large number of herbaceous species. In the common desert, vegetation coverage is only 20–30%, soil organic matter content is low, and there are few or no herbaceous species. In the extremely arid desert, vegetation coverage is <10%. According to the life form of the dominant plants, Chinese desert areas can be divided into four categories (Editing Committee of China vegetation 1980; Hou 1982): (1) Small tree desert dominated by *Haloxylon ammodendron* and *Haloxylon persicum*, (2) Shrubby desert dominated by *Ephedra przewalskii*, *Sarcozygium xanthoxylon*, *Nitraria sphaerocarpa*, *Gymnocarpos przewalskii*, *Ammopiptanthus mongolicus*, *Potaninia mongolica*, *Caragana tibetica*, *Caragana korshinskii*, *Calligonum arborescens*, and *Ammodendron argenteum*, (3) Subshrub, half shrub desert with *Reaumuria songarica*, *Salsola passerine*, *Sympegma regelii*, *Nanophyton erinaceum*, *Anabasis elatior*, and *Artemisia saposhnikovii*, and (4) Mat-forming half shrub desert mainly with *Ceratoides compacta* and *Ajania tibetica*.

5.1.2 Characteristics of Oases and Grasslands in Chinese Desert Areas

We need to pay attention to oases and grasslands in the desert because they often are heavily impacted by human activities such as farming and grazing, thereby creating open niches for the establishment of invasive species. An oasis is a unique geographical landscape feature in arid, semi-arid areas with sufficient water to support plant growth, and it is one of the three major geographical landscapes in the desert (Gao et al. 2002; Lü et al. 2004). It has an "island" shape (Fan et al. 2004; Sun et al. 2005), and due to the presence of water the natural or cultivated vegetation forms a strong contrast with the surrounding desert environment. Biomass production of an oasis is significantly higher than that of the surrounding deserts (Wang 2010). Water is the basic element for formation of an oasis, and the vegetation mainly depends on water from outside the system, i.e., water comes from springs, wells or an irrigation system (Wang 2000).

Oases can be divided into different types, according to the level of human intervention (semi-artificial oasis, natural oasis, and artificial oasis), different climate zones (tropical, subtropical, temperate zone and temperate desert oasis), history or age (ancient oasis, old oasis and new oasis), geomorphology (valley oasis, fan-shaped alluvial plain and delta oasis type), source of water (irrigation oasis, and natural oasis with well or spring water) and economic function (agricultural, town and industrial or mining oasis) (Chen 1995; Zhang and Wang 1995).

Oases in Chinese deserts are mainly distributed in the west of the Helan, Wushaoling, Qilian and Altun Mountains, as well as vast areas of the northern Kunlun Mountains, all of which are part of the vast Eurasia Steppe Desert Grassland Region (Comprehensive Survey Team of Xinjiang Resources Development of China Academy 1989). In China, about 69,646 km^2 of the arid land is oasis, accounting for 4.33% of the Chinese desert land area (Lü et al. 2004).

Although oases occupy less than 5% of the Chinese desert land area, they support 95% of the human population and 90% of the social wealth of the desert. Therefore, the rise and fall of the oasis economy is closely related to the lives of people in the desert (Lü et al. 2004). In recent decades, the scale of development and utilization of water and soil resources has increased rapidly in desert regions of China for industrial, agricultural and economic development of oases. However, along with human population growth, the oasis ecosystem in China continues to deteriorate. Signs of this damage include: (1) river shortening, lake atrophy or drying, and water pollution including increased salinity; (2) serious soil erosion and expansion of desertified land; (3) increased surface salt accumulation, i.e., salinization; (4) farmland soil pollution; (5) increases in number of invasive species; (6) increases in plant diseases and insect pests; and (7) increases in amount of cultivated land, all of which seriously affect the survival and development of the oasis (Lü et al. 2004; Li et al. 2006).

Grasslands, which are dominated by herbaceous plant species in arid and semi-arid regions, are the most widely distributed type of vegetation globally (Liu et al.

2005), and they need to be given attention in terms of being potential sites for the establishment of invasive species. The intensive use of grasslands for grazing often results in overgrazing and thus destruction of the herbaceous canopy, thereby creating open sites for the establishment for invasive species. The herbaceous plant community is dominated by perennial tufted or rhizomatous grasses and various forbs that have the ability to tolerate drought. According to their biological and ecological characteristics, the grasslands of China can be divided into four types, meadow steppe, plain grassland (steppe), desert steppe (desert grassland) and alpine grassland, with the desert grassland being the most widespread.

The desert grassland region of China has a continental climate and receives ≤200 mm of precipitation annually. It is mainly distributed in central and northern Inner Mongolia Autonomous Region, central and eastern Ordos Plateau of Inner Mongolia Autonomous Region, central part of Ningxia Hui Autonomous Region, eastern Gansu province, and the low slopes of mountains in Xinjiang Uyghur Autonomous Region. Plants are normally 23–30 cm in height, and cover 30–40% of the soil surface. Various species of grasses can grow in desert grassland, including *Stipa tianschanica var. klemenzii*, *Stipa glareosa*, *Stipa tianschanica* var. *gobica*, and *Cleistogenes songorica*. Although biomass production is relatively low, desert grasslands are suitable for grazing of sheep and horses. Desert grasslands have pale chestnut, brown and light sierozem soils, and the humus layer is thin. At present, with global climate change and increased grassland desertification, desert grasslands in China are facing problems such as decline in productivity. It is fragile and may be subject to vegetation and soil degradation, increased desertification and invasion by alien species.

5.1.3 Characteristics of Agriculture and Stockbreeding in Chinese Desert Areas

Most of the agriculture occurs in the oases, and it mainly relies on river water that comes from melted snow in the mountains. Oasis agriculture is mainly distributed on the alluvial plain, and crop production is greatly influenced by the elevational gradient, ground water depth and soil texture, salinity, and fertility. In addition, oasis agricultural systems are surrounded by large tracts of Gobi-type desert; thus, crop land can be influenced by wind, drought and salinity. The agricultural oasis has low species diversity and thus low ecological stability; therefore, it is has little ecological resistance against the invasion of alien species.

The climate of the desert oasis is well suited for crop production. Light and heat resources of the desert oasis during the warm season are particularly rich with long days of clear skies; i.e., little rain or clouds. Therefore, high utilization of light energy results in good fruit production and coloring and growth of crops such as cotton, corn and tomato, sunflower and melon. Thus, the good light, heat and water resources in the desert oasis provide a reliable guarantee of high yield and quality of crops. Application of plastic film technology to fields to reduce loss of the water applied to crops during irrigation further strengthens the agricultural productivity of the oasis, especially for crops such as cotton and melons. Thus, oasis agriculture is an important aspect of the economy of the arid region, and is known for its ancient history (more than 1000 years), advanced technology (irrigation agriculture), and high and stable yields.

Facility agriculture, i.e., crop production in structures where light, water and temperature are maintained at optimal levels to maximize growth, is an important part of agriculture in Chinese desert areas. At the end of 2011, the total facility agriculture area in Ningxia, Gansu, Xinjiang and Qinghai was 805 km^2, 880 km^2, 760 km^2 and 133 km^2, respectively. Some of main crops are planted in agriculture facilities in autumn, winter and early spring, including tomato, pepper, melon, watermelon and cucumber.

The natural grasslands in the desert region of China have a large number of forage species that have high nutritional value and good yield, i.e., *Festuca ovina*, *Stipa capillata*, *Agropyron cristatum*, *Phleum pretense*, *Bromus inermis*, *Dactylis glomerata*, *Poa pratensis*, *Roegneria kamoji*, *Alopecurus pratensis*, *Trifolium pretense*, *T. repens*, *Onobrychis viciuefolia* and *Medicago sativa*. Also, these forage species are well adapted to the natural seasonal conditions of the grasslands and have different feeding values and utilization modes for the various kinds of livestock. According to the management and feeding mode, animal husbandry can be divided into three grazing systems: mountain, plain oasis, and plain oasis confined. The most important/abundant livestock are sheep and goats, and the secondarily important animals are cattle, horse, goat, donkey, camel, mule and yak. Due to the influence of vertical climate belts in the mountains, forage growth and thus grazing have a strong seasonal character. Thus, depending on the region, grasslands can be grazed in (1) summer, spring and winter, (2) summer and winter, and (3) in all four seasons (in the plains region). Only 10.18% of the grassland area in China can be grazed in all seasons; therefore, animals in most areas have to be moved/shifted to different pastures.

5.2 Current Status of Alien Biological Invasion in Chinese Desert Areas

There are 165 alien invasive species in the desert region of China, of which 88, 57 and 20 species are plants, animals and microorganisms, respectively (Tables 5.1, 5.2 and 5.3; Fig. 5.2). The invasive plants belong to 23 families and 64 genera, and the most common families are Asteraceae, Poaceae and Brassicaceae. The invasive animals are insects which mainly belong to Coleoptera (24 species), Plecoptera (12 species), Diptera (13 species) and Lepidoptera (8 species). Fungi, viruses and bacteria account for 44.3%, 36.1% and 19.6%, respectively, of the invasive microbes (Table 5.3). Richness of invasive species differs among provinces. There are more invasive species in Xinjiang, Gansu and Inner Mongolia than in Qinghai and

Table 5.1 Exotic plants in Chinese desert areas

Family	Species	Distribution					Life form	Origin	Reported time
		Xinjiang	Inner Mongolia	Ningxia	Gansu	Qinghai			
Amaranthaceae	*Alternanthera pungens*					√	Annual	South America	1950s
Amaranthaceae	*Amaranthus blitoides*		√				Annual	North America	1857
Amaranthaceae	*Amaranthus caudatus*	√	√	√	√	√	Annual	Pantropical distribution	1850–1919
Amaranthaceae	*Amaranthus lividus*	√	√	√	√	√	Annual	South America	?
Amaranthaceae	*Amaranthus retroflexus*	√	√	√	√	√	Annual	South America	?
Amaranthaceae	*Amaranthus spinosus*	√			√		Annual	South America	?
Amaranthaceae	*Amaranthus tricolor*	√	√	√	√	√	Annual	Asia	?
Amaranthaceae	*Amaranthus viridis*	√	√	√	√	√	Annual	Africa	1864
Amaranthaceae	*Amaranthus albus*	√	√	√	√	√	Perennial	North America	1929
Anacardiaceae	*Rhus typhina*	√			√		Perennial	Europe and North America	1959
Apocynaceae	*Catharanthus roseus*		√		√		Perennial	Mediterranean, Asia, South America	?
Asteraceae	*Achillea millefolium*	√	√				Perennial	Europe, Africa	?
Asteraceae	*Ambrosia artemisiifolia*	√					Annual	North America	?
Asteraceae	*Ambrosia trifida*		√				Annual	North America	1960s
Asteraceae	*Centaurea diffusa*	√	√	√	√	√	Biennial	Europe	?
Asteraceae	*Chrysanthemum carinatum*		√				Annual	Mediterranean	Ancient
Asteraceae	*Cichorium intybus*	√					Perennial	Europe	Colonial period
Asteraceae	*Conyza bonariensis*				√		Annual or biennial	South America	1857

(continued)

Table 5.1 (continued)

Family	Species	Distribution					Life form	Origin	Reported time
		Xinjiang	Inner Mongolia	Ningxia	Gansu	Qinghai			
Asteraceae	*Conyza canadensis*	√	√				Annual or biennial	North America	1934
Asteraceae	*Erechtites valerianaefolia*				√		Annual	South America	?
Asteraceae	*Erigeron annuus*	√	√	√			Annual or biennial	North America	1886
Asteraceae	*Galinsoga parviflora*		√		√	√	Annual	South America	1915
Asteraceae	*Gynura crepidioides*				√		Annual	Africa	1930s
Asteraceae	*Leucanthemum vulgare*				√		Perennial	Europe and North America	1919–Now
Asteraceae	*Mikania micrantha*	√			√		Perennial	South America	1919
Asteraceae	*Sanvitalia procumbens*	√					Annual	North America	?
Asteraceae	*Senecio vulgaris*	√					Annual	Europe	19 Century
Asteraceae	*Solidago canadensis*	√					Perennial	North America	1935
Asteraceae	*Sonchus oleraceus*	√	√	√	√	√	Annual or biennial	Europe	Ancient
Asteraceae	*Xanthium italicum*	√					Annual	North America and Europe	1991
Asteraceae	*Xanthium spinosum*	√	√	√			Annual	South America	?
Asteraceae	*Zinnia peruviana*				√		Annual	North America	1964
Asteraceae	*Sonchus asper*	√		√	√	√	Annual	Europe	Ancient
Boraginaceae	*Heliotropium europaeum*	√			√		Annual	Europe	1919–Now
Boraginaceae	*Symphytum officinale*	√	√	√	√	√	Perennial	Europe	1963
Brassicaceae	*Brassica juncea*	√					Annual	Europe, Mediterranean, Asia	?

Brassicaceae	*Brassica kaber*	√	√	√	√		Biennial	Europe	1998
Brassicaceae	*Lepidium campestre*		√		√		Annual or biennial	Europe and Asia	1919–Now
Brassicaceae	*Lepidium virginicum*	√	√	√	√	√	Annual or biennual	North America	1933
Cannabaceae	*Cannabis sativa*	√	√	√	√	√	Annual	Asia	?
Caryophyllaceae	*Agrostemma githago*	√	√				Annual	Mediterranean	19 Century
Caryophyllaceae	*Stellaria apetala*	√					Biennial	Mediterranean	1996
Caryophyllaceae	*Stellaria pallida*	√					Annual or biennial	Europe, Asia and North America	Ancient
Caryophyllaceae	*Vaccaria segetalis*	√	√	√	√	√	Annual	Europe	Ancient
Chenopodiaceae	*Chenopodium hybridum*	√	√	√	√	√	Annual	Europe, Asia	1864
Convolvulaceae	*Cuscuta europaea*	√	√		√	√	Annual	Europe, Africa and Asia	?
Convolvulaceae	*Ipomoea nil*	√	√	√	√	√	Annual	South America	220–450
Convolvulaceae	*Calonyction aculeatum*	√	√	√	√	√	Annual	South America	?
Convolvulaceae	*Ipomoea purpurea*	√					Annual	?	?
Crassulaceae	*Bryophyllum tubiflora*	√		√	√	√	Perennial	Africa	?
Euphorbiaceae	*Ricinus communis*	√					Annual	Africa	?
Euphorbiaceae	*Euphorbia helioscopiae*		√	√	√	√	Annual	Europe, Africa and Asia	?
Fabaceae	*Lotus corniculatus*	√			√		Perennial	North America	1980s
Fabaceae	*Medicago sativa*	√	√	√	√	√	Perennial	Asia	126 BC
Fabaceae	*Melilotus albus*	√			√	√	Biennual	Europe	19 Century
Fabaceae	*Trifolium repens*	√	√	√	√		Perennial	Europe	19 Century
Fabaceae	*Vicia villosa*	√			√		Annual	Europe, Asia	?

(continued)

Table 5.1 (continued)

Family	Species	Distribution					Life form	Origin	Reported time
		Xinjiang	Inner Mongolia	Ningxia	Gansu	Qinghai			
Malvaceae	*Abutilon theophrasti*	√	√	√	√	√	Annual	Europe, North America and Asia	?
Malvaceae	*Hibiscus trionum*	√	√	√	√	√	Annual	Africa	1406
Malvaceae	*Malvastrum coromandelianum*				√		Perennial	North America	1861
Oxalidaceae	*Oxalis corymbosa*	√	√	√	√	√	Perennial	South America	19 Century
Papaveraceae	*Papaver nudicaule*	√	√	√			Perennial	North America, Europe	?
Phytolaccaceae	*Phytolacca americana*			√	√	√	Perennial	North America	1935
Plantaginaceae	*Plantaga lanceolata*	√			√		Perennial	Europe	1959
Poaceae	*Avena fatua*	√	√	√	√	√	Annual	Europe and Mediterranean	19 Century
Poaceae	*Bromus cartharticus*	√					Annual or biennial	South America	~1935
Poaceae	*Lolium multiflorum*				√	√	Annual	Africa, Europe, Asia	?
Poaceae	*Lolium perenne*	√	√	√	√	√	Perennial	Europe	1959
Poaceae	*Lolium persicum*					√	Annual	Asia	?
Poaceae	*Lolium temulentum*	√	√	√	√	√	Annual or biennial	Europe	1954
Poaceae	*Lolium temulentum var. longiaristatum*					√	Annual or biennial	Europe	~1960
Poaceae	*Cenchrus incertus*		√				Annual	North America	1990
Poaceae	*Festuca arundinacea*	√					Perennial	Europe and Asia	?
Pontederiaceae	*Eichhornia crassipes*		√				Perennial	South America	1901

Scrophulariaceae	*Veronica didyma*	√					Annual or biennial	Asia	?
Scrophulariaceae	*Veronica hederaefolia*				√		Annual or biennial	Europe, Africa	1980s
Scrophulariaceae	*Veronica persica*	√					Annual or biennial	Europe	1933
Scrophulariaceae	*Veronica polita*	√			√	√	Annual or biennial	Asia	1406
Solanaceae	*Datura metel*	√			√	√	Annual	Pantropical distribution	Ancient
Solanaceae	*Datura stramonium*	√	√	√	√	√	Annual	North America	1578
Solanaceae	*Solanum rostratum*	√	√				Annual	North America	1982
Solanaceae	*Solanum sarachoides*			√	√	√	Annual	South America	?
Solanaceae	*Nicandra physaloides*				√		Annual	South America	1964
Solanaceae	*Physalis angulata*	√	√	√	√	√	Annual	?	?
Umbelliferae	*Coriandrum sativum*	√			√		Annual or biennial	Europe, Mediterranean	Ancient
Umbelliferae	*Daucus carota*	√	√	√	√	√	Biennial	Europe	1406
Umbelliferae	*Hydrocotyle vulgaris*	√					Perennial	Europe	?

Table 5.2 Exotic animals in Chinese desert areas

Family	Species	Distribution					Origin	Reported time
		Xinjiang	Inner Mongolia	Ningxia	Gansu	Qinghai		
Agromyzidae	*Liriomyza huidobrensis*	√				√	South America	1993
Agromyzidae	*Liriomyza sativae*	√	√	√	√		South America	1994
Aleyrodidae	*Bemisia tabaci*	√					Asia	1995
Aleyrodidae	*Trialeurodes vaporariorum*		√				South America	?
Anatidae	*Branta canadensis*	√					Asia and Oceania	1998
Anguinidae	*Anguina agrostis*		√				?	?
Anthribidae	*Araecerus fasciculatus*	√		√	√	√	Asia	?
Aphididae	*Eriosoma lanigerum*	√				√	North America	1914
Apidae	*Apis mellifera ligustica*	√	√	√	√	√	Oceania	1917
Blattellidae	*Blattella germanica*	√	√	√			Africa	1953
Bostrychidae	*Sinoxylon japonicum*					√	?	?
Bruchidae	*Acanthoscelides pallidipennis*		√				North America	1980
Carposinidae	*Carposina nipponensis*	√	√	√	√	√	Asia	1930
Cecidomyiidae	*Mayetiola destructor*	√					Europe	1980
Chloropidae	*Meromyza saltatrix*	√	√	√	√	√	Europe	1955
Cholodkovsky	*Aphanostigma piri*	√			√	√	?	1981
Chrysomelidae	*Bruchus pisorum*	√	√	√	√	√	Mediterranean	1950s
Chrysomelidae	*Leptinotarsa decemlineata*	√					North America	1993
Cichlidae	*Oreochromis niloticus*				√		Africa	1994
Cionidae	*Ciona intestinalis*	√					?	?
Circetidae	*Ondatra zibethicus*		√	√	√	√	North America/Europe/ Asia	1927
Clariidae	*Clarias lazera*	√	√	√	√	√	Africa	1981
Curculionidae	*Cosmopolites sordidus*	√	√	√	√	√	Asia	?

Curculionidae	*Cryptorrhynchus lapathi*	√	√		√		?	1953
Curculionidae	*Hypera postica*	√					?	?
Curculionidae	*Ips sexdentatus*				√		Palearctic	?
Curculionidae	*Sitophilus granarius*	√			√		Europe, Mediterranean and Africa	1954
Cyprinidae	*Cyprinus carpio var. specularis*	√	√				Europe	1982
Dermestidae	*Anthrenus picturatus*	√	√	√	√	√	?	1984
Dermestidae	*Anthrenus verbasci*	√	√	√	√	√	Europe	?
Eurytomidae	*Bruchophagus gibbus*	√	√		√		North America	?
Gelechiidae	*Anarsia lineatella*	√	√	√	√	√	Mediterranean	?
Gelechiidae	*Phthorimaea operculella*				√		America	1937
Limacidae	*Lehmannia valentiana*		√				Europe and Africa	?
Molgulidae	*Molgula manhattensis*	√	√				North America	?
muridae	*Mus musculus*		√				Europe	?
Muridae	*Rattus exulans*			√	√	√	Asia	1999
Muridae	*Rattus norvegicus*			√	√	√	Asia	18 Century
Myocastoridae	*Myocastor coypus*			√			North America	1953
Palaemonida	*Macrobrachium rosenbergii*	√	√	√	√	√	Asia and Oceania	1976
Percidae	*Perca flavescens*	√	√	√	√	√	North America	2003
Percoidea	*Lucioperca lucioperca*	√					Euasia	1960s
Phylloxeridae	*Moritziella castaneivora*			√			Asia	1997
Pieridae	*Pieris rapae crucivora*	√	√	√	√	√	Asia	?
Psittaculidae	*Psittacula krameri*	√					Africa and Asia	1900
Psittaculidae	*Trichoglossus haematodus*	√					Asia	1998
Pyralidae	*Ephestia kuehniella*				√		Europe	1982
Pyuridae	*Halocynthia roretzi*		√				Asia	?

(continued)

Table 5.2 (continued)

Family	Species	Distribution					Origin	Reported time
		Xinjiang	Inner Mongolia	Ningxia	Gansu	Qinghai		
Ranidae	*Rana catesbeiana*			√	√		North America	1959
Salmonidae	*Salmo aguabonita*	√		√	√	√	Asia	1996
Sciuridae	*Callosciurus erythraeus thai*			√			Asia	17 Century
Sciuridae	*Sciurus vulgaris exalbidus*		√				Asia, Europe	1950s
Sphaeromatidae	*Styela canopus*	√					North America	?
Tephritidae	*Carpomyia vesuviana*	√					Asia	2007
Tetranychidae	*Tetranychus urticae*	√		√	√		?	1983
Tortricidae	*Cydia inopinata*	√	√	√	√	√	Asia	?
Tortricidae	*Cydia pomonella*	√	√	√	√		Europe	1953

Table 5.3 Exotic microorganisms in Chinese desert areas

Family	Species	Distribution					Origin	Reported time
		Xinjiang	Inner Mongolia	Ningxia	Gansu	Qinghai		
Acholeplasmataceae	*Paulownia witches broom phytoplasma*	√		√			?	?
Ascomycetes	*Botryosphaeria laricina*				√	√	Asia	1970
Ascomycetes	*Ceratocystis fimbriata*			√			North America	1937
Bunyaviridae	*Tomato spotted wilt virus*	√	√	√	√	√	Oceania	?
Closteroviridae	*Beet necrotic yellow vein virus*				√	√	Europe	1970s
Closteroviridae	*Poplar mosaic virus*				√	√	North America	1979
Cronartiaceae	*Cronartium ribicola*	√	√	√	√	√	?	?
Hyaloscyphaceae	*Lachnellula willkommii*		√				North America	1975
Microbacteriaceae	*Clavibacter michiganensis* subsp. *sepedonicus*	√	√	√	√	√	Europe	?
Microbacteriaceae	*Clavibater michiganensis*. Subsp. *michiganensis*	√	√				North America	1981
Moniliaceae	*Verticillium albo-atrum*	√					Europe	?
Moniliaceae	*Verticillium dahliae*	√	√				North America	1935
Mycosphaerellaceae	*Mycosphaerella pini*		√				North America	1982
Nectriaceae	*Fusarium oxysporum* f. sp. *vasinfectum*	√			√		North America	1892
Pseudomonadaceae	*Acidovorax avenae* subsp. *citrulli*		√		√		?	1988
Pseudomonadaceae	*Xanthomonas campestris* pv. *citri*	√	√	√		√	?	Ancient
Pseudomonadaceae	*Xanthomonas oryzae* pv. *oryzae*		√	√	√	√	?	1950
Pythiaceae	*Phytophthora sojae*		√				North America	1991
Sclerosporaceae	*Peronosclerospora sacchari*	√		√			North America	1937
Taphrinaceae	*Taphrina deformans*	√	√	√	√	√	Europe	?

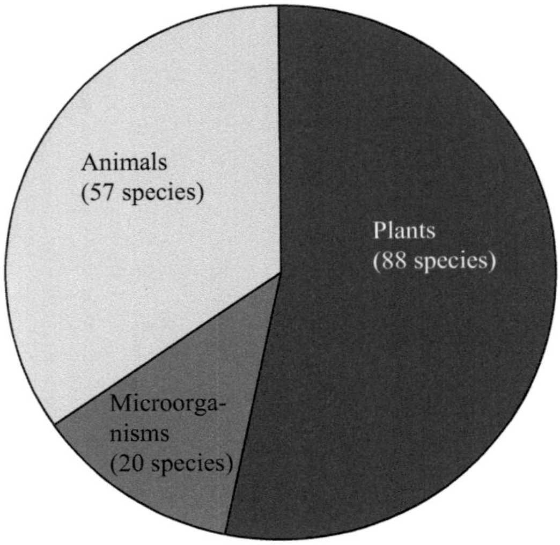

Fig. 5.2 Percentages of invasive alien species of different taxa in Chinese desert areas

Ningxia. About 120 invasive species have been found in Xinjiang, while the number is only about 70 in Ningxia and Qinghai (Fig. 5.3).

About 50% of the alien invasive species in the desert region of China comes from North America and Europe. This may be due to similar climates and environments in North America, Europe and China, which allows alien species to quickly adapt to habitat conditions in China and successfully become invasive within a relatively short period of time. Since China is one of the largest trading partners with North America and Europe, frequent trade and personnel exchanges many provide many opportunities for alien species to enter the Chinese mainland (Fig. 5.4).

Some of the invasive species in the Chinese desert area were intentionally introduced (e.g. plants propagated for forage, fodder, industrial materials, medicines, vegetables, lawn plants, and ornamentals) or unintentionally with the importation of logs or plants growing in soil. However, some invasive species have spread by natural ways, such as along with water and/or wind, or by insects flying from one site to another. Unintentional introductions are the main reason for the increase in number of invasive species in the Chinese desert region in the past 20 years. It should be noted that over 80% of the invasive species in this region are found in disturbed areas such as fields.

The invasion history of alien species into the China can be divided into four stages, depending on number of species introduced. The first stage was from sixteenth to nineteenth century, when only <20 foreign species were introduced because during this time there was little domestic and foreign exchanges in the desert area. In the second stage, from early-mid nineteenth century to 1949, there was a rapid increase in the number of introduced alien species due to increased trade

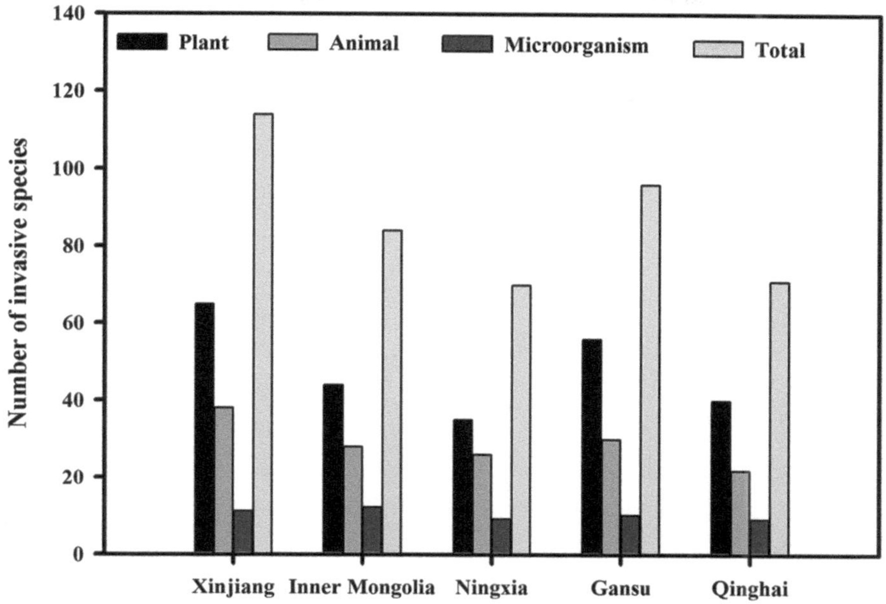

Fig. 5.3 Number of alien species in each province and autonomous region of the Chinese desert area

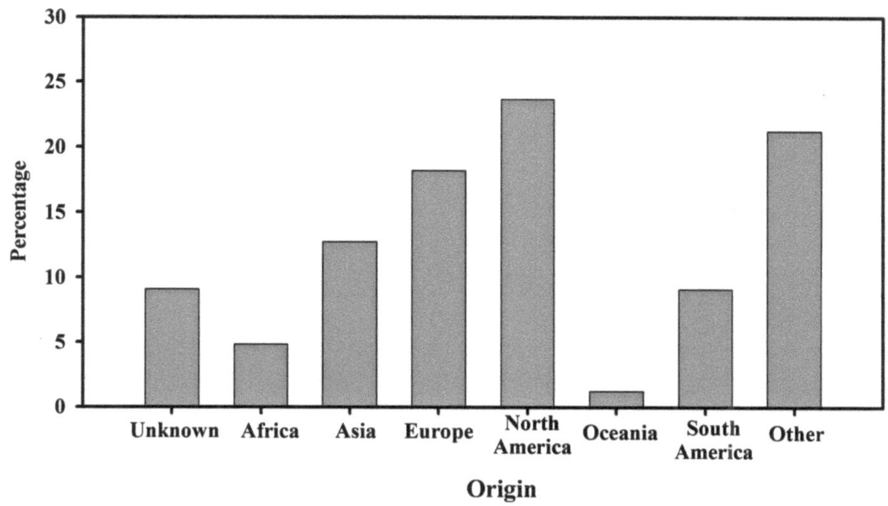

Fig. 5.4 Origin of alien plants in Chinese desert areas

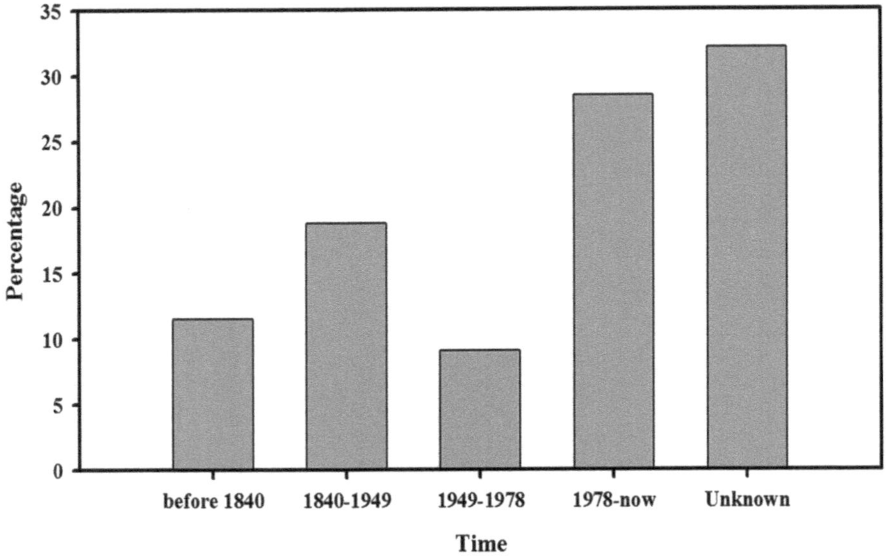

Fig. 5.5 Invasion trends of alien plants in Chinese desert areas

between China and surrounding countries. The third stage, from 1949 to 1978, corresponds with the development of much exchange between China and surrounding countries. The fourth stage, from 1978 to present, is the outbreak stage, which corresponds with China's rapid economic development and exchange with countries all over the world. Thus, the great increase in number of alien species has contributed to outbreaks of alien species and led to an increase of quarantines to help control the spread of pests (Fig. 5.5).

5.3 Analysis of Primary Cause of Biological Invasion in Chinese Desert Areas

1. **The eco-geographical environment is diverse**. The desert area of China is characterized by complex landforms, diverse climate, and mountains surrounded by basins in which there is a mosaic of oases and various kinds of deserts, including desert grassland, and some lakes and wetlands. This complex geographical environment creates a diverse ecological system in the desert area of China, in which alien species from various habitats can find a suitable habitat for growth and reproduction. Also, the ecological environment in the desert area of China increasingly is being disturbed by human activities such as destruction of the natural vegetation by farming and overgrazing, which promote the invasion of

alien species by removing species that are potential competitors, thereby increasing the speed of range expansion.

2. **The ecosystem is fragile**. Excessive grazing and land reclamation for irrigated farming may cause serious disturbance and/or destruction of the original ecological system, and potentially result in rapid expansion of desertified land and atrophy of natural lakes. These disturbances of the natural environment can lead to the formation of a large number of vacant niches in the ecological community that increase the susceptibility of ecosystems to biological invasions. Due to human-related disturbances, desert areas have become one of the most vulnerable regions to biological invasions in China.
3. **The desert area of China is a connection to western Asia and Europe**. The desert area of northwestern China is bordered by the People's Republic of Mongolia, Russia, Kazakhstan, Kyrgyzstan, Tajikistan, Afghanistan, Pakistan and India. In ancient times, the "Silk Road", which was the main route for trade between China and western Asia and Europe, passed through the desert area of northwest China. Thus, the Chinese desert region has played an important role in the communication between the east and the west, and since the reform and opening up of China, many roads and airports have been built in these regions. The "Eurasian Continental Bridge railroad", which crosses the desert area, is a good foundation for China to further strengthen the development of foreign exchange and no doubt promotes some additional biological invasions. Further, a lot of foreign agricultural products have been introduced into the desert area of China.
4. **Chinese desert area has superior and unique natural conditions**. Climates of the Chinese desert area differ greatly. The Yili River Valley in Xinjiang has a mild and humid climate, and mean annual average temperature is 10.4 °C, sunshine is 2870 h, precipitation is 417.6 mm (600 in mountainous areas), making this the most humid desert area in China. Not surprisingly, the Yili River Valley is one of the main regions invaded by alien species in the Chinese desert area. In contrast, the Turpan Basin has a dry and hot climate, and the mean annual average temperature is ~13 °C, precipitation is about 16 mm, and evapotranspiration is up to 3000 mm. In Toksun County in the Turpan Basin, the highest recorded temperature is 49.6 °C, and average annual precipitation is only 5.9 mm, making it the most arid place in China. In the Turpan Basin, the average maximum temperature is over 38 °C in June to August, the annual amount of sunshine is 3200 h, annual accumulated temperature above 5300 °C, and frost free period is up to 270 days. These unique climatic conditions in the Turpan Basin are suitable for the successful invasion of harmful alien organisms that require a high temperature environment, such as the insects *Aphis gossypii* and *Carpomya vesuviana*.
5. **Economic development promotes invasion of alien species in the Chinese desert area**. When areas of the desert are disturbed during the construction of economic or tourism projects, they can serve as centers for the establishment of invasive species and for the dispersal of invasive organisms to other areas. For

example, in Ürümqi city, one of the fastest growing economic centers in China, the GDP reached 15.8 billion USD in 2010 and economic growth has increased nearly 200 times in the past 20 years. At the same time, Ürümqi city is a domestic and foreign transportation hub, and it is the biggest commodity distribution center in the Chinese desert area with eight national-level ports. Thus, the city is the largest entry point of manufactured goods and alien species in northwestern China. Clearly, the special geographical and economic attributes of Ürümqi city have made it the area with the highest incidence of invasion of alien species in the desert of northwest China.

6. **Climate change has increased the risk of biological invasion**. There is a close relationship between global climate change and biological invasions, and global climate warming in particular may aggravate the occurrence of and damage by alien species. Studies on climate change for more than 40 years in the Altai area of northern Xinjiang and the invasions of the insect *Locusta migratoria* have shown that temperature increases have favored the establishment of this insect (Li et al. 2008). Increased temperatures allow eggs to overwinter and healthy individuals to hatch in spring, thereby favoring the growth of populations.
7. **Facility agricultural production provides a bridge for invasion of alien species**. Facility agriculture has developed rapidly in almost all the prefectures, cities and counties in the Chinese desert area, and it has become a feature of local agricultural production and an important way for farmers to increase their income. However, facility agriculture may create a situation that benefits invasions of alien diseases, insects, and plants. This is mainly because the environmental conditions of winter (greenhouse) agricultural production not only create extensive shelter in which alien invasive species can safely overwinter, but they provide abundant food sources and habitats. Thus, the environment that promotes the growth of vegetables in winter may also promote establishment and growth of pest populations. For example, *Liriomyza sativae* was first found in greenhouses in Xinjiang in 1996, and then it spread all over Xinjiang in a few years. Invasions of the whitefly *Bemisia tabaci*, which is a major vector of viruses, also underwent the same process. *B. tabaci* was first found in 1998 in Ürümqi city in Xinjiang, and then it was found in Turpan City in 1999 in a local greenhouse used for growing vegetables. Later, the population expanded greatly, and *B. tabaci* migrated from greenhouse vegetables to cotton fields, where it caused great damage. In the late autumn, *B. tabaci* transferred from cotton fields back to greenhouses during the winter and caused harm to vegetable plants in greenhouses. This whitefly has spread rapidly to most oases in the Chinese desert area, causing significant damages to oasis ecosystems. It was first found on cotton in Akesu city and Awati County in 2011, and in the same year *B. tabaci* was found in an area of 175 km^2 in Kashkgar and Hotan.

5.4 Effect of Invasive Species in the Chinese Desert Area on Economy, Agriculture, Human Health and Social Stability

5.4.1 *Economy*

Spread and extensive damage by alien invasive species in Chinese desert areas have directly caused economic losses. For example, the invasions of *Locusta migratoria* from Kazakistan into Xinjiang in 1999 covered an area of 4180 km^2 and caused total crop failure in an area of 1330 km^2 (~32% of invaded area) (Yu and Hou 2004). The cotton aphid *Aphis gossypii* invaded in the 1980s, and its sustained outbreaks in cotton growing areas in Chinese desert areas have caused severe economic losses (Wang and Jiao 2000; Yao 2004). Although *Verticillium dahliae* was reported early in the 1950s and 1960s in the desert region of China, it only occurred sporadically in some cotton growing areas and caused no obvious harm. However, this fungus has spread rapidly and has caused serious damage since the 1980s, along with the rapid development of cotton production in the desert region of China, and extensive introduction of new cotton varieties from other parts of China where *V. dahliae* occurs. At present, it is widely distributed in the desert area in China. *Leptinotarsa decemlineata* is a major invasive insect, and it has spread to all parts of Xinjiang, causing major damage to many crops such as potato and eggplant and thus severe economic losses (Guo et al. 2010). The alien invasive insect *Carpomya vesuviana* was first reported in Xinjiang in 2007, and during that year it occurred in an area of 54.7 km^2. This insect attacked red jujubes, causing very heavy economic losses for this fruit growing industry.

In the desert area of China, great efforts have been made in prevention and control programs. Direct economic losses caused by the 16 main invasive species, including *L. migratoria*, *A. gossypii*, *V. dahliae*, *L. decemlineata*, *C. vesuviana*, on agriculture production in Xinjiang in 2010 was 183 million USD and control costs were 4.95 billion USD (Guo et al. 2010).

5.4.2 *Agriculture*

Some alien species compete with native species for resources and space and may reduce species diversity of natural ecosystems. In the agricultural landscape, invasive plants such as *Solanum rostratum*, *Xanthium spinosum*, and *Xanthium italicum* that grow in oases, grasslands, pastures, roadsides, ditches, abandoned farmlands and landfills can outcompete native plants for water, light and nutrients. Due to the loss of native species, the biological diversity in invaded areas has been reduced greatly, and ecological equilibrium has been destroyed (McCormick 1977; Lin and Tan 2007).

Alien species can largely effect growth of crop species. For example, water extracts of *X. spinosum* roots, stems and leaves can inhibit seed germination of wheat and alfalfa. These extracts not only reduce the final germination percentage of seeds, but they also extend the seed germination period (Dong et al. 2014). Water extracts of *S. rostratum* stems and leaves can significantly inhibit the germination of tomato seeds (Ping et al. 2012). *Ambrosia artemisiifolia*, which is one of the world's most harmful weeds, can quickly form a single-dominant community, thereby inhibiting the growth of crops and wild plants and causing decline and extinction of the native plant community (Zhang and Liu 2000).

Some alien species can hybridize with crop species, thereby causing changes in genetic structure and diversity (e.g. genetic pollution). For example, the invasive plant *S. rostratum* has crossed with tomato and potato, and such genetic introgression can facilitate its invasiveness (Lin and Tan 2007).

Invasions of poisonous species can cause death or decline in health of livestock. The desert area of northwest China is one of the important livestock producing areas in the country, and the native vegetation of the desert steppe is vital for the animal husbandry. However, the geographical position and fragile ecological environment make the desert area vulnerable to invasions by alien species. In recent years, with an increase in international trade, the risk of invasion of alien species into the desert area of China has increased greatly. At the end of 1980s, there were 81 poisonous weeds such as *Aconitum carmichaeli*, *Achnatherum inebrians*, *Oxytropis glabra*, and *Pedicularis abrotanifolia* in desert grassland of China, and they occurred in a limited area of only 1296 km^2. However, in 2007 there were 104 species that occurred in an area of 187,280 km^2.

The invasive plants *S. rostratum* and *X. spinosum* have spread to Xinjiang and the Inner Mongolia Autonomous Region in the Chinese desert area. Plants of both species have numerous thorns, which can cause serious skin irritation of livestock and reduce the value of wool (Lin and Tan 2007). Further, the leaves, berries and roots of the *S. rostratum* contain alkaloids that act as neurotoxins, and thus are highly toxic (Orgell et al. 1958). If cattle and sheep eat *S. rostratum*, they can develop severe enteritis and bleeding, and even finally die (Simic 1943; Bah et al. 2004).

5.4.3 Human Health

Sweet potato long beak shell (*Ceratocystis fimbriata*) is one of the dangerous pathogens that can damage crops in China. It can cause a variety of diseases such as sweet potato blight. This fungus can cause potato pieces used for planting to decay, and it can kill plants and destroy the sweet potatoes after they are harvested and placed in storage. In addition, toxic substances produced by the fungus often cause human and livestock poisoning (Liu et al. 2003). Another threat to human health in the desert area is due to activities of the invasive potato beetle. This insect feeds on leaves, which causes the concentration of glycoalkaloid to increase significantly in the tubers. If humans eat these tubers, they can become ill (Dinkins et al. 2008;

Dinkins and Peterson 2008). Further, pollen of *Ambrosia artemisiifolia* can cause allergic reactions in susceptible people when pollen is dispersed from June to September. People who lack antibodies to the pollen may experience asthma, chest tightness, dizziness and other symptoms, and they may have other complications that result in death. If *A. artemisiifolia* becomes mixed with cow feed, the milk and milk products have an unpleasant smell and taste (Zhang and Liu 2000).

5.4.4 Social Stability and Development

Invasive species can have a profound detrimental impact on the lives of local farmers and herdsmen. The plant and animal resources of the agricultural and grassland ecosystems are an indispensable part of the lives of the local people. Thus, when invasive alien species cause a serious negative impact on the productive of these systems, people will suffer. That is, invasions of alien species usually result in a reduction of the income of farmers and herdsmen, which directly lowers their quality of life. For example, in the Inner Mongolia grasslands, invasive rats heavily impact an area of about 373,300 km^2 each year (i.e. 14% of the available pastoral lands) resulting in an annual loss of nearly 20 million tons of forage grass. *Ondatra zibethicus* is one of the most harmful species among the invasive rats, and its numbers can reach 80 thousand to120 thousand each year. This species digs a 20 m long hole, which causes serious damage to the grassland vegetation, and it not only eats forage grasses but also feeds on grass seeds. Thus, this species causes major economic losses in Inner Mongolia grasslands (Tian et al. 2015).

5.4.5 Minority Ethnic Group Culture

National culture is the sum of the material and spiritual wealth that is produced and created by the people. The deserts of northwest China have many ethnic groups, and each has its own special features and characteristics that contribute to the formation of a special regional culture. Ethnic groups such as Uyghur and Hui mainly live in the oases, and they are engaged in agriculture. The Kazak and Mongol mainly live in the desert grasslands, and they are involved in animal husbandry. Alien invasive species can have detrimental effects on both crop production systems and natural ecosystems of the desert grasslands, which would endanger the stability of regional ethnic cultures. *Agrostemma githago* is the main weed in wheat crops, and its seeds are poisonous to humans and livestock if eaten, along with the grain (Pang et al. 2015). *Cenchrus pauciflorus* is one of the main weeds in Inner Mongolia grasslands. When its fruits mature, the spiny involucre became hard and sharp, and they can injure livestock. Thus, this species has a serious negative impact on grazing and utilization of natural grassland (Pang et al. 2015).

5.5 Problems and Countermeasures for Prevention and Control of Alien Species Invasions in the Chinese Desert Area

5.5.1 The Existing Problems

Although people working in the Chinese desert area have made great efforts in preventing establishment and spread of alien species and reduced their harmful effects on agricultural production, many unsolved problems still exist. For example, there is a great need to establish regulatory policies, improve coordination between departments, and increase technical research and application.

1. **Lack of knowledge**. Overall, the public in desert areas of China has not yet learned fully about the problems associated with biological invasions. Thus, traveling people may carry plants and their products out of epidemic areas, and relevant organizations or individuals may not conscientiously perform quarantines. In some cases, people introduce new species without conducting tests to determine if they have the potential to become invasive. Since 1995, about 68% of the new alien invasive species in the Chinese desert area were intentionally introduced by humans *via* importation of nursery stock, flowers and germplasm resources, etc.
2. **Control is not strict**. In the Chinese desert area, programs/procedures for risk assessment, early warning, eradication, and control of biological invasions have not been established. Further, in areas where biological invasions have occurred, no plans for ecological restoration have been developed. Prevention and management responsibilities for invasive species are dispersed among various departments, and there is lack of coordination. Thus, it is very difficult to have an early detection program and to do the early eradication treatments that are required to stop the spread of a new invasive species in an area. For the majority of alien invasive species, if they are not controlled shortly after they invade an area, then we can expect further dispersal and ecological harm. In general, the present management system of alien species does not meet the needs of modern agricultural production and cannot safeguard agriculture or natural desert ecosystems.
3. **Research is weak**. There is a poor understanding of the reasons why alien species spread in Chinese desert areas, and research is needed to determine the dispersal mechanisms and to develop the methods of preventing dispersal.
4. **Investment is insufficient**. In recent years, the government has invested some money in monitoring and in prevention and control of invasive species in the Chinese desert area. However, funding needs to be increased to develop research programs devoted to solving the problems of how to control invasive species.

5.5.2 Countermeasures

The countermeasures on prevention and control of alien species invasions in Chinese desert area are to (1) strengthen the training of government employees to recognize the species that are invasive, (2) increase public awareness and develop a central office where people can report locations where they have seen invasive species, (3) strengthen enforcement of the existing laws relevant to alien invasive species to ensure that these species remain inside quarantine areas, (4) establish additional special laws and regulations, (5) unify and coordinate management operations for control of invasive species, (6) formulate emergency prevention and control plans and response mechanisms by relevant departments, (7) increase investment in science and technology and thus improve research ability, (8) enhance invasion risk assessment, (9) learn more about the ecology and biology, (10) establish a monitoring, risk assessment and early warning system for regional major alien species, (11) establish an information-sharing system for alien species between local regions and departments, (12) establish a mechanism for collection of information about major alien species in neighboring countries, then develop a system of risk analysis, information release and joint action, and (13) strengthen comprehensive control of regional major invasive species.

Acknowledgments This work was partially supported by National Natural Science Foundation of China (31560067) and the National Basic Research Program of China (973 Program) (2010CB134510), and the Specimen Platform of China, Teaching Specimens Sub-platform (2005DKA21403-JK, http://mnh.scu.edu.cn/), and the Graduate Student Scientific Research Innovation Project of Xinjiang, China (XJGRI2013104).

References

Bah M, Gutierrez DM, Escobedo C, Mendoza S, Rojas JI, Rojas A (2004) Methylprotodioscin from the Mexican medical plant *Solanum rostratum* (Solanaceae). Biochem Syst Ecol 32:197–202

Chen CD (1987) Types of deserts in China and their economic exploitation. Acta Phytoecol Geobot Sin 11(2):81–91. [in Chinese with English abstract]

Chen LH (1995) Formation conditions and processes of desert oases. J Arid Land Res Environ 9(3):49–57. [in Chinese with English abstract]

Comprehensive Survey Team of Xinjiang Resources Development of China Academy (1989) Study on the ecological environment of Xinjiang. Science Press, Beijing. [in Chinese]

Dinkins CLP, Peterson RKD (2008) A human dietary risk assessment associated with glycoalkaloid responses of potato to Colorado potato beetle defoliation. Food Chem Toxicol 46:2837–2840

Dinkins CLP, Peterson RKD, Gibson JE, Hu Q, Weaver DK (2008) Glycoalkaloid responses of potato to Colorado potato beetle defoliation. Food Chem Toxicol 46:2832–2836

Dong FH, Liu Y, Jing MJ, Li HB, Hu YX (2014) Allelopathic effect of *Xanthium spinosum* on seeds of *Triticum aestivum* and *Medicago sativa*. Arid Zone Res 3:530–535. [in Chinese with English abstract]

Editing Committee of China Vegetation (1980) Vegetation in China. Science Press, Beijing. [in Chinese]
Fan ZL, Ma YJ, Alishir K, Shen YL (2004) On formation, evolution and sustainable development of artificial oasis ecosystem in desert area in China. J Desert Res 24(1):10–16. [in Chinese with English abstract]
Gao YH, Chen YC, Lu SH (2002) Role of irrigation in maintenance and development of modern oasis. J Desert Res 22(4):383–386. [in Chinese with English abstract]
Guo WC, Tuerxun XJJ, Liu J, He J, Li J, Ma DC, Wan J (2010) Research on the identification of colorado potato beetle & its distribution dispersal and damage in Xinjiang. Xinjiang Agric Sci 47(4):906–909. [in Chinese with English abstract]
Hou XH (1982) Vegetation geography of China and chemical composition of its dominent plants. Science Press, Beijing. [in Chinese]
Li YS, Jia XH, Lu WY (2006) The utilization of water resource and its influence on eco-envirornnent in arid area of Northwest China. Res Soil Water Conserv 13(1):217–219. [in Chinese with English abstract]
Li Q, Wang SZ, Feng CH, Zhang M, Jiang F, Yang G, Luo LM (2008) The physiological and biochemical indexes of the cold-hardiness of *Locusta migratoria tibetensis* Chen. Acta Ecol Sin 28(3):1314–1320. [in Chinese with English abstract]
Lin Y, Tan DY (2007) The potential and exotic invasive plant: *Solanum rostratum*. Acta Phytotaxon Sin 45(5):675–685. [in Chinese with English abstract]
Liu YL, He YH, Ruan XY (2003) *Ceratocystis fimbriata* – a plant pathogen with wide distribution and many plants attacked. J Yunnan Agric Univ 18(4):408–412. [In Chinese with English abstract]
Liu YX (1982) Observations on the formation of Chinese desert floras. J Syst Evol 20(2):131–142. [in Chinese with English abstract]
Liu YY, Liu HF, Liu M (2005) Study on biodiversity conservation in the karamori mountain nature reserve, Xinjiang. Arid Environ Monit 19(3):131–135. [in Chinese with English abstract]
Lü X, Yang L, Zhang FH, Gu XL, Zhang W, Zhang WF, Wang SM (2004) Agricultural characteristics and its sustainable development strategies of desert oasis region. J Desert Res 1(1):10–16. [in Chinese with English abstract]
McCormick LL (1977) Category I—Weed survey. Southern States research reports. Southern Weed Sci Soc 30:184–215
National Bureau of Statistics (2012) China statistical yearbook 2012. China Statistics Press, Beijing. [in Chinese]
Natural Division Working Committee of China Academy of Sciences (1959) China natural division. Science Press, Beijing. [in Chinese]
Orgell WH, Vaidya KA, Dahm PA (1958) Inhibition of human plasma cholinesterase in vitro by extracts of solanceous plants. Science 13:1136–1137
Pang LD, Eminniyaz A, Liu GX (2015) Problems and countermeasures of exotic invasive plants in Inner Mongolia Autonomous Region. Pratacul Sci 32(12):2037–2046. [in Chinese with English abstract]
Ping YF, Zhu JW, Zhang ZG (2012) Effects of *Solanum rostratum* on seed germination and seedling growth of tomato. Agric Res Arid Areas 03:176–179. [in Chinese with English abstract]
Simic WJ (1943) Solanine poisoning in swine. Vet Med 38:353
Sun HB, Wang RH, Zhang HZ, Huang JF, Zhao ZY (2005) Mountain-oasis-desert system and characteristics of climate in Xinjiang. Arid Land Geogr 28(2):199–204. [in Chinese with English abstract]
The Central Weather Bureau (1978) Climatic Atlas of People's Republic of China. Map Publishing House Press, Beijing. [in Chinese]
Tian WT, Liu Y, Wang SY, Han B (2015) Alien invasive species in Inner Mongolia and their influence on the grassland. Pratacul Sci 32(11):1781–1788. [in Chinese with English abstract]
Wang YX (2000) Oasis ecosystem and its environmental characteristics. Arid Land Geogr 23(1):7–12. [in Chinese with English abstract]

Wang T (2010) Some issues on oasification study in China. J Desert Res 5:995–998. [in Chinese with English abstract]
Wang YJ, Jiao L (2000) Oasis divisions and basic types in China. Arid Land Geogr 23(4):344–349. [in Chinese with English abstract]
Wu ZY (1980) Vegetation in China. Science Press, Beijing, pp 956–960. [in Chinese]
Yao YS (2004) The theory and practice of high efficiency and high quality of cotton in Xinjiang. Xinjiang Science and Technology Book Press, Ürümqi, pp 160–240. [in Chinese with English abstract]
Yin LK (1997) Diversity and *ex-situ* conservation of plants in the desert region of temperate zone in China. Chinese Biodiv 5(1):40–48. [in Chinese with English abstract]
Yu LH, Hou H (2004) Occurrence and control measures of *Locusta migratoria* in Xinjiang. China Plant Prot 9:25–26. [in Chinese]
Zhang FC, Liu XT (2000) Relation of Xinjiang water resources exploitation and use with natural environment variation and sustainable development. Hydropower Station Des 16(3):13–18. [in Chinese with English abstract]
Zhang LY, Wang NA (1995) On the oases genesis types and evolution in time and space. J Arid Land Res Environ 9(3):32–42. [in Chinese with English abstract]

Chapter 6
Biological Invasions in Nature Reserves in China

Hui Guo, Susan J. Mazer, Xinyu Xu, Xi Luo, Kailing Huang, and Xiaohong Xu

Abstract One consequence of the rapid economic development in China is the rapid increase in the ecological threats posed by alien species. Nature reserves serve as barriers to species invasions, but there is now no reserve in the world that is free from introduced alien species. In this chapter, we review studies of alien species invasions in nature reserves of China, analyze the invasion pattern across reserves, and propose suggestions for the management of invasive species in nature reserves in China. By searching available databases, we found a total of 37 studies focusing on biological invasions in 24 nature reserves in China. The Dinghushan Nature Reserve has the largest number of invasive species, followed by Taohongling, Tianmushan and Ganshiling reserves, whereas Yiwulvshan, Yalujiangkou and Dayudao have the fewest invasive plant species. *Alternanthera philoxeroides, Amaranthus spinosus* and *Euphorbia hirta* are the species most frequently reported to occur in nature reserves, while Compositae, Amaranthaceae and Gramineae are the three most frequently reported families. The number of invasive species reported declines with increasing latitude, but is not significantly correlated with reserve age, elevation, or area. We analyzed an index of invasion (the ratio of invasive to total plant species), which increases with nature reserve age, and decreases with elevation and the number of plant species in reserves; the effect of elevation, however, is not detected when controlling for the number of plant species or reserve area. When controlling for the effect of reserve area, the number of invasive species is positively correlated with the total number of plant species. Although manual, chemical,

H. Guo (✉) • X. Xu • X. Luo • K. Huang
College of Resources and Environmental Sciences, Nanjing Agricultural University, Nanjing, Jiangsu 210095, China
e-mail: hui.guo@njau.edu.cn

S.J. Mazer
Department of Ecology, Evolution & Marine Biology, University of California, Santa Barbara, CA 93106, USA

X. Xu
Virtual Simulation Experiment Teaching center of Agricultural Biology, Nanjing Agricultural University, Nanjing, Jiangsu 210095, China

F. Wan et al. (eds.), *Biological Invasions and Its Management in China*, Invading Nature - Springer Series in Invasion Ecology 11,
DOI 10.1007/978-94-024-0948-2_6

mechanical, and biological methods have been suggested for the control or eradication of invasive species, each method has its own limitation. We suggest that global changes and disturbances (e.g., N deposition) should be taken into account when assessing the risk of alien species and designing management strategies.

Keywords Biological invasion • Degree of invasion • Invisibility • Nature reserve

6.1 Introduction

Since the implementation of the "Reform and opening" policy in the 1970s, China has achieved and maintained a high growth rate in gross domestic product and international trade. Economic development has induced habitat destruction and/or fragmentation as well as air and water pollution, and their combined effects have caused a dramatic decline of plant biodiversity; as many as 5000 species are currently endangered in China (López-Pujol et al. 2006). One of the consequences of rapid economic development in China is the dramatic increase in the threats posed by alien species (Ding et al. 2008). For example, the total numbers of harmful alien animals, plants, and other pest organisms intercepted at Chinese borders grew at least tenfold from 1990 to 2005 (Ding et al. 2008). The invasion of alien species may not only displace native and agricultural species, but it may also cause damage to habitats and ecosystems. In sum, in addition to affecting ecosystems and contributing to the extinction of native species, invasive species can cause major socioeconomic damage.

Nature reserves can help to prevent invasions by alien species by providing a buffer of relatively undisturbed vegetation; invasions appear to be most likely in disturbed habitats. For example, in an analysis of more than 27,000 non-native plant presence records in South Africa's Kruger National Park, Foxcroft et al. (2010) reported that the number of records of non-native invasive plants inside the park declined rapidly beyond 1500 m from the edge.

In China, the first modern nature reserve was founded in 1956 and new reserves were regularly established until 1964 (Zhujing 1989; Fig. 6.1). After a six-year 'retreat' stage from 1965 to 1972, the establishment of natural conservation areas in China continued again. In the 1990s, as public awareness of conservation grew, even with growing tension between activities that promoted economic development and practices that promoted conservation of natural resources, the number of national nature reserves constructed per year increased, reaching a peak around 2000 (Fig. 6.1). After 2000, many new nature reserves were set up by provinces or counties rather than by the nation. By the end of 2014, China had established a total of 2718 national and local reserves, primarily forests (52.4%), wildlife refuges (24.5%) and inland wetlands (13.91%) (Fig. 6.2). Compared with the natural conservation areas in developed countries such as United States, those in China are apparently much younger (Appendix I).

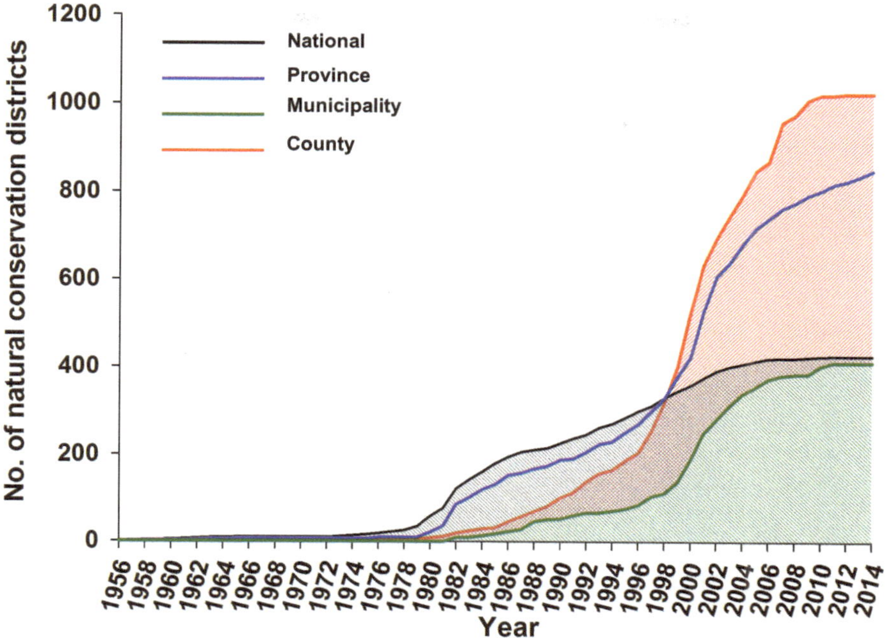

Fig. 6.1 The cumulative number of natural conservation districts set up since 1956

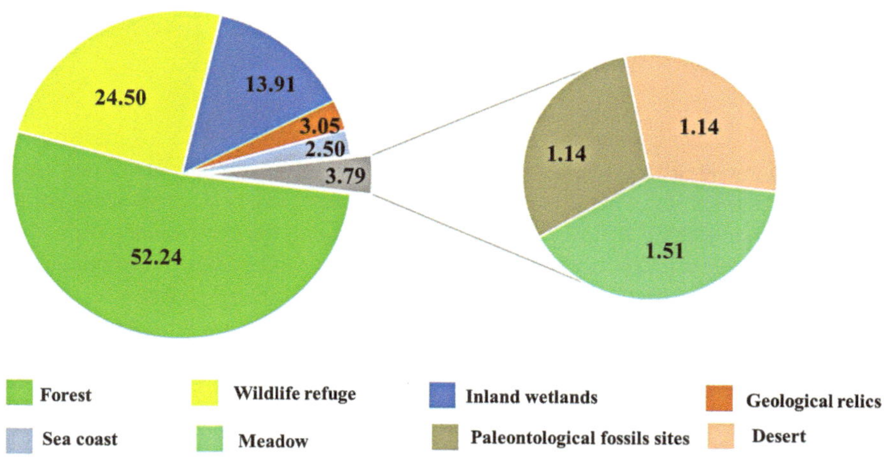

Fig. 6.2 The percentage of different types of natural conservation districts

Because of rugged mountains in the west and vast desert areas in central China, the population is concentrated within a surprisingly small area (Riley 2004). Previous studies in other countries have reported a positive relationship between the extent of exotic species' invasion and human disturbance or visitation. For example, in a survey of 234 New Zealand reserves, Timmins and Williams (1991) found that the proximity of towns, distance from roads/railway lines, human use, reserve shape and habitat diversity were the most important factors determining the degree of invasion. Similarly, others have reported that an increase in the number of human visitors led to an increase in the number of exotic species (Usher 1988; Macdonald et al. 1989; Lonsdale 1999). Given that the introduction of plant species from one region to another is closely related to human activities, China's high population density poses a greater risk of exotic species' invasions into reserves than that faced by countries with small or low-density populations.

High population density and economic growth have also resulted in elevated atmospheric nitrogen deposition in terrestrial and aquatic ecosystems (Liu et al. 2013). Over the past 30 years, China's emissions have increased to the point where it has become by far the largest creator and emitter of nitrogen globally (Liu et al. 2011). N deposition levels have commonly exceeded 20 kg N ha−1 year−1 in central and east China, a level above which ecosystems health is seriously threatened (Bobbink et al. 2010). In ecosystems with predominantly nutrient-poor soils, the addition of nutrients can constitute a major disturbance, which has been shown in many examples to facilitate invasions by non-native species (Hobbs 1992).

In this chapter, we review studies of alien species invasions in nature reserves, analyze the invasion pattern across reserves, and offer several suggestions for the management of invasive species in China.

6.2 Brief Review of the Studies on Alien Species Invasions in Nature Reserves in China

Nature reserves are a suitable "laboratory" for studying the effect of undisturbed habitats on the likelihood of alien species invasions. Indeed, studies in various kinds of nature reserves (e.g., national parks, biosphere reserves, and small-scale nature reserves) have yielded a few generalizations that apply at both global (Lonsdale 1999) and national levels (Pyšek et al. 2002), but few studies have been conducted in China. Here, we first summarize those studies that directly address the process and outcome of biological invasions in nature reserves and conservation areas in China.

By searching the available databases (CNKI, WANFANG DATA and Web of Science) using the keywords, "invasion", "exotic" or "alien" or "introduced" "invasive species", "nature reserve" or "natural conservation area" or "protected area", and "China", we found only 37 studies focusing on biological invasions in 24 nature reserves (Appendix II). In most of these studies, the authors simply listed the names of the invasive plants that had been reported in focal reserves, and briefly introduced their invasion histories or inferred the possible paths of their invasions. For example, invasive alien plant species in the Dinghushan National Nature Reserve were investigated using transects and plot surveys, and a total of 14 invasive alien plant species were found in the experimental, buffer, and core zones (Song et al. 2009). In the Shiwandashan National Natural Reserve, 23 invasive species representing 21 genera and 12 families were reported (Wei et al. 2006). Similar investigations were conducted in the Dashahe nature reserve in Guizhou province; Bawangling and Ganshiling nature reserves in Hainan province; Hengshuihu nature reserves in Hebai province; Jinfoshan nature reserve in Chongqing city; Jinhuacha nature reserve in Guangxi Zhuang Autonomous Region; and Lishan nature reserve in Shanxi province.

A few studies focused on single invasive plant species. For example, *Solidago canadensis* (Asteraceae), a noxious invasive plant in the Lushan Natural Reserve of Jiangxi Province, was studied to reveal its pattern of spread and to identify its potential effects on biodiversity in a woodland (Tang et al. 2012). While low sunlight and the predominance of foggy weather of the Lushan Nature Reserve were not suitable for the spread of *S. canadensis,* the herb layer was nevertheless dominated by this species, which caused a significant decline in the diversity of co-occurring species (Tang et al. 2012). Qin et al. (2004) reported the effects of the invasive *Spartina alterniflora* on the flora, fauna, and ecosystem of the Chongming island and Jiuduansha nature reserves, and they proposed strategies for controlling *S. alterniflora* in coastal nature reserves. Bai et al. (2013) reported the consequences of *Phyllostachys edulis* invasions for plant community composition by examining the changes in plant diversity in different communities over a seven-year period (2005–2011) in the Tianmushan Nature Reserve. *P. edulis* negatively affected plant communities by causing significant declines in two measures of diversity: Simpson's Diversity Index and Pielou's Eveness Index (Bai et al. 2013).

We identified only five studies that reported invasions of animals or insects into nature reserves in China (Jiang et al. 2007, 2010; Luo et al. 2007, 2012; Zeng et al. 2013). We found no experimental studies that identified the factors that promote invasive species and no comparative studies examining patterns of invasions across nature reserves. Similarly, we found no studies that identified the process by which any alien species has invaded a nature reserve.

The shortage of relevant studies could be due to several factors. First, the importance and value of nature reserves in resisting biological invasions has not received sufficient attention by plant community ecologists and conservation biologists. Second, there might not be adequate funding available to support the research of plant ecologists engaged in basic scientific research. Our understanding of the effects of invasive species on the structure, composition, and function of natural ecosystems would be improved if more financial resources were allocated to researchers to conduct experimental and comparative studies of the effects of invasive species on species diversity and ecosystem function. Third, the management policies of government-supported nature reserves do not place a high priority on the study and control of invasive species. Fourth, public awareness of the ecological risks posed by introduced species needs to be increased. The preservation of China's biological diversity for future generations can only be ensured if government, media, and educational and research institutions collectively take the responsibility for educating and informing the public of its value.

6.3 Invasive Species in Nature Reserves in China

Our survey of the basic attributes of nature reserves in China is presented in Appendix III. The number of invasive plant species per reserve ranges from 3 to 51, wherein Dinghushan has the largest number of invasive species, followed by Taohongling, Tianmushan and Ganshiling reserves; by contrast, only three invasive species have been reported in Yiwulvshan, Yalujiangkou and Dayudao (Fig. 6.3).

Alternanthera philoxeroides, Amaranthus spinosus and *Euphorbia hirta* are the three exotic species that have been reported most frequently in the 24 well-studied nature reserves in China (these species have been reported in 17,17 and 15 reserves, respectively), followed by *Erigeron annuus*, *Bidens pilosa*, *Chromolaena odorata* and *Ipomoea purpurea* (Table 6.1). Compositae, Amaranthaceae and Gramineae are the three families most frequently reported in the 24 nature reserves, followed by Euphorbiaceae, Solanaceae, Convolvulaceae and Leguminosae (Fig. 6.4). Exotic species representing other families are relatively rare and only occasionally found in these reserves (Fig. 6.4).

Annual invasive plant species have the highest frequency across all nature reserves, accounting for 388 occurrences (Fig. 6.5a; Table 6.1). Perennial invasive plant species also are very high in frequency, accounting for 266 occurrences, whereas biennial invasive plant species are comparatively rare (Fig. 6.5a). When invasive plant species are classified by growth form, herbaceous species have the highest frequency across reserves, while shrubs, trees and climbing vines are rarely recorded (Fig. 6.5b).

Only 11 invasive animal and insect species have been reported in nature reserves in China. Specifically, *Ondatra zibethica* (muskrat), *Neovison vison* (American mink), *Oncorhynchus mykiss* (Steelhead Trout) and *Amphiesma vibakari* (Asian keelback) were reported to invade the Changbaishan reserve and to have large effects on the populations of native species (e.g., wild duck) (Luo et al. 2012); *Procambarus clarkia* (Red Swamp crayfish), *Rana catesbeiana* (Bullfrog) and *Pomacea canaliculata* (Channeled applesnail) were found in the Poyanghu reserve (Zeng et al. 2013); *Pomacea canaliculata, Hemiberlesia pitysophila* (Pine needle scale) and *Oracella acuta* (Mealybug) were reported to invade the Jiulianshan reserve (Jiang et al. 2010); *Rattus norvegicus* were found in the Shedaolaotieshan reserve (Luo et al. 2007) ; and *Hyphantria cunea* (Fall webworm) were found in the Kunyushan reserve (Jiang et al. 2007). *Pomacea canaliculata* was the only species to have been reported in two reserves (Poyanghu and Jiulianshan reserves).

Although there are no invasive species reported in other reserves, this does not mean they are absent. On the contrary, the invasive species listed above have a high probability of having invaded reserves with similar habitats or environments. Additional investigations are urgently needed in other nature reserves, so as to avoid or limit large losses of ecosystem function and economic values that may result from unreported and uncontrolled biological invasions.

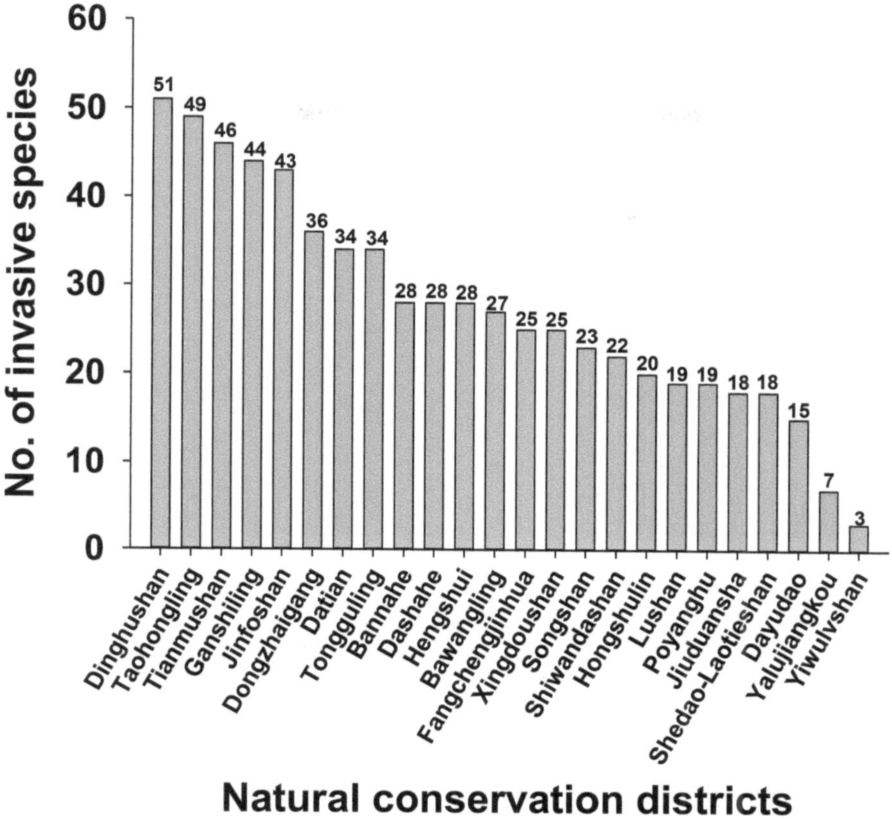

Fig. 6.3 The number of invasive species recorded in each of the 24 natural conservation districts

6.4 Patterns of Plant Invasions in Natural Conservation Districts in China

Although it seems that there is now no nature reserve in the world that is free of introduced alien species, nature reserves were reported to be invaded about half as often as sites outside reserves (Lonsdale 1999). The degree to which a nature reserve is invaded depends on a variety of factors, including the attributes of the alien species, the diversity of native species, the area of a reserve, the degree of human disturbance, ongoing environmental change, etc. (Usher 1988; Lonsdale 1999). As these factors change in the future, the importance of alien species in nature reserves may increase. Here, we synthesize data across reserves in order to identify the primary factors that appear to determine the extent of invasions by alien plant species into China's nature reserves. We compiled and then analyzed a data set based on the published literature; given that there are limited data available for animals and insects, we focus on invasive plants here.

Table 6.1 The frequency of the 20 most commonly reported invasive plant species among 24 nature reserves

Species	Count	Probability
Alternanthera philoxeroides	17	0.025
Amaranthus spinosus	17	0.025
Euphorbia hirta	15	0.022
Erigeron annuus	14	0.021
Bidens pilosa	13	0.019
Chromolaena odorata	13	0.019
Ipomoea purpurea	13	0.019
Ageratum conyzoides	12	0.018
Amaranthus viridis	12	0.018
Chenopodium ambrosioides	12	0.018
Amaranthus retroflexus	11	0.016
Axonopus compressus	10	0.015
Mimosa pudica	10	0.015
Crassocephalum crepidioides	9	0.013
Ipomoea cairica	9	0.013
Oxalis corymbosa	9	0.013
Ricinus communis	9	0.013
Avena fatua	8	0.012
Conyza canadensis	8	0.012
Datura stramonium	8	0.012

Count refers to the number of reserves in which each species has been reported; probability refers to the proportion of occurrence of each species to the overall occurrence of all invasive species

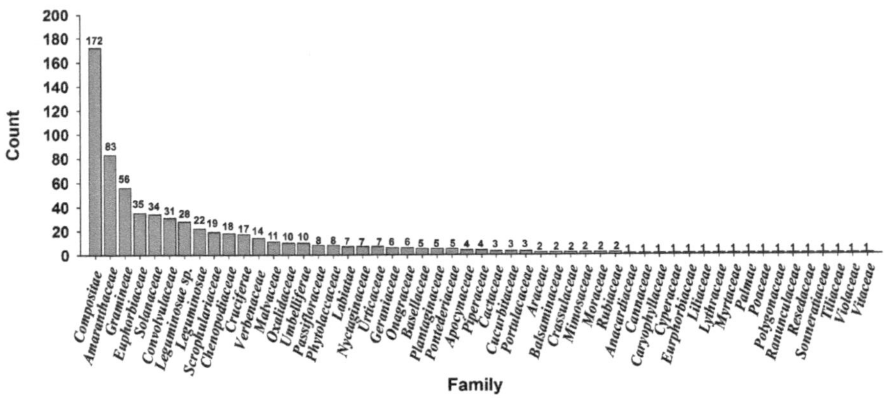

Fig. 6.4 The count of species classified by family reported in natural conservation districts (Count refers to the number of reserves in which each family has been reported)

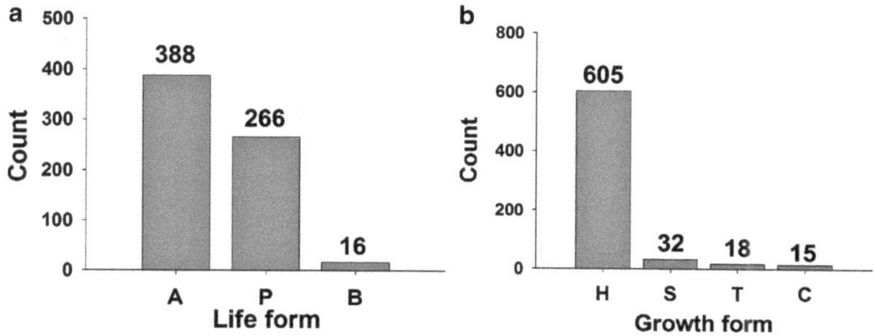

Fig. 6.5 The count of invasive species classified by life form (**a**) and growth form (**b**). *A* annual, *P* perennial, *B* biennial, *H* herb, *S* shrub, *T* tree, *C* climbing vine (Count refers to the number of reserves in which each life form or growth form has been reported)

6.4.1 Data Set

The data used in this section were extracted primarily from the primary scientific literature and from the Ministry of Environmental Protection of the People's Republic of China. Species lists were obtained for each of the 24 nature reserves (Appendix III). The total number of all plant species and the number of invasive plant species were recorded for each reserve. For each nature reserve, the following attributes were also obtained:

Year Established We aimed to determine whether the number of invasive species increases with the age of a reserve. If older nature reserves were established prior to any major disturbance, they may exhibit greater resistance to alien species invasions than younger ones.

Mean Elevation The mean elevation of each nature reserve was estimated as the midpoint between minimum and maximum elevation recorded (Pyšek et al. 2002). Given that habitats and vegetation change with increasing elevation, different elevations may be differentially invasible to alien species. We tentatively hypothesized that nature reserves at higher elevations may be less invasible due to lower temperatures and more stressful abiotic environments.

Latitude We recorded the boundary latitudes including both highest and lowest values across the range of each reserve. Temperate ecosystems are predicted to be more invasible than tropical ecosystems (Lonsdale 1999).

Nature Reserve Area One controversy is whether several small reserves will contain more species than would a single reserve of equal total area. Large reserves are able to harbour larger populations and contain greater habitat diversity but well-designed and carefully managed small reserves can effectively protect more populations of rare species than a single large area that is not well-designed (Primack

2000). By analyzing the relationship between reserve area and the number of invasive species, we tried to evaluate whether small nature reserves are more vulnerable to invasions than large ones. This question has been repeatedly raised but not rigorously answered (Usher 1988).

Native Plant Species Native species richness may predict the degree of exotic species richness because it reflects not only the area of the site but also its habitat diversity (Lonsdale 1999). Here, we examined the correlation between the total number of plant species and the number of invasive species across reserves in order to investigate the effect of the flora's diversity on the degree of invasions. Our primary hypothesis is that – all else being equal – relatively species rich communities are more stable, have fewer vacant niches and are less invasible than relatively depauperate communities; an alternative hypothesis is that communities with more plant species also have higher habitat diversity, which may facilitate the introduction of a greater number of alien species.

Mean Annual Temperature & Mean Annual Precipitation These variables were obtained from the China Meteorological Administration and used to detect whether the extent of plant invasion is correlated with climatic factors.

6.4.2 Data Analyses

Linear bivariate regressions were conducted to detect whether the focal dependent variable (the number of invasive plant species) was correlated with each independent variable, i.e., the attributes of nature reserves represented by year, latitude, reserve area, the number of native plant species, mean annual temperature and precipitation. In addition to using the number of invasive species as the dependent variable, we conducted a separate set of linear bivariate regressions in which we used an index of invasion, V, as the dependent variable, where V is defined as:

$$V = 100 * I / S$$

where I is the number of invasive species, and S is the number of plant species (Usher 1988). Because the Dongzhaigang, Jiuduansha and Hongshulin reserves belong to typical tidal ecosystems characterized by relatively low plant species richness compared to other ecosystems, we excluded these three reserves when analyzing the relationship between V and the independent variables above. We also examined the relationship, among reserves, between the number of invasive species and the residuals of the linear regression of the number of plant species on the reserve area. Using these residuals allowed us to control statistically for the effect of reserve area when examining the relationship between the number of invasive species and the number of plant species per reserve.

6.4.3 *Factors Correlated with the Degree of Plant Invasion in Nature Reserves*

We estimated Pearson pairwise correlations between all pairs of independent variables (Table 6.2). The number of invasive species is significantly correlated only with latitude; reserves at higher latitudes have a lower number of invasive species. The index of invasion, however, is negatively correlated with elevation, the number of native plant species and reserve age (Table 6.2).

6.4.3.1 Established Time

There is no significant relationship between the number of invasive species and the year in which a reserve was established (Fig. 6.6a). However, the index of invasion was positively related to the year of establishment ($r^2 = 0.21$, $P = 0.030$, $n = 21$; Fig. 6.7a). In younger reserves, the ratio of invasive species to total plant species was significantly higher than in older ones. This pattern suggests that older nature reserves show higher resistance to alien species invasions than younger ones, perhaps because the older reserves were established at locations with less disturbance.

6.4.3.2 Elevation

No significant correlation was found between the number of invasive species and elevation (Fig. 6.6b), whereas the index of invasion significantly decreased with increasing elevation ($r^2 = 0.41$, $P = 0.0012$, $n = 21$; Fig. 6.7b and Table 6.2). The reason is that the total number of plant species increases with increasing elevation, but the number of invasive species remains constant (the correlation is not significant between the number of invasive species and elevation). This result is consistent with the hypothesis that nature reserves at higher elevations may have lower invasibility due to their lower temperatures and more stressful abiotic environments. Consequently, as native species expand their ranges into higher elevations in response to global warming, invasive species may not have the same potential to migrate to higher elevations.

6.4.3.3 Latitude, Annual Temperature and Annual Precipitation

A significant negative relationship was found between the number of invasive species and latitude ($r^2 = 0.18$, $p < 0.0324$, $n = 24$; Fig. 6.6c), but not between the index of invasion and latitude (Fig. 6.7c). These results suggested that the reserves in lower latitudes may be more susceptible to invasions than those in high latitudes. The independence between the index of invasion and latitude indicates that the ratio of invasive to total plant species does not change with increasing latitude; however, we do not know whether other

Table 6.2 The Pearson pairwise correlation coefficients between all focal variables

	Year established	Elevation (m)	Latitude	Area (hm²)	# of plant species	Average annual Temp. (°C)	Annual precipitation (mm)	# of invasive species	Index of invasion
Year established	1.00	−0.12	0.18	0.20	−0.40	−0.057	−0.43*	−0.26	0.46*
Elevation (m)		1.00	−0.12	0.18	0.67***	−0.12	0.33	0.16	-0.64**
Latitude			1.00	0.24	−0.24	−0.91***	−0.66***	−0.43*	0.016
Area (hm²)				1.00	0.27	−0.31	−0.02	−0.33	−0.38
# of plant species					1.00	−0.10	0.39	0.37	−0.63**
Average annual Temp. (°C)						1.00	0.56**	0.26	0.17
Annual Precipitation (mm)							1.00	0.25	−0.37
# of invasive species								1.00	0.019
Index of invasion									1.00

Significance levels:*** $P<0.001$; ** $P<0.01$; * $P<0.05$

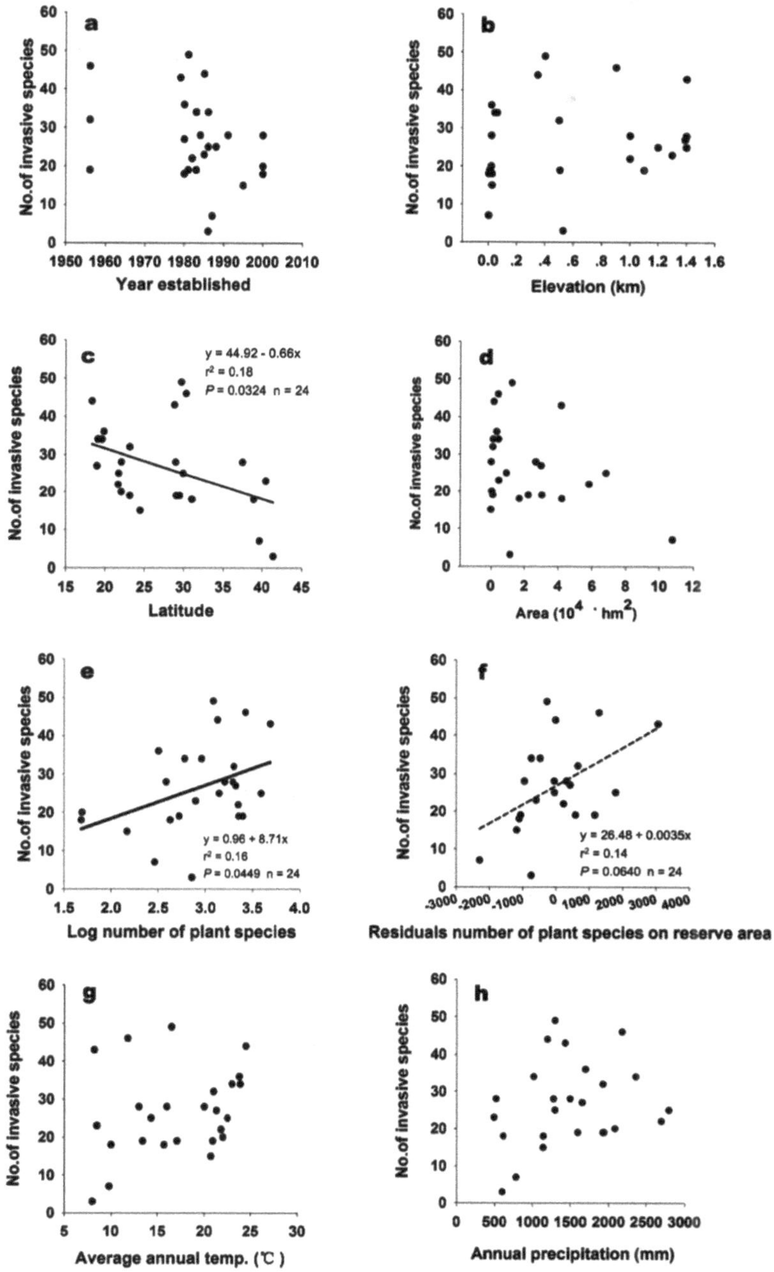

Fig. 6.6 The linear regression relationships between the number of invasive species and (**a**) the established year; (**b**) the elevation; (**c**) the latitude; (**d**) the reserve area; (**e**) the log number of plant species; (**f**) the residuals of the total number of plant species on the reserve area; (**g**) the average temperature and (**h**) the annual precipitation of natural reserves

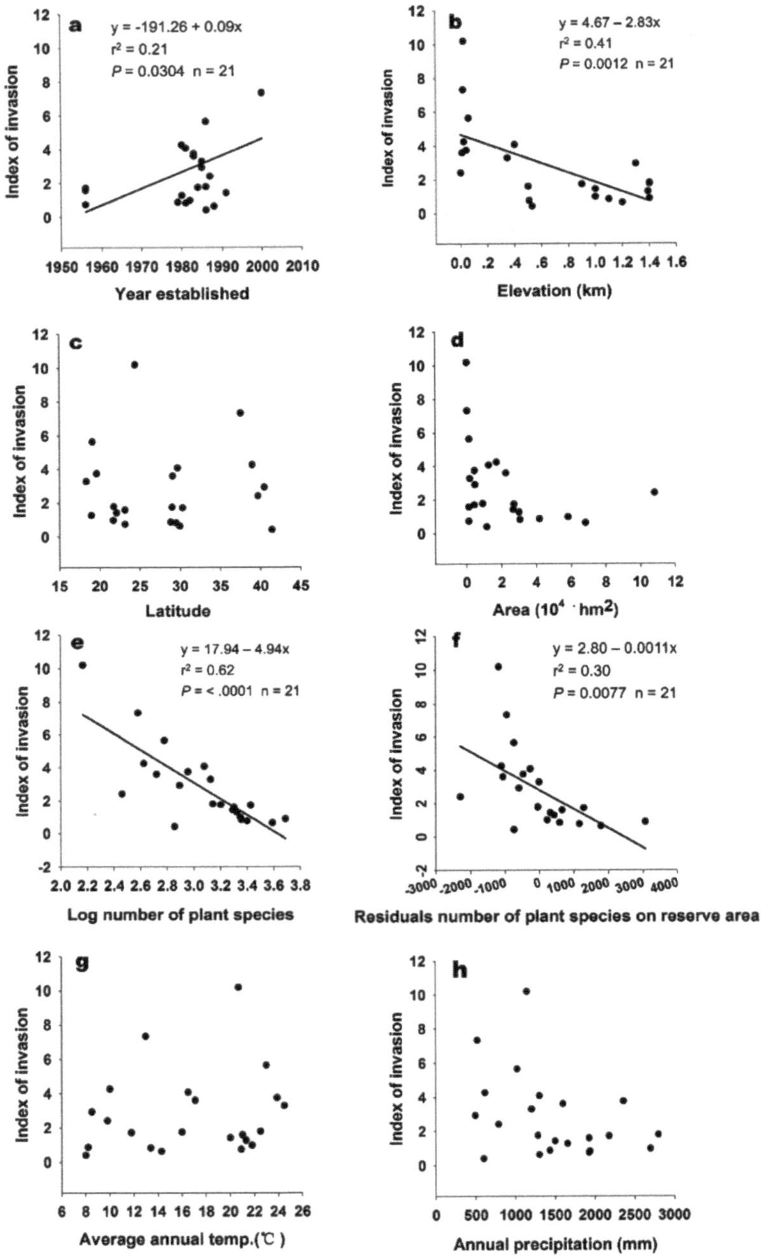

Fig. 6.7 The linear regression relationships between the index of invasive and (**a**) the established year; (**b**) the elevation; (**c**) the latitude; (**d**) the reserve area; (**e**) the log number of plant species; (**f**) the residuals of the total number of plant species on the reserve area; (**g**) the average temperature and (**h**) the annual precipitation of natural reserves

characteristics of native or invasive species may change with latitude. Neither the number of invasive species nor the index of invasion was correlated with mean annual temperature or precipitation (Figs. 6.6g, h and 6.7g, h), but these climatic variables provide only a rough approximation of a large or topographically complex reserve's climate.

6.4.3.4 Nature Reserve Area

The results of this study with respect to reserve area and invasibility of reserves could potentially inform strategies for reserve establishment and design. We found that there was no significant linear relationship between the number of invasive species or index of invasion and reserve area (Figs. 6.6d and 6.7d). However, a non-linear relationship existed between the index of invasion and reserve area ($y = 4.58–0.01x^{1/2}$; $p < 0.015$, n = 21; Fig. 6.7d), indicating that the largest reserves have a dramatically lower number of invasive species and index of invasion than smaller ones.

6.4.3.5 Native Plant Species

We detected a significant relationship between the number of invasive species and the total number of plant species ($r^2 = 0.16$, P = 0.0449, n = 24; Fig. 6.6e). When controlling for the effect of reserve area, a nearly significant positive linear relationship was detected ($r^2 = 0.14$, P = 0.0640, n = 24; Fig. 6.6f). The positive relationship between invasive and plant species richness found here is consistent with the hypothesis that richer plant communities are evidence of greater habitat diversity; if so, then both native and alien plant species respond to the greater habitat diversity in a similarly positive way. However, among reserves, the index of invasion was negatively correlated with the number of plant species ($r^2 = 0.62$, $p < 0.0001$, n = 21; Fig. 6.7e), which suggests that reserves with higher species richness may be more stable and less susceptible to invasions than those with less plant species. The negative correlation between index of invasion and the number of plant species was independent of reserve area ($r^2 = 0.30$, $p < 0.0077$, n = 21; Fig. 6.7f).

The multivariate general linear models also indicated that the number of native plant species was the most important factor in controlling the number of invasive species and index of invasion, independent of variation in elevation or area (Table 6.3).

6.4.3.6 Coastal Reserves

The three coastal reserves – Dongzhaigang, Hongshulin and Jiuduansha – have very high indices of invasion (v = 11.36, 40.82 and 37.5, respectively: almost 20 times higher than other reserves (which have a median value of 1.78). These results suggest that the plant communities in reserves located on coasts are characterized by a much larger fraction of invasive plant species. One interpretation is that the coastal

Table 6.3 Multivariable analyses using general linear models

		Number of invasive species			Index of invasion	
Model 1 $R^2 = 0.3622$	**Source**	**DF**	**F Ratio**	**Prob > F**	**F Ratio**	**Prob > F**
	Number of plant species	1	5.82	**0.0267**	1.90	0.1841
	Elevation (m)	1	0.48	0.4960	3.37	0.0826
	Area (hm²)	1	5.51	**0.0305**	1.85	0.1903
Model 2 $R^2 = 0.1669$	**Number of plant species**	1	3.21	0.0890	2.69	0.1171
	Elevation (m)	1	0.40	0.5341	3.25	0.0872
Model 3 $R^2 = 0.3451$	**Number of plant species**	1	6.93	**0.0164**	10.51	**0.0043**
	Area (hm²)	1	5.68	**0.0277**	1.66	0.2125

Model 1: Dependent variable: Number of invasive species and Index of invasion; Factors: Number of plant species, Elevation (m) and Area (hm^2); Model 2: Dependent variable: Number of invasive species and Index of invasion; Factors: Number of plant species, Elevation (m). Model 3: Dependent variable: Number of invasive species and Index of invasion; Factors: Number of plant species, Area (hm^2)

reserves are more invasible due to their higher vulnerability and diffusibility relative to inland sites; another interpretation is that coastal reserves contain so few native species that even a few invasive species result in a high index of invasion (the average number of native species is 1581 in inland reserves, and 138 in coastal reserves). Coastal and island reserves should receive more attention in preventing and managing invasive species.

6.5 Risk Assessment and Management of Invasive Species in Nature Reserves in China

A central goal of protecting natural areas is to preserve biodiversity and to sustain ecosystem function. While nature reserves or protected areas provide barriers to invasive species (e.g., Foxcroft et al. 2010), the management of invasive vegetation in natural reserves is a significant challenge. Therefore, to assess the risk of alien invasive species and to identify and implement effective methods for managing invaded vegetation are important and urgent objectives. In this section, we review the management strategies previously proposed in studies of nature reserves in China and offer our suggestions based on the most recent publications.

6.5.1 The Management Strategies Proposed by Chinese Government and Researchers

The Ministry of Environmental Protection of the People's Republic of China provides little general advice for controlling alien invasive species. For example, it's suggested that relevant people or departments should investigate alien species regularly; draw up a plan for invasive species prevention and monitoring; replace invasive species by local species when restoring the vegetation; and enhance public education to increase awareness of alien species, etc. While the government is paying more attention and providing more funding to manage alien invasive species, there are few specific policies, regulations, and implementation of the above suggestions in specific reserves.

By reviewing studies of biological invasions in nature reserves in China, we found that most authors proposed methods and strategies for invasive species management, and even discussed the potential effectiveness of each method. For example, in studies of Tongguling, Dongzhaigang, Datian, Shiwandashan, Xingdoushan, Fangchengjinhuacha and Dinghushan reserves, it has been suggested to build databases of alien invasive species and to promote early warning and emergency response systems (Qin et al. 2008; Wei et al. 2006; Lu et al. 2005; Wu and Li 2012; Song et al. 2009). In addition, the authors encourage using manual, mechanical and biological methods to control invasive plants. Manual removal is time-consuming but can be a highly effective method for controlling invasive plants. One of the negative effects is that pulling plants from the ground may cause unintended soil disturbance, and subsequently result in further invasions. Although mechanical removal was always suggested, this method may not be appropriate in natural areas because of the serious disturbance to soils and non-target vegetation caused by heavy equipment. If it's essential to use heavy equipment to remove invasive species, then plans for replanting and monitoring native plants following mechanical removal should be carefully developed prior to mechanical removal. There was no successful case of biological control reported so far, even though it's appealing and may be implemented in the future.

Chemical control of invasive plants was suggested by several authors. For example, herbicides were thought to be effective for the control of invasive plants, as reported in studies of the Dashahe (Lin et al. 2008), Huacha (Wu et al. 2009) and Tianmushan reserves (Chen et al. 2011). While herbicides or pesticides were supposed to be effective, we have not seen relevant laws or regulations that control the training and certification of those who apply these chemicals. Anyone who applies herbicides in natural areas should have basic training in herbicide application technology and be able to comply with all the instructions and directions for use.

Some researchers suggested that the government should enhance public education (Lu et al. 2005; Qin et al. 2008; Wu and Li 2012). The importation and spread of invasive plants can be significantly reduced by public education. People who are aware of the problems caused by invasive species have the responsibility to educate others about their identity, impacts, and control so as to reduce further ecological destruction or degradation of native ecosystems.

Another method suggested by Chinese researchers is to explore the use of invasive species as natural resources (Liu et al. 2008; Chen et al. 2011). It has been proposed, for example, that some invasive plants may potentially be used as Chinese medicine (Chen et al. 2011).

6.5.2 Suggestions: Consider Global Changes and Disturbance

Global changes and biological invasions are primary threats to global biodiversity that may interact in the future (Dukes and Mooney 1999). For example, a meta-analysis revealed that in terrestrial (primarily plant) systems, native and non-native species responded similarly to environmental changes, but in aquatic (primarily animal) systems, increases in temperature and CO_2 largely inhibited native species (Sorte et al. 2013). As climate change proceeds, aquatic systems may be particularly vulnerable to invasions. Across systems, there could be a higher risk of invasions at sites that become more climatically hospitable, whereas sites shifting towards harsher conditions may become more resistant to invasions (Sorte et al. 2013). Therefore, climate change and its impacts on native vs. alien species should be taken into account when assessing the risk of alien species and drawing up strategies for their management.

In addition, conservation efforts must include explicit consideration of disturbance to preserve biodiversity and ecosystem function. While moderate frequency or intensity of disturbance fosters high species richness, major disturbances in plant communities can promote alien species invasions. For example, input of atmospheric nitrogen was apparently to blame for the increasing dominance of grass species and the loss of many forbs, regardless of management (mowing, grazing and burning) in grassland communities (Lake and Leishman 2004; Bai et al. 2010).

6.6 Take Advantage of Ecology Theories

6.6.1 Allee Effect

The Allee effect refers to a positive relationship between individual fitness and population size or density. The concept of Allee effects can be integral to risk assessments and to the prioritization of resources allocated to manage alien species, as some species with strong Allee effects may be less successful as invaders. While conservation biologists or restoration ecologists may attempt to minimize Allee effects (by maximizing genetic variation) in native species so that their local extinction is less likely, invasion biologists should consider the Allee effect as a factor that can be used to inhibit the establishment success or the spread of an invading species (Tobin et al. 2011). To reduce population size or density below an Allee threshold could be an effective strategy to control invasive species. The reduction of population size or density could simply be induced by the application of herbicide or pesticides or by manual removal. In this way, invasive species with strong Allee effects might be effectively controlled. Other methods could also be chosen to give rise to an Allee effect in invasive species. For example, the disruption of mating can prevent population growth, potentially contributing to population extinction. The release of natural enemies can also lead to widespread population reduction or extinction of an alien species, thereby causing a dramatic population collapse and range retraction of invasive species (Elkinton et al. 2004). Moreover, habitat loss or fragmentation could decrease population size and be exploited to slow down the rates of range extension and population spread, and even promote species eradication (Tobin et al. 2011). For example, propagules that arrive ahead of the expanding front could be strategically fragmented through control tactics or selectively targeted based on their density and spatial extent so that only fragmented colonies remain (Tobin et al. 2011).

6.6.2 Corridors

Habitat corridors that increase species dispersal and richness for many native taxa may have unintended negative effects, potentially increasing the spread of invasive species. In addition, corridors may increase the spread of unwanted disturbances such as fire. Haddad et al. (2014), however, conducted a literature review and meta-analysis to evaluate the prevalence of each of these negative effects, and found no evidence that corridors increase disturbance or non-native species invasions. Another study focused on river corridors, which provide crucial links to the surrounding landscape but are also major conduits for invasion of alien species (Foxcroft et al. 2007). Foxcroft et al. (2007) developed a framework to assess the risk that alien plants in watersheds adjacent to a protected area would invade the protected area along rivers; this framework was then applied to Kruger National Park (KNP) in South Africa. They reported that KNP was facing increasing pressure from alien species in the upper regions of the drainages of neighboring watersheds (Foxcroft et al. 2007). Their framework might be applicable to plants and other passively dispersed species that invade protected areas located in the lower regions of drainage basins. The inconsistency among results calls for more studies to seek evidence of corridor-mediated effects on invasive species or disturbance, and to provide theoretical support for invasive species management.

6.6.3 Invasibility

The success of a given species' invasions into a reserve depends on the invasibility of the focal plant community and on the invasiveness of the alien species. Invasibility can be defined as "the susceptibility of biological communities to colonization and dominance by introduced organisms" (Lonsdale 1999; Fridley 2011), and can be strongly influenced by community features such as species composition, diversity, and biomass. The degree of invasion in any given community is the outcome of previous interactions between its invasibility and the invasiveness of its alien species (e.g., propagule or invader attributes or traits), and depends on the proximity to exotic species sources such as ports and large cities, on disturbance, and on the time since invasions (Guo et al. 2015). To quantify the invasibility and degree of invasion in a given reserve based on unified indices of invasibility and degree of invasion can be used to uncover basic ecological patterns across systems, and to inform land management and ecological restoration (Guo et al. 2015). In future studies, reserve managers or conservation biologists need to measure community features (e.g., species richness, biomass, and area), the degree of invasion, and the invasiveness of alien invasive species in order to evaluate the potential for continued invasions and to generate effective management strategies.

6.7 Other Issues

6.7.1 Encouragement to Establish Non-profit Organizations for the Control of Invasive Species

The Florida Exotic Pest Plant Council (FLEPPC) is a nonprofit professional organization founded in 1984 to increase public awareness of the significant threat that alien invasive plant species pose to native species, communities, and ecosystems, and to develop integrated management and control strategies to halt the spread of exotic species in natural areas. FLEPPC maintains a list of plant species considered by a committee of botanists, ecologists, and land managers to be invasive in Florida. This list is available on the FLEPPC website (http://www.fleppc.org). Similar organizations or councils should be initiated in each province in China, and then integrated into a nationwide network that provides accessible databases of invasive species; such a resource would facilitate studies of invasion mechanisms and invasive species distributions across different spatial scales.

6.7.2 Policy-Making

Policy-making by government or managers should involve scientists (local scientists in particular) from different research disciplines. Local scientists often know most about their region's reserves and invasive species, and scientists with different backgrounds may identify different problems and propose alternative and complementary (or synergistic) solutions. In addition, specific reserves should develop customized strategies for controlling alien species invasion based on their specific attributes.

6.8 Conclusion and Perspectives

Nature reserves represent suitable laboratories for studying the factors that serve as barriers to rapid invasions. By searching the available peer-reviewed literature, we found only 37 studies that focused on biological invasions in 24 nature reserves in China. In particular, there was shortage of studies that identify mechanisms or patterns of biological invasions. In the present study, we found that the age of nature reserves, latitude, and the number of plant species influence the number of invasive species or the index of invasion in nature reserves. The positive relationship between the number of invasive species and total number of plant species found in the present study (Fig. 6.6e) provides indirect support for the hypothesis that richer plant communities contain greater habitat diversity. Moreover, coastal and island reserves should receive greater attention and protection against further invasions due to their relatively high indices of invasion.

Although China's government is paying more attention and providing more funding to manage alien invasive species, the Ministry of Environmental Protection of the People's Republic of China provides little guidance for the control of alien invasive species, and there is a lack of specific policies, regulations, and implementation of management strategies or methods in China's nature reserves.

We suggest that both global change and disturbance (e.g., N deposition) should be taken into account when assessing the risk of alien species and developing management strategies. Meanwhile, ecological concepts such as the Allee effect may be important and potentially applied to controlling alien species in nature reserves.

In this chapter, we first conducted a brief survey of studies regarding biological invasions and nature reserves in China. Many nature reserves that have not been studied yet are likely to have been invaded. More investigations and studies should be conducted in other nature reserves, so as to avoid large losses of ecosystem function and economically important resources from biological invasions that have not yet been reported. To assess the risk of alien invasive species and find viable methods to manage invaded vegetation is important and urgent for the sustainability of China's wild species.

Acknowledgements The authors thank S.H. Zhong, F. LUO, Z.J. Bai, Fei Yang and F.Y. Yuan for assistance in searching publications and compiling data. We are grateful to Prof. M.X. Jiang and Prof. A.B. Zhan for valuable comments and suggestions. The research was supported by NSFC program (no. 31100298), Jiangsu Natural Science Foundation (Bk20161445) and Key Project of NAU (0306 J0887 & KYTZ201404).

References

Bai YF, Wu JG, Clark CM, Naeem S, Pan QM, Huang JH, Zhang LX, Han XG (2010) Tradeoffs and thresholds in the effects of nitrogen addition on biodiversity and ecosystem functioning: evidence from Inner Mongolia grasslands. Glob Chang Biol 16:e358–e372

Bai SB, Zhou GM, Wang YX, Liang QQ, Chen J, Chen YY, Shen R (2013) Plant species diversity and dynamics in forests invaded by Moso bamboo (*Phyllostachys edulis*) in Tianmu Mountain Nature Reserve. Biodivers Sci 21(3):288–295. (in Chinese with English abstract)

Bobbink R, Hicks K, Galloway J et al (2010) Global assessment of nitrogen deposition effects on terrestrial plant diversity: a synthesis. Ecol Appl 20(1):30–59

Chen J, Xu P, Yao ZS (2011) Analysis and control of invasive plants of Tianmushan Nature Reserve. China Medical Herald 8(9):137–139. (in Chinese with English abstract)

Ding JQ, Mack RN, Lu P, Ren MX, Huang HW (2008) China's booming economy is sparking and accelerating biological invasions. Bioscience 58:317–324

Dukes JS, Mooney HA (1999) Does global change increase the success of biological invaders? Trends Ecol Evol 14:135–139

Elkinton JS, Liebhold AM, Muzika RM (2004) Effects of alternative prey on predation by small mammals on gypsy moth pupae. Popul Ecol 46:171–178

Foxcroft LC, Rouget M, Richardson DM (2007) Risk assessment of Riparian plant invasions into protected areas. Conserv Biol 21:412–421

Foxcroft LC, Jarŏs'lk V, Pÿsek P, Richardson DM, Rouget M (2010) Protected-area boundaries as filters of plant invasions. Conserv Biol 25:400–405

Fridley JD (2011) Invasibility of communities and ecosystems. In: Simberloff D, Rejmanek M (eds) Encyclopedia of biological invasions. University of California Press, Chicago, pp 356–360

Guo QF, Fei SL, Dukes JS, Oswalt CM, Iannone BV, Potter KM (2015) A unified approach for quantifying invasibility and degree of invasion. Ecology 96:2613–2621

Haddad NM, Brudvig LA, Damschen EI, Evens DM, Johnson BL, Levey DJ, Orrock JL, Resasco J, Sullivan LL, Tewksbury WSA (2014) Potential negative ecological effects of corridors. Conserv Biol 28:1178–1187

Hobbs RJ (1992) Disturbance, diversity, and invasion: implications for conservation. Conserv Biol 6:324–337

Jiang MK, Qin WG, Wang Z, Xu WG, Cai L, Zhang LR, Li J, Qiang S (2007) An investigation of exotic species in costal nature reserves in China. 13:37–43 (in Chinese)

Jiang Y, Zeng ZY, Zhang ZB (2010) The effect of alien species invasion on. Jiulianshan Nat Reserve of Jiangxi province 38(17):9127–9128

Lake JC, Leishman MR (2004) Invasion success of exotic plants in natural ecosystems: the role of disturbance, plant attributes and freedom from herbivores. Biol Conserv 117:215–226

Lin MX, Liu ZY, Han F, Ren MB, Zhang J (2008) Weed Sci 1:31–32. (in Chinese)

Liu F, Tao GD, Wang DS (2008) Investigation on the alien invasive plants in Nabanhe Nature Reserve and precautionary measures. For Invent Plan 33(6):112–117. (in Chines with English abstract)

Liu X, Duan L, Mo J, Du E, Shen J, Lu X, Zhang F (2011) Nitrogen deposition and its ecological impact in China: an overview. Environ Pollut 159(10):2251–2264

Liu X, Zhang Y, Han W, Tang A, Shen J, Cui Z, Vitousek P, Erisman JW, Goulding K, Christie P, Fangmeier A, Zhang F (2013) Enhanced nitrogen deposition over China. Nature 494:459–462

Lonsdale WML (1999) Global patterns of plant invasions and the concept of invasibility. Ecology 80:1522–1536

López-Pujol J, Zhang FM, Ge S (2006) Plant biodiversity in China: richly varied, endangered, and in need of conservation. Biodivers Conserv 15(12):3983–4026

Lu SF, Liu SX, Fang YP (2005) A preliminary research of invasive species around Xingdoushan National Nature Reserve, Hubei. J Huanggang Normal Univ 25(3):48–52. (in Chinese with English abstract)

Luo Q, Qin WH, Zhou SB, Wang Z, Wang L, Jiang MK (2007) Status of invasion of exotic species into snake island and their ecological influence. Chin J Wildlife 28(3):29–32

Luo YM, Piao ZJ, Wang H, Wang C, Sui YC, Zeng FW, Luo YJ, Wang ZC (2012) Effect of invasive animals on rivers biodiversity in the Changbai Mountain Nature Reserve. J Beihua Univ (Nat Sci) 13(5):578–580. (in Chinese with English abstract)

Macdonald IAW, Loope LL, Usher MB, Hamann O (1989) Wildlife conservation and the invasion of nature reserves by introduced species: a global perspective. In: Drake JA et al (eds) Biological invasions: a global perspective. Wiley, Chichester, pp 215–255

Primack RB (2000) A primer of conservation biology. Sinauer Publication, Sunderland

Pyšek P, Jarošík V, Kucera T (2002) Patterns of invasion in temperate nature reserves. Biol Conserv 104:13–24

Qin WH, Wang Z, Jiang MK, (2004) Invasion of *Spartina alterniflora* to Dongtan and Jiuduansha Nature Reserves. Weed Sci 04–0015-02

Qin WH, Wang Z, Xu WG, Jiang MK (2008) Investigation and analysis on alien invasive plants in three national nature reserves in Hainan. J Plant Resour Environ 17(2):44–49

Riley NE (2004) China's population: new trends and challenges. Popul Bull 59(2):1–36

Song L, Cao F, He YH, Qiang S, Qin WH, Jiang MK (2009) A survey of invasive alien plant species in Dinghushan National Nature Reserve. J Zhejiang For Coll 26(4):538–543. (in Chinese with English abstract)

Sorte CJB, Ibáñez I, Blumenthal DM, Molinari NA, Miller LP, Grosholz ED, Diez JM, D'Antonio CM, Olden JD, Jones SJ, Dukes JS (2013) Poised to prosper? A cross-system comparison of climate change effects on native and non-native species performance. Ecol Lett 16:261–270

Tang MZ, Ren MX, Zheng JM, Ding JQ (2012) Seasonal dynamics of plant diversity in woodland invaded by *Solidago canadensis* (Asteraceae) in Lushan Nature Reserve. Plant Sci J 30(4):366–373. (in Chinese with English abstract)

Timmins SM, Williams PA (1991) Weed numbers in New Zealand's forest and scrub reserves. N Z J Ecol 15:153–162

Tobin PC, Berec L, Liebhold AM (2011) Exploiting Allee effects for managing biological invasions. Ecol Lett 14:615–624

Usher MB (1988) Biological invasions of nature reserves – a search for generalizations. Biol Conserv 44:119–135

Wei YL, Ye D, Wen YG, Men YY (2006) Invasive plant species in Shiwandashan Mountain National Natural Reserve, Guangxi, China. China For Sci Technol 20(6):23–26. (in Chinese with English abstract)

Wu RH, Li FY (2012) J Green Sci Technol 11:47–49 (in Chinese)

Wu RH, Pan ZP, Huang KT (2009) Pop Sci Technol 2009(8):151–152. (in Chinese)

Zeng ZY, Wu HP, Jiang Y, Peng GH (2013) Dynamics and impacts of invasion by exotic species to Poyang Lake National Nature Reserve. Energy Res Manag 4(15–18):35. (in Chinese with English abstract)

Zhujing Z (1989) Nature Conservation in China. British Ecol Soc 26(3):825–833

Chapter 7
Roles of Chinese Government on Prevention and Management of Invasive Alien Species

Rui Wang, Fanghao Wan, and Bo Li

Abstract The invasion and expansion of alien species in China is rising sharply in recent decades, due to increased transport, trade, and travel. China has become one of the countries most seriously affected by invasive species. In response to the severe threat of invasive alien species in China, the national strategies for managing invasive species have been developed, which defined the roles and responsibilities of government. In the past decades, Chinese government has initiated a series of activities to build administrative capacity, strengthen research capacity, promote international cooperation and raise public awareness. In addition, Chinese government promoted the implementation of coordinated measures and cooperative efforts throughout China and development of new prevention and control technology to prevent or minimize adverse impacts of invasive alien species (IAS) on biodiversity, as well as their consequences for the economy and human health and well-beings. But to meet the challenge of IAS in the new era, continued support of both existing and new initiatives should be reinforced or established.

Keywords Biological invasions • China • Invasive alien species • National management strategy • National activities

R. Wang • F. Wan (✉)
State Key Laboratory for Biology of Plant Disease and Insect Pests, Institute of Plant Protection, Chinese Academy of Agricultural Sciences, Beijing 100193, China
e-mail: wanfanghao@caas.cn

B. Li
Department of Science, Technology and Education, Ministry of Agriculture, Beijing 100125, China

F. Wan et al. (eds.), *Biological Invasions and Its Management in China*, Invading Nature - Springer Series in Invasion Ecology 11,
DOI 10.1007/978-94-024-0948-2_7

7.1 Introduction

The introduction of species beyond their natural range is rising sharply due to the unprecedented rate and magnitude of globalization. Increased international trade, transport and travel have enormously promoted the biological invasions in China, which now has become one of the countries most seriously affected by invasive species (Wan et al. 2002; Lin et al. 2007; Ding et al. 2008). In China, invasive alien species cost the forestry and agriculture industries billions of dollars annually in increased costs of production, harvest losses and losses to international markets, threatening the sustainable development of Chinese forestry and agriculture (Xu et al. 2006; Wan et al. 2009). For example, the outbreak of tobacco whitefly (*Bemisia tabaci* Genndius) has caused fetal losses to vegetable production (Hu et al. 2011). Moreover, invasive alien species have severe and often irreversible impacts on native ecosystems, severely threatening China's biodiversity and genetic resources (Wan et al. 2002, 2005, 2008). For example, Crofton weed (*Argentina adenophora*), an invasive alien weed has formed mono-dominant colonies in the invaded area by its strong reproductive capacity and rapid dispersal as well as its competition and allelopathy effects to repel surrounding native plants (Yang et al. 2006). The weed spreads rapidly in large areas of southwestern China, excluding a large number of native plants and simplifying flora composition as well as corresponding fauna (Niu et al. 2007; Wan et al. 2009). Additionally, some invasive species are affecting public health. For example, important fire ant is ferocious, and it often builds its nest near the residential area and attacks the people interfering with nest (Lu et al. 2008; Ju et al. 2012). As international trade, travel and other factors are increasing the frequency of new species introductions to China, the threats posed by invasive alien species would much severer in the future.

As a response to biological invasion, China has developed the national strategies for managing invasive species. In the strategies, the roles and responsibilities of government in were clearly defined, and the government has initiated a series of activities.

7.2 National Management Strategies for IAS

The IAS problem is not only induced by IAS themselves, but also related to human activities and ecosystems (Wan et al. 2011a, b). Thus, biological invasions are not only a scientific topic to be addressed by academia, but also a management issue that government and the public has to face, both at regional and international levels. Accordingly, the China national management strategies of IAS focus on three aspects: building administrative capacity, strengthening research capacity, promoting international cooperation and raising public awareness.

Managing biological invasion is a systematic project which requires collaborating of different administrative sectors. Therefore, the development of administrative capacity mainly focuses on the establishment of policy and legislation to clarify leadership and coordination roles between different levels of governments and define their duties and responsibilities for managing IAS, including establishing early detection and warning systems, initiating rapid responses, building information sharing platforms, and designing national plans and priority actions to prevent new invasions, etc.

The strengthening of research capacity mainly focused on, by providing sufficient financial support, conducting in-deep fundamental and applied research (Wan et al. 2009). Different emphases should be put for fundamental and applied research. The fundamental research is mainly to explore key scientific issues of invasion process, including invisibility of alien species, introduction mechanism, IAS population establishment and eco-adaption evolution, IAS competitive, exclusion mechanisms and displacement effect on native species, invasion impacts on structure and function of ecosystem. While applied research mainly focuses on developing prevention and control technologies, including risk assessment and early warning techniques for potential IAS, real-time inspection techniques and rapid and precise detection system for invaded and potential invasive species, field monitoring, emergency and eradication of worst IAS, biological control and ecological restoration of worst IAS in specific ecosystem.

As invasive alien species are an international problem, international cooperation and extensive information exchange should be established by making national cooperation plan for joint issues with common interests. Similarly, within the country, public education should also be promoted via various media, plans for public campaign to support management of biological invasions as raising public awareness is key to changing public perceptions and values and creating support for programs directed at controlling or preventing invasive alien species.

7.3 National Activities on Biological Invasions in China

In accordance with the national strategies as described above for managing IAS, Chinese government has initiated a series of activities to realize the objectives.

7.3.1 Activities for Developing Administrative Capacity

Clear leadership or appropriate coordination is now in place in China for IAS prevention and mitigation, which involve relevant sectors and different levels of government. Since 2003, Chinese government implemented series of activities for managing biological invasions. Firstly, central government (State Council) assigned Ministry of Agriculture (MOA) to be in charge of the leadership and coordination among different ministries and departments of Central and Local governments in 2003. MOA's role and responsibility include: establishing a coordinating platform

Table 7.1 Partial information on eradication activities carried out in China in the past 15 years

Year	Target species number	Number of provinces	Number of counties
2003	3	3	15
2005	5	10	100
2015	20	21	>1000

for dealing IAS problem among different ministries and departments, coordinating the IAS actions of different departments, making national strategies and plans for prevention and management of IAS, and developing international cooperation.

To fulfill its duty and responsibility, MOA has established three new organizations to meet the requirements of the administrative, research and monitoring of IAS. MOA firstly established Office for Alien Species Management in 2003 for coordinating and dealing with IAS problems among different governmental ministries and departments. In the same year, Center for Management of Invasive Alien Species (CMIAS) was established to advance IAS research on basic and applied sciences and technologies. In 2012, MOA established Rural Energy & Environment Agency to improve its ability in warning and monitoring of IAS.

In addition to establishment of new organizations, MOA has also carried out several other activities. They are: (i) drafted a 10-year work plan to identify specific initiatives with other governmental ministries, (ii) conducted an evaluation on the current legal and regulatory authorities relevant to invasive species, (iii) worked on IAS risk assessment criteria and procedure and screening 52 most important invasive alien species for priority actions, and (iv) launched a nation-wide integrated eradication and control activities on 15 IAS pests since 2003. In 2003, these activities focused on only three major invasive alien species in 15 counties of three provinces. Until 2015, these activities had covered 20 major species in more than 1000 counties of 21 provinces (Table 7.1).

Additionally, several regulations, contingency plans and technical specifications have been drafted by national department, province or local government in China. Hunan province issued provincial regulation for management of IAS in 2011. Ministry of Environmental Protection issued technical guideline for assessment on environmental risk of alien species in 2012. Almost all the provinces and more than 100 counties released contingency plans for management of important IAS.

7.3.2 *Activities for Developing Research Capacity*

Research plays a critical role in the management of invasive alien species as it is the way in which new information is obtained and communicated. Complying with strategy framework for research capacity building, a series of research programs were launched by different ministries including Ministry of Agriculture (MOA), Ministry of Science and Technology (MOST), and National Natural Science

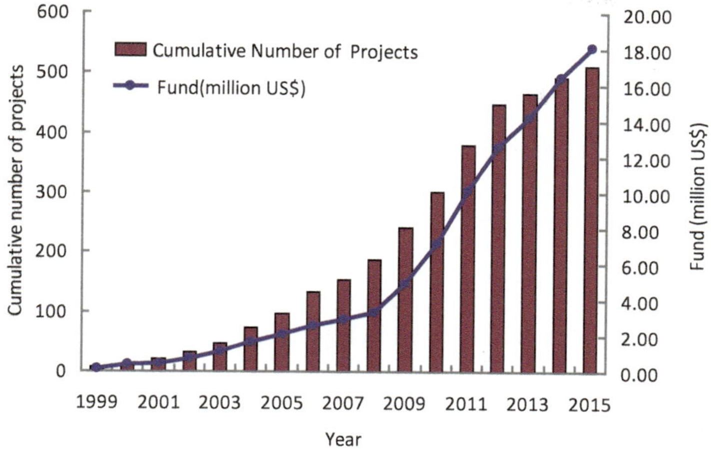

Fig. 7.1 Projects sponsored by NSFC from 1999 to 2015

Foundation of China (NSFC), etc. Generally, these programs can be classified into two types, fundamental research and applied research.

Since twenty-first century, China has started to put more attention on fundamental research of IAS. Fundamental research projects are mainly funded by the MOST and NSFC. Until 2015, MOST and NSFC have funded approximately US$ 30 million for fundamental research projects. MOST has supported two successive projects under national basic research and development program ("973" program) in the field of fundamental research on invasive mechanisms of IAS from 2003 to 2013. Generally, these projects focus issues in invasion biology, including population establishment and range expansion of IAS, ecological adaption and evolution of IAS, and structure breakdown and function recession of invaded ecosystems.

Since 1999, NSFC have funded more than 500 projects with accumulative fund exceeding 18 US$ million and both number of projects and funds have increased steadily (https://isisn.nsfc.gov.cn/egrantindex/funcindex/prjsearch-list) (Fig. 7.1). Compared to the national basic research and development program, NSFC funded projects cover more scientific questions including risk assessment and early warning, detection and surveillance, genetic diversification and evolution, ecological adaptation mechanisms, interactions between invasive species and native species, the process of expansion and outbreak, resistance of ecosystems to biological invasions, ecological and economic impacts of biological invasions, and the impacts of human activities on biological invasions and control and management. And the scientific questions covered by the NSFC funded projects has broadened gradually since 1999 and 10 main scientific questions of invasion biology have been covered after 2007 (Table 7.2).

In addition to fundamental research, a series of applied research programs have been sponsored by different ministries (MOA, MOST, etc). Since 2006, seven projects were supported by Ministry of science and technology focusing on exploring

Table 7.2 NSFC sponsoring range of general projects in Invasion Biology

	Scientific questions									
Year	RA	D/S	GV	EA	IBS	PE	RE	EEI	HAI	CM
1999	√					√		√		√
2000								√		√
2001	√			√	√			√		
2002	√					√				√
2003			√		√			√		
2004		√	√	√		√				
2005		√	√		√	√	√	√	√	√
2006		√	√	√	√	√	√	√	√	√
2007	√	√	√	√	√	√	√	√	√	√
2008–2015	√	√	√	√	√	√	√	√	√	√

RA risk assessment, *D/S* detection/surveillance, *GV* genetic variation, *EA* ecological adaptation, *IBS* interactions between species, *PE* population expansion, *RE* resistance of ecosystem, *EEI* ecological and economic impacts, *HAI* human activity impacts, *CM* control & management

risk assessment and early warning techniques for potential IAS, developing real-time inspection techniques and rapid and precise detection system for invaded and potential invasive species, setting up bio-control of IAS in agriculture and forestry, IAS survey and bio-safety evaluation system in China. Since 2007, MOA and Ministry of environment protection funded more than 10 projects for development of new control technology for major invasive species e.g. Crofton weed, coastal plain yellow-tops, the western flower thrips, oriental fruit fly, tobacco whitefly, spiraling whitefly, etc.

7.3.3 International Cooperation and Public Awareness

According the national strategy, we carried out widely international cooperation. Firstly, Chinese government cooperated and harmonized international agreements related to IAS (such as CBD, IPPC, etc). Secondly regional strategies were developed for adjacent countries, bi-lateral & multi-lateral cooperation. For example by sponsoring the APEC workshop on IAS, the APEC IAS Strategy Framework was submitted to and endorsed by APEC senior officials and ministers.

A series of public awareness education activities were conducted through various media, such as TV and training course. For example since 2003, Center for Management of Invasive Alien Species (CMIAS) and Rural Energy & Environment Agency have organized more than 20 training courses for administrative staffs, local officials, technicians and the public to improve their technical skills and administrative capacity on control and management of IAS and public's understanding and preventive awareness of IAS.

7.4 Future Perspectives for Managing IAS in China

In the past decades, Chinese government has put great efforts on managing the IAS in China and made great progresses in administrative and research capacity building in managing IAS. Generally, achieved progress promoted the development and implementation of coordinated measures and cooperative efforts throughout China to prevent or minimize adverse impacts of invasive alien species (IAS) on biodiversity, as well as their consequences for the economy and human health. But to meet the challenge of IAS in the new era, the following should be reinforced or established.

Firstly, new national legislation specific to management and control of invasive alien species should be issued and existing regulations related to management and control of alien species should be enforced. While current regulations in China provide a foundation for protecting plants and habitats from invasive alien species, a complete legal analysis should be conducted to evaluate all existing legislative and regulatory coverage relating to invasive alien species in China, identify weakness or gaps, assess operational capabilities and recommend means by which a comprehensive, integrated national/provincial legal framework can be established. Additionally, the regulations and supervision systems should be standardized for isolation and quarantine of introduced species. Implementing these measures should help change the current incomplete supervision of introduced species. Strict inspection should be conducted on various biological resources introduced intentionally.

Secondly, national guideline for research and management of biological invasions should be developed to promote the practical implementations in relevant national and provincial departments. According to the current conditions of administrative management and scientific research, the guideline should comprise the medium and long-term development plans on prevention, risk evaluation, early warning, fast detection, epidemic monitoring, and communication, effective control technique, ecological restoration or repair of ecosystems, etc.

Thirdly, based on the current monitoring and early warning systems for agricultural, forestry and marine ecosystems, centers for IAS monitoring and early warning should be established at national and regional levels to form a monitoring and early warning network across the country. A reporting mechanism should be established for IAS monitoring and early warning. A national rapid response system between different departments as well as rapid response task force within each department should be set up to make standard management plans for first discovery of IAS to later dispose and to form an effectively coordinated operation mechanism from the central and local governments. National emergency management plan for IAS and general principles of emergency management procedures should be set as well as corresponding provincial management system.

Additionally, dissemination of scientific knowledge and public education on biological invasions should be enhanced. A complete database system and information sharing platform should be set up through integrating all the information of established and potential IAS in China. A national technique training base should be established for prevention and control of biological invasions. Public awareness education should be improved through various vectors such as lectures on internet, TV and radios or training courses.

Finally, international cooperation and extensive information exchange should be reinforced by making national cooperation plan for joint issues with common interests. The comprehensive cooperation should be conducted on risk warning and management, classic biological control technology, ecological repair technology, area-wide integrated management technology, control technology and strategy, etc.

References

Ding JQ, Mack RN, Lu P, Ren MX, Huang HW (2008) China's booming economy is sparking and accelerating biological invasions. Bioscience 58:317–324

Hu J, De Barro P, Zhao H, Wang J, Nardi F, Liu SS (2011) An extensive field survey combined with a phylogenetic analysis reveals rapid and widespread invasion of two Alien Whiteflies in China. PLoS One 6(1):e16061. doi:10.1371/journal.pone.0016061

Ju RT, Li H, Shih CJ, Li B (2012) Progress of biological invasions research in China over the last decade. Biodivers Sci 20(5):581–611

Lin W, Zhou GF, Cheng XY, Xu RM (2007) Fast economic development accelerates biological invasions in China. PLoS One 2

Lu YY, Liang GW, Zeng L (2008) Study on xxpansion pattern of red imported fire ant, *Solenopsis invicta* Buren, in South China. Sci Agric Sin 41(4):1053–1063

Niu HB, Liu WX, Wan FH, Liu B (2007) An invasive aster (*Ageratina adenophora*) invades and dominates forest understories in China: altered soil microbial communities facilitate the invader and inhibit natives. Plant Soil 294:73–85

Wan FH, Guo JY, Wang DH (2002) Alien invasive species in China: their damages and management strategies. Biodivers Sci 10(1):119–125

Wan FH, Zheng XB, Guo JY (2005) Biology and management of alien invasive species in agriculture and forestry. Science Press, Beijing

Wan FH, Guo JY, Zhang F (2009) Research on biological invasions in China. Science Press, Beijing

Wan FH, Xie BY, Chu D (2008) Biological invasions: legislations and management strategies. Science Press, Beijing, p 316

Wan FH, Xie BY, Yang GQ (2011a) Invasions biology. Science Press, Beijing

Wan FH, Yan Y, Wang R, Yang GQ (2011b) Invasion biology: development and perspective in China. J Biosafety 20(1):1–19

Xu HG, Ding H, Li MY, Qiang S, Guo JY, Han ZM, Huang ZG, Sun HY, He SP, Wu HR, Wan FH (2006) The distribution and economic losses of alien species invasion to China. Biol Invasions 8:1495–1500

Yang GQ, Wan FH, Liu WX, Zhang XW (2006) Physiological effects of allelochemicals from leachates of Ageratina adenophora (Spreng.) on rice seedlings. J Allelopathy 18(2):237–246

Part II
Invasion and Management of Major Alien Insects in China

Chapter 8
The Whitefly *Bemisia tabaci* (Gennadius)

Xiaowei Wang and Nianwan Yang

Abstract The whitefly *Bemisia tabaci* (Gennadius) is a species complex containing at least 35 morphologically indistinguishable cryptic species. Some members of the complex are pests of agricultural and horticultural crops in temperate and tropical regions. During the past 20 years, two species of the complex, Middle East-Asia Minor 1 (hereafter MEAM1) and Mediterranean (hereafter MED), which have been commonly referred to as the B and Q 'biotypes', have risen to international prominence due to their global invasions. In the middle-1990s, the MEAM1 species invaded China most probably with the import of infested plants and seedlings and has become a pest since the late 1990s. In 2003, the MED species of the *B. tabaci* complex was first recorded in China and it is now the dominant species in the Yangtze River Valley and eastern coastal areas. In this chapter, we first reviewed the invasion histories of MEAM1 and MED whiteflies in China and their negative effects. Then, the research progresses on behavior, biotic, environmental and molecular mechanisms of MEAM1 and MED whitefly invasions and replacement of native whitefly species were discussed. Finally, the strategies for whitefly management in China were summarized. These research efforts have provided solid foundation for future investigations on the molecular mechanisms of whitefly invasions and are expected to open important avenues for the discovery of novel strategies for whitefly management in China.

Keywords *Bemisia tabaci* • China • Invasion • Management • Whitefly

X. Wang (✉)
Ministry of Agriculture Key Laboratory of Agricultural Entomology, Institute of Insect Sciences, Zhejiang University, Hangzhou 310058, China
e-mail: xwwang@zju.edu.cn

N. Yang (✉)
State Key Laboratory for Biology of Plant Diseases and Insect Pests, Institute of Plant Protection, Chinese Academy of Agricultural Sciences, Beijing 100193, China
e-mail: yangnianwan@caas.cn

F. Wan et al. (eds.), *Biological Invasions and Its Management in China*,
Invading Nature - Springer Series in Invasion Ecology 11,
DOI 10.1007/978-94-024-0948-2_8

8.1 Introduction

The whitefly *Bemisia tabaci* (Gennadius) is a phloem-feeding insect that lives predominantly in temperate and tropical regions (Byrne and Bellows 1991; Naranjo et al. 2010). It is highly polyphagous and attack more than 600 plant species including food, vegetable, fiber and ornamental host plants under field and greenhouse conditions (Brown 1995). *Bemisia tabaci* is now recognized as a species complex containing at least 35 cryptic species (Boykin and De Barro 2014; De Barro et al. 2011; Liu et al. 2012; Xu et al. 2010a). Some members of the complex are considerable pests of ornamental, vegetable and cotton production, causing damages directly through feeding and indirectly through the transmission of plant viruses (Fig. 8.1) (Naranjo et al. 2010). During the past 20 years, two species of the complex, Middle East-Asia Minor 1 (hereafter MEAM1) and Mediterranean (hereafter MED), which have been commonly referred to as B and Q whiteflies, have risen to international prominence due to their global invasions (Dalton 2006; De Barro et al. 2011; Liu et al. 2007). The invasive MEAM1 has been considered as one of the world's top 100 invasive species according to the International Union for the Conservation of Nature and Natural Resources (IUCN) (http://www.issg.org) and are regulated species in a number of countries, e.g., Australia, Africa, China, EU, and United States. Similarly, the MED species has also begun to invade a large number of countries in Africa, Asia, Europe and America from the Mediterranean region.

Fig. 8.1 Whiteflies cause damage directly through feeding and indirectly through the transmission of begomoviruses

Interestingly, despite MEAM1 and MED species of the *B. tabaci* complex are highly invasive, other members of the complex apparently lack the capacity to invade and establish in neighboring regions. It suggests that the invasive ability for MEAM1 and MED is actually uncommon. Therefore, unraveling the process and mechanism of whitefly invasions may be crucial to advance studies on their biology and introduce new methods to control them. In this review, we will outline the invasion processes and mechanisms of MEAM1 and MED whiteflies in China and then summarize the recent developments in the management of *B. tabaci*. These research efforts have provided valuable information for future investigations on the molecular mechanisms of whitefly invasions and are expected to open important avenues for the discovery of novel strategies for whitefly management.

8.2 The Invasion of *B. tabaci* in China

Although the invasions of *B. tabaci* have caused devastating consequences worldwide, the invasion history and processes of MEAM1 and MED have been poorly documented in most instances. The whitefly *B. tabaci* was first recorded in China in 1949. However, for several decades, this insect was not considered as a severe agricultural pest. In the middle-1990s, the MEAM1 species invaded China most probably with the import of infested plants and seedlings (Luo et al. 2002; Zhang 2000). Subsequently *B. tabaci* has gradually become a severe agricultural pest and significantly threatened the vegetable production in China since the late 1990s. The exotic MEAM1 whitefly species was found to be the dominant species in many agricultural production regions in 1990s (Luo et al. 2002). For example, among the 23 *B. tabaci* populations collected from 1999 to 2001 from 14 provinces, 17 populations were MEAM1. Interestingly, MEAM1 whiteflies mainly occur in regions near the coast or with convenient transportation while only six populations are other species distributed in the inland or mountainous regions (Qiu et al. 2003a). It was also found that during the invasion of *B. tabaci* MEAM1, a number of native whitefly species have been completely replaced in many regions, such as in Zhejiang Province, China during 2004–2006 (Liu et al. 2007).

In 2003, one population collected from poinsettia in Yunnan Province was identified as *B. tabaci* MED based on the mitochondrial cytochrome oxidase subunit I (mtCOI) gene (Chu et al. 2005). Later on, Chu et al. (2012) examined the whitefly populations collected from ten provinces and found that MED was only occasionally detected in Yunnan, Beijing and Henan and MEAM1 was still the dominant species in China. However, in the subsequent years, *B. tabaci* MED was detected in many regions in China and has replaced the previously well-established MEAM1. For instance, in 2001, among the 22 whitefly populations collected from 15 provinces, MED were found in 19 populations (Teng et al. 2010) . Of the 14 samples collected from 2008 to 2009, nine populations belonged to MED, three were MEAM1, and two were a mixture of MED and MEAM1 (Wang et al. 2010b). Hu et al. (2011) found that during June 2009 to March 2010 among the 16 provinces in south and east China, 12 provinces had MED. Pan et al. (2012) found that, for the 55 populations collected from 18 provinces, 43 populations were composed of MED. These data demonstrate that though the exotic MEAM1 was predominant across China in the past, the MED is now the dominant species in the Yangtze River and eastern coastal areas, and MEAM1 is the predominant species only in the south and south eastern coastal areas.

8.3 Invasion Mechanisms

The population dynamics of *B. tabaci* is governed entirely by four processes – birth, death, immigration and emigration. Each of them is governed by a wide array of interacting biotic and abiotic factors that determine whitefly occurrence and abundance over time and space (Naranjo et al. 2010). During the last decade, a number of factors that contribute to the invasion and spread of MEAM1 and MED in China have been investigated, such as mating interactions (Liu et al. 2007), host plant adaptation (Xu et al. 2015a), insecticide resistance (Pan et al. 2015), virus-vector-host mutualism (Luan et al. 2013a) and the roles of endosymbionts (Bing et al. 2013; Himler et al. 2011). These studies suggest that the successful invasions of MEAM1 and MED in China is largely due to their intrinsic biological traits, such as high insecticide resistance, wide host range and strong reproduction capacity. These findings will help us to develop novel strategies to manage the invasion and spread of MEAM1 and MED in China.

8.3.1 Asymmetric Mating Interactions

Reproductive interference is one of the major factors mediating species exclusion among insects. Liu et al. (2007) found that asymmetric mating interactions between different cryptic species of *B. tabaci* species complex was a driving force contributing to widespread invasions of the MEAM1 species in China and Australia. The authors conducted long-term field surveys, caged population experiments, and detailed behavioral observations in Zhejiang, China, and Queensland, Australia, to investigate the invasion process and its underlying behavioral mechanisms. During invasion and displacement, they found that the increased frequency of copulation led to increased production of female progeny among invaders, as well as reduced copulation and female production in indigenous genetic groups. Such asymmetric mating interactions may be critical to determining the capacity of a haplodiploid invader and the consequences for its closely related indigenous organisms.

The cryptic species MEAM1 and MED of *B. tabaci* complex have invaded many parts of the world and often exhibit niche overlap and reproductive interference. However, contrasting patterns of competitive displacement between the two invaders have been observed. Sun et al. (2014) investigated the role of reproductive interference on species exclusion between MEAM1 and MED in China. In mixed cohorts of the two species, MEAM1 always excluded MED in a few generations when the initial proportion of MEAM1 was ⩾0.25. Importantly, as MEAM1 increased in relative abundance, MED populations became increasingly male-biased. All these results suggest that MEAM1 should have a stronger reproductive interference than MED, leading to reduced frequency of copulation and female progeny in MED. These findings not only reveal the importance of reproductive interference in the competitive interactions between the two invasive whiteflies, but also provide a valuable framework against which the effects of other factors mediating species exclusion can be explored (Sun et al. 2014).

8.3.2 Host Plant Adaptation

The capacity of the MEAM1 and MED species of *B. tabaci* species complex to invade has often been linked to their presumed wider host range than the indigenous competitors. To determine whether the invasive MEAM1 species and the indigenous Asia II 1 whitefly species, commonly referred to as the 'ZHJ2 biotype', differ in their ability to use different host plants, Xu et al. (2011) compared their development, survival and reproduction on eight crop species/cultivars that are commonly cultivated in Zhejiang, China. Of the eight host plants tested, MEAM1 performed substantially better than ZHJ2 on squash, tomato and tobacco; MEAM1 and ZHJ2 preformed equally well on cotton and sweet potato, while ZHJ2 performed better than MEAM1 on kidney bean and pepper. These results indicate that while MEAM1 generally has a wider host range than many indigenous *B. tabaci*, an indigenous *B. tabaci* can perform as well as or better on some host plants. This suggests that the differential capacity to use various host plants between whitefly species is important in mediating the process of invasions by an alien whitefly species. Sun et al. (2013a) also examined the effects of host plants on the competitive interactions between the two species in the laboratory. When cohorts with equal abundance of MEAM1 and MED were set up on different host plants, MEAM1 displaced MED on cabbage and tomato in five and seven generations, respectively, but MED displaced MEAM1 on pepper in two generations.

Despite its importance, few studies have addressed the molecular mechanisms of *B. tabaci* against major plant secondary defense compounds. Alon et al. (2012) compared the gene expression between *B. tabaci* adults on wild-type *Nicotiana tabacum* plants or transgenic plants constitutively activating the phenylpropanoid/flavonoids biosynthetic pathway. Both the SSH and cDNA microarray analyses indicated a complex interaction between *B. tabaci* and secondary defense metabolites produced by the phenylpropanoids/flavonoids pathway, such as expression of detoxification, immunity, oxidative stress and general stress related genes. However, the elevated transcriptional activity was not accompanied by reduction of whitefly reproductive performance, indicating high adaptability of *B. tabaci* to this large group of plant secondary defense metabolites (Alon et al. 2012). The same research group also investigated the effects of aliphatic and indolic glucosinolates on the expression of detoxification genes in MEAM1 and MED whiteflies (Elbaz et al. 2012). The result suggests that the two species use rather different strategies to cope with plant defense responses. While MEAM1 utilizes inducible defenses, MED invests significant resources in being always 'ready' for a challenge (Elbaz et al. 2012). Xu et al. (2015a) compared the performance of the invasive MEAM1 and the indigenous Asia II 3 whitefly species following host plant transfer from a suitable host (cotton) to an unsuitable host (tobacco) and analyzed their transcriptional responses. Transcriptional analysis showed that the patterns of gene regulation were very different with most of the genes up-regulated in MEAM1 but down-regulated in Asia II 3. Compared to the constitutive high expression of detoxification genes in MEAM1, most of the detoxification genes were down-regulated in Asia II 3. Enzymatic activities of P450, GST and esterase further verified that the detoxification of MEAM1 was much higher than that of Asia II 3. These results reveal obvious differences in response to host transfer in MEAM1 and Asia II 3.

8.3.3 Roles of Vector–Virus–Plant Interactions in Biological Invasions

Begomoviruses are a group of single stranded DNA viruses exclusively transmitted by the whitefly *B. tabaci* in a persistent, circulative manner (Czosnek and Ghanim 2002). During this process, virions are acquired by the stylet of whitefly vectors from the plant phloem, moving along the esophagus to the midgut, then crossing the gut epithelial cells to the hemocoel, circulating with the hemolymph and reaching the salivary glands, and finally were secreted with saliva (Gray et al. 2014). During the circulation, viruses have engaged host factors or unique strategies for replication, movement, transmission and pathogenesis (Wei et al. 2014). Meanwhile, the insect vectors have evolved immunologic surveillance system against viral invasions (Luan et al. 2011; Wang et al. 2016). Interactions among vector insects, plant viruses and host plants are complex and diverse. These interactions may affect the process of biological invasions and the displacement of indigenous species by invaders when the invasive and indigenous organisms occur with niche overlap. Jiu et al. (2007) compared the performance of the invasive MEAM1 and indigenous Asia II 3 whiteflies on healthy, TbCSV-infected and TYLCCNV-infected tobacco plants. Compared to its performance on healthy plants, the invasive MEAM1 increased its fecundity and longevity significantly when feeding on virus-infected plants. In contrast, the indigenous Asia II 3 species performed similarly on healthy and virus-infected plants. The indirect mutualism between the MEAM1 whitefly and these viruses via their host plants, and the apparent lack of such mutualism for the indigenous whitefly, may contribute to the ability of the MEAM1 whitefly to invade and displace indigenous whiteflies. Interestingly, transcriptional analyses of MEAM1 whiteflies feeding on TYLCCNV-infected and uninfected tobacco plants indicated that the genes involved in the oxidative phosphorylation pathway and detoxification enzyme were down-regulated in whiteflies feeding on virus-infected plants (Luan et al. 2013a). The reduced detoxification activity is likely to attenuate energy costs, thus, enhancing the performance of whiteflies on virus-infected plants (Luan et al. 2013a).

Zhang et al. (2012) found that TYLCCNV and betasatellite coinfection suppresses jasmonic acid defences in the plant. Impairing or enhancing defences mediated by jasmonic acid in the plant enhances or depresses the performance of the whitefly. The result indicate that TYLCCNV can benefit its whitefly vector indirectly, through suppression of jasmonic acid-mediated plant defense (Zhang et al. 2012). To reveal why begomovirus infection can benefit its whitefly vector, Luan et al. (2013b) used a next-generation sequencing technology to identify defense genes differentially regulated in whitefly-infested and/or virus-infected tobacco. The authors found that many of terpenoid synthesis genes were up-regulated in whitefly-infested plants. In contrast, in TYLCCNV-infected leaves, most terpenoid genes were unchanged and five genes were declined. Interestingly, in co-infested plants, most terpenoid genes were unchanged and only three terpenoid genes were up-regulated. This study demonstrates that virus infection can deplete the terpenoid-mediated plant defense against whiteflies, thereby favoring vector–virus mutualism (Luan et al. 2013b). Furthermore, Li et al. (2014b) found that βC1 of TYLCCNV can suppress plant terpene biosynthesis. βC1 directly interacts with the basic helix-loop-helix transcription factor MYC2 to compromise the activation of MYC2-regulated terpene synthase genes, thereby reducing whitefly resistance.

Shi et al. (2013) showed that TYLCV directly and indirectly modified the feeding behavior of *B. tabaci* in favor of MED rather than MEAM1. They further quantified the salicylic acid (SA) titers and relative gene expression of SA in tomato leaves that were infested with viruliferous or non-viruliferous MEAM1 and MED. The authors found that SA titer was always higher in leaves that were infested with viruliferous MEAM1 than with viruliferous MED, whereas the SA titer did not differ between leaves infested with non-viruliferous MEAM1 and MED. Their results also indicate MED may have a mutualistic relationship with TYLCV that results in the reduction of the plant's defense response (Shi et al. 2013). Interestingly, He et al. (2015) found that when feeding on either cotton, a non-host of TYLCCNV, or uninfected tobacco, a host of TYLCCNV, virus-infection of the whiteflies impeded their feeding. Interestingly, when viruliferous whiteflies fed on virus-infected tobacco, their feeding activities were no longer negatively affected; instead, the virus promoted whitefly behaviour related to rapid and effective sap ingestion.

8.3.4 Roles of Endosymbionts in **B. tabaci** *Invasions*

The life history traits of multicellular organisms are often influenced by interactions with symbiotic species. The whitefly *B. tabaci* harbors a primary symbiont "*Candidatus* Portiera aleyrodidarum", which compensates for the deficient nutritional composition of its food sources and a variety of secondary symbionts. Interestingly, all of these secondary symbionts are found in co-localization with the primary symbiont within the same bacteriocytes, which should favor the evolution of strong interactions between symbionts. Recent study showed that *Rickettsia* sp. nr. bellii swept into a population of the invasive MEAM1 in just 6 years. Compared with uninfected whiteflies, Rickettsia-infected whiteflies produced more offspring, had higher survival to adulthood, developed faster, and produced a higher proportion of females (Himler et al. 2011). The *Rickettsia* thus functions as both mutualist and reproductive manipulator. The observed increased performance and sex-ratio bias of infected whiteflies are sufficient to explain the spread of Rickettsia across the southwestern United States.

Bing et al. (2013) tested five of the six S-endosymbiont lineages (excluding *Fritschea*) from 340 whitely individuals representing six putative species from China. *Hamiltonella* was detected only in the two exotic invaders, MEAM1 and MED. *Rickettsia* was absent in Asia II 1 and MED, scarce in Asia II 3 (13%), but abundant in Asia II 7 (63.2%), China 1 (84.7%) and MEAM1 (100%). *Wolbachia*, *Cardinium* and *Arsenophonus* were absent in the invasive MEAM1 and MED but mostly abundant in the native putative species. Furthermore, phylogenetic analyses revealed that some S-endosymbionts have several clades and different *B. tabaci* putative species can harbor different clades of a given S-endosymbiont, demonstrating further the complexity of S-endosymbionts in *B. tabaci*. These results demonstrate the variation and diversity of S-endosymbionts in different putative species of *B. tabaci*, especially between invasive and native whiteflies.

Owing to the importance of endosymbionts in whitefly biology, a number of studies have been carried out to reveal the function of whitefly endosymbionts. In order to gain insight into the metabolic role of each symbiont, Rao et al. (2015) analyzed the genome sequences of the primary symbiont *Portiera* and of the secondary symbiont *Hamiltonella* in MED. Sequencing results shows that the genome of *Portiera* is highly reduced (357 kb). It has kept a number of genes encoding most essential amino-acids and carotenoids, but lacks almost all the genes involved in the synthesis of vitamins and cofactors. Interestingly, *Hamiltonella* can not only provide vitamins and cofactors, but also complete the missing steps of some of the pathways of *Portiera*. The data suggests that *Portiera* and *Hamiltonella* are not only complementary but could also be mutually dependent to provide a full complement of nutrients to their hosts. On the other hand, genome analysis revealed that another secondary symbiont *Rickettsia* was unable to synthesize amino acids required for complementing the whitefly nutrition (Zhu et al. 2016). Parallel genomic and transcriptomic analysis further revealed that the host genome contributes multiple metabolic reactions that complement or duplicate *Portiera* function, and that *Hamiltonella* may contribute multiple cofactors and one essential amino acid, lysine. These results shows that bacteria with genomic decay enable host acquisition of complex metabolic pathways by multiple independent horizontal gene transfers from exogenous bacteria (Luan et al. 2015).

8.3.5 Roles of Pesticides Application in B. tabaci *Invasions*

At present, more than 50 insecticides have been employed to control the growth of *B. tabaci* populations and viral transmission (Horowitz et al. 2011). However, due to the rapidly rising resistance to insecticides, utilizing chemical agents to control *B. tabaci* is facing ever-increasing difficulties (Dennehy et al. 2010; Wang et al. 2009). During the last 20 years, a number of studies have been carried out to reveal the molecular mechanisms of whitefly resistance to insecticides. Pyriproxyfen is one of the major insecticides used to control the whitefly, however, whitefly resistance to pyriproxyfen has been observed in many regions (Crowder et al. 2007). To investigate the molecular basis underlying this resistance, a cDNA microarray was used to monitor changes in gene expression in a resistant *B. tabaci* population (Ghanim and Kontsedalov 2007). Functional analysis showed that many of the up-regulated ESTs were associated with resistance and xenobiotic detoxification, protein, lipid and carbohydrate metabolism and JH-associated processes (Ghanim and Kontsedalov 2007). Yang et al. (2013b) analyzed the differences between resistant and susceptible stains at both transcriptional and translational levels. Among the 1338 differentially expressed genes, 118 were putatively linked to insecticide resistance. The same research group also compared gene expression in the egg, nymph and adult stages of a thiamethoxam-resistant strain with a susceptible strain using a custom whitefly microarray (Yang et al. 2013a). Gene ontology and bioinformatic analyses revealed that at all life stages many of the DEGs encoded enzymes were involved in metabolic processes and/or metabolism of xenobiotics. In addition, several ATP-binding cassette transporters were highly over-expressed at the adult stage of the TH-R strain and may play a role in resistance by active efflux (Yang et al. 2013a).

The wide application of insecticides in China may be the key factor driving the rapid displacement of MEAM1 by MED. Under laboratory conditions, the MEAM1 can displace MED without the selection of insecticides (Wu et al. 2010). However, field surveys in China after 2003 indicate that in many regions MED has been replacing the earlier invader MEAM1. Sun et al. (2013a) conducted laboratory experiments and field sampling to examine the effects of insecticide application on the competitive interactions between MEAM1 and MED. In the laboratory, MEAM1 displaced MED in five generations on cotton when initial populations of the two species were equal and no insecticide was applied. In contrast, MED displaced MEAM1 in seven and two generations, respectively, when imidacloprid was applied. Field sampling indicated that in a single season MED displaced MEAM1 on crops heavily sprayed with neonicotinoid insecticides but the relative abundance of the two species changed little on crops without insecticide spray. As field populations of MED have lower susceptibility than those of MEAM1 to commonly used insecticides, insecticide application seems to have played a major role in shifting the species competitive interaction in favor of MED in the field. Later on, Pan et al. (2015) confirmed that the rapid replacement of the MEAM1 by MED throughout China is because MED is more tolerant of insecticides. The field monitoring also revealed that the insecticide resistance of MED generally is higher than MEAM1.

For instance, Luo et al. (2010) found that MEAM1 remained largely susceptible to acetamiprid, imidacloprid, and thiamethoxam, whereas MED expressed 20–170 folds resistance to these insecticides. These studies strongly support the hypothesis that insecticide use reverses the MEAM1-MED competition in China and allows MED to displace MEAM1 in different regions.

8.3.6 *Adaptation of* **B. tabaci** *to Heat Stress*

In nature, whiteflies are continuously exposed to abiotic stresses. The great adaptability of MEAM1 to harsh temperature conditions plays a major role in its rapid invasions and spread. Results of the comparative studies demonstrated that MEAM1 whiteflies possess greater tolerance to harsh temperatures than the greenhouse whitefly *T. vaporariorum* and some of the native whitefly species (Cui et al. 2008; Gao et al. 2015). The MEAM1 adults showed greater ability to adapt to higher temperature than *T. vaporariorum* under laboratory conditions. The thermal thresholds for survival of MEAM1 and *T. vaporariorum* were 45 °C and 43 °C, respectively. Fecundity of the MEAM1 was not significantly affected after 1 h heat shock, whereas that of *T. vaporariorum* was decreased (Cui et al. 2008). Mahadav et al. (2009) compared the expression patterns of MEAM1 and MED under 25 °C and 40 °C heat stress using microarray. The authors found that compared to the treatment of MED, exposure of MEAM1 to heat stress was accompanied by rapid alteration of gene expression. These differences might be due to better adaptation of one species over another and might eventually lead to change of MEAM1 and MED distribution (Mahadav et al. 2009). To reveal why females are more heat resistant than males, Lu and Wan (2008) identified the DEGs in male and female whiteflies, respectively. The authors found that difference of heat-resistance under heat-shock condition was associated with DEGs between *B. tabaci* sexes.

8.3.7 *Genetic Base of* **B. tabaci** *Invasions*

The recent advancement in genomic technologies offers great opportunities for a better understanding of the complex mechanisms underlying whitefly invasions (Edwards and Papanicolaou 2012). With the development of high performance sequencing technology, transcript profiling techniques allow the simultaneous examination of thousands of genes, and can be utilized to study changes in gene expression (Gibbons et al. 2009). In 2010, a new short read sequencing technology (Illumina) was employed to analyze the transcriptome of the MED. A total of 168,900 distinct sequences were assembled. Based on similarity search with known proteins, 27,290 sequences with a cut-off E-value of 10^{-5} were identified (Wang et al. 2010a). In 2011, Wang et al. (2011) re-sequenced the transcriptome of the invasive MEAM1 whitefly using Illumina and compared it with the MED transcriptome. The comparison revealed that the level of sequence divergence in coding region was 0.83%, strongly supporting to the previous proposition that MEAM1 and MED whiteflies are two species. This study further showed that 24 genes, which have evolved in response to positive selection, were involved in metabolism and insecticide resistance. These genes might contribute to the divergence of the two whitefly species (Wang et al. 2011). To reveal the possible mechanism of whitefly invasions, 52,535 transcriptome sequences were identified from the native Asia II 3 species (Wang et al. 2012). Comparison of the sequence divergence between the transcriptomes of Asia II 3 and the invasive species MEAM1 and MED indicated that the overall divergence of coding sequences between the orthologous gene pairs of Asia II 3 and MEAM1, and that between Asia II 3 and MED, was 1.73 and 1.84%, respectively, much higher than that between MEAM1 and MED (0.83%). The data also demonstrated that the most divergent gene classes between the native and invasive species were related to cytochrome P450, glutathione metabolism and oxidative phosphorylation, which seemed relevant to the invasion, displacement and speciation of the species in the *B. tabaci* complex (Wang et al. 2012).

Wang et al. (2013) further examined the transcriptional difference between the two invasive whitefly species, MEAM1 and MED, and one indigenous whitefly species Asia II 3. The results showed that 2422 genes between MEAM1 and MED; 3073 genes between MEAM1 and Asia II 3; and 3644 genes between MED and Asia II 3 were differentially expressed. Carbohydrate, amino acid and glycerolipid metabolisms were more active in MEAM1 and MED than in Asia II 3. Furthermore, the majority of genes involved in basic metabolism and detoxification were expressed at a higher level in MEAM1 and MED than in Asia II 3, which might be responsible for their higher resistance to insecticides and environmental stresses (Wang et al. 2013). Sequencing and comparison of the gut transcriptomes of MEAM1 and MED revealed that many genes related to detoxification were expressed at an elevated level in the gut of MED compared to MEAM1, which might be responsible for the MED's higher resistance to insecticides and environmental stresses (Ye et al. 2014). Xu et al. (2015a) compared the transcriptional responses of MEAM1 and Asia II 3 whitefly species following host plant transfer from a suitable host (cotton) to an unsuitable host (tobacco). Transcriptional analysis showed that compared to the constitutive high expression of detoxification genes in MEAM1, most of the detoxification genes were down-regulated in Asia II 3. Enzymatic activities of P450, GST and esterase further verified that the detoxification of MEAM1 was much higher than that of Asia II 3. These results reveal obvious differences in responses of MEAM1 and Asia II 3 to host transfer.

8.4 Management of *B. tabaci*

8.4.1 Monitoring and Treatment Decisions

By collecting information of several factors (e.g., longitude, latitude, altitude, average annual rainfall, average annual temperature and its range, average temperature in January and July, annual highest and lowest temperature) at whitefly occurred spots in China and analyzing with Maxent prediction model, Ren et al. (2011) reported that South and East China, as well as the southern part of North China were suitable for *B. tabaci* distribution. Using life table data of *B. tabaci* and 10 years' climate information in different regions in China, Ren et al. (2011) warned that *B. tabaci* may have 11–15 generations in Guangdong, Guangxi and Hainan provinces, 7–15 generations in Shandong, Henan and Chongqing areas, and one generation in Tibet and Qinhai provinces. Based on this result, a monitoring and early warning platform has been established (http://www.ipm.ioz.ac.cn/them_gefeng/fenshi/index.as).

The use of yellow sticky cards (24 cm × 20 cm) is an important method of monitoring whitefly populations in the wild. Five yellow sticky cards were put in every 667 m^2 field using 5-points or "Z"-shaped methods, and replaced with new ones every 10 days. The upper edge of the card was kept at an equal level to or slightly above the plants. If the number of adult whitefly captured on the cards reached 0.25–0.5 adult/cm^2, biological control measures should be taken; while it reached 3–4 adult/cm^2, chemical control measures should be taken to suppress the outbreak of *B. tabaci* (Ren et al. 2014). Sampling via leaf-turn method was also adopted. For Solanaceous and Cucurbitaceae plants, the third and fourth leaf from the top was sampled; while for vegetables such as cabbage and Chinese kale, the young leaves in plant heart part were sampled. If the number of adult whitefly reached 0.5–2 adult/leaf, biological control measures should be taken; while it reached 5–6 adult/leaf, environmental friendly insecticides should be sprayed (Tian et al. 2015). Such action threshold has been defined as the level of pest populations at which control should be implemented to avoid significant damages to crops (Dik and Albajes 1999), while an "economic injury level" (EIL) requires more rigorous criteria employing economic consideration. Related to market and environmental conditions in China, an economic injury level of 18 adult *B. tabaci* per cucumber plant at the four-leaf stage was determined (Shen et al. 2005); for greenhouse tomato, a level of 13.6 adults per 100 leaves was established (Cao and Cao 2011); while for cabbage, an economic threshold of 6.0–6.8 adults per plant in the seedling stage and 2.1–3.8 adults per plant in the rosette stage were determined (Wang 2007).

8.4.2 Cultural Control

Cultural control is to improve crop system to make the environment less favorable to pest reproduction, dispersal, survival and damage, while more suitable to natural enemies. Cultural control options for *B. tabaci* include the use of physical barriers or other barriers to prevent the pest from reaching the crop, adjusting planting dates, planting in low infestation areas, rotation with non-susceptible crops, destroying crop residues and selecting resistant crops or cultivars (Ren et al. 2001).

Planting Dates One way to reduce whitefly infestations is to adjust planting dates to avoid the heaviest insect migration periods or crop overlap. Early planting in spring or delayed planting in fall can be an effective way to avoid whiteflies, since they reproduce more rapidly under hot and dry conditions. Highly susceptible crops such as cucurbits, crucifers and tomatoes should avoid to be planted when whitefly migration is expected.

Crop Termination Whitefly infestations most often occur in fields where damages ever occurred. Susceptible crops should not be planted near infestation sources. Godfrey et al. (2008) suggested that cotton should be planted at least one-half mile upwind from other key host crops, such as melons, tomatoes, from key ornamental plants, and from key weed species that harbor *B. tabaci*. Infected plants should be removed and destroyed, and susceptible crops should not be grown continuously. It is reported that *B. tabaci* can continue to increase up to 6 weeks after final cotton irrigation even following defoliation since red eye pupa were able to continue development to the adult stage on cotton leaves that abscised and fell from the plants (Nuessly et al. 1994). Thus crop residues, which can harbor whitefly and virus inoculum, should be rapidly and completely destroyed after the final harvest.

Water and Fertility Management Water and fertility management play important roles as cultural control methods in whitefly management. Overuse of nitrogen fertilizer will greatly increase whitefly numbers and honeydew production thus exacerbate damages from *B. tabaci* infestations (Bi et al. 2001, 2005). Higher populations of *B. tabaci* were observed on water-stressed cotton as compared to well-watered plants (Flint et al. 1996). However, *B. tabaci* feeding on well-watered plants produced more honeydew and sugars per gram of honeydew than on water stressed cotton (Henneberry et al. 2002).

8.4.3 Host Plant Resistance

Planting varieties that are resistant to *B. tabaci* and associated viruses (e.g., tomato yellow leaf curl virus, TYLCV) is a preferred solution for minimizing damages caused by whitefly at the early stage. The *Mi-1* gene, present in many varieties of cultivated tomato (*Solanum lycopersicum* L.) and introduced into this plant from its wild relative, *S. peruvianum*, regulates resistance to *B. tabaci*, aphid *Macrosiphum euphorbiae* and root-knot nematodes *Meloidogyne* spp. (Roberts and Thomason 1986; Rossi et al. 1998; Nombela et al. 2001). However, tomato varieties bearing *Mi-1* gene has not been widely used in China yet. A number of studies on different plant varieties resistant to this pest have been carried out in China during the last decade. Studies have compared the biotic potential of whiteflies in different cultivated plants such as tomato, cucumber, beans, eggplant, peanut, squash and pepper in China (Ren et al. 2001; Xu et al. 2010b; Sun et al. 2013b; Wu 2013; Ji 2015).

8.4.4 Physical Control

Fine-mesh nylon screen (normally pore diameter 0.125 mm) can be used in greenhouse production to reduce the potential for infestation of greenhouse pests including whitefly, flower thrips, aphid and leafminer. Under field conditions, several types of barriers can reduce whitefly problems, including reflective mulches that tend to repel whiteflies, oil-coated yellow mulches that act as a trap for whiteflies, floating row covers that exclude whiteflies during the vegetative growth of crops, as well as barrier and trap crops (Ren et al. 2001; Tian et al. 2015).

Yellow sticky cards were also used to trap whitefly adults, especially in greenhouse condition. It is suggested that yellow sticky card (24 cm × 20 cm) could trap whitefly efficiently when placed vertically among plants and one card per 3–5 m, kept the upper edge of the card at an equal level or slightly above the plants (Qiu and Ren 2006; Chen et al. 2012).

8.4.5 Biological Control

In China, a number of natural enemies of *B. tabaci* have been recorded including 56 species of parasitoids, mainly belonging to the genera *Eretmocerus* and *Encarsia*; 54 species of arthropod predators, dominated by lady beetles and lacewings in Coleoptera and Neuroptera; and seven species of entomopathogenic fungi (Li et al. 2011). More than ten species of parasitoids and predators are commercially available in China now (Table 8.1).

Parasitoids *Encarsia bimaculata* and *Eretmocerus* sp. nr. *furuhashii* are the two dominate species parasitizing *B. tabaci* in South China, while the population of *Encarsia formosa* is abundant in north China in protected fields, and *Eretmocerus hayati* is abundant in Xinjiang province, located in northwest China, on cotton in fields (Li et al. 2011; Abuduhani et al. 2013; Zhang et al. 2015).

Among the four Aphelinid species which are commercially available in China (Table 8.1), *Encarsia sophia* (formerly known as *En. transvena*) is a solitary, arrhenotokous, heteronomous autoparasitoid. For this species, fertilized eggs are laid in whitefly nymphs and developed into female progeny, while the unfertilized eggs are

Table 8.1 Biological control parasitoids and predators commercially available in China

Biocontrol insects	Company/commercial facility	Location
Parasitoids		
Encarsia formosa	Quentian ™ Bio-Tech Co., LTD	Beijing
	NewLand™ BioControl Service Co., Ltd	Changchun, Jilin Province
	LandGreen® BioTech Service	Beijing
	TIANYI® Biologicaol Control	Hengshui, Hebei Province
	Beijing Academy of Agriculture and Forestry Sciences	Beijing
Encarsia sophia	Quentian ™ Bio-Tech Co., LTD	Beijing
	LandGreen® BioTech Service	Beijing
	TIANYI® Biologicaol Control	Hengshui, Hebei Province
	Institute of Plant Protection, Chinese Academy of Agricultural Sciences	Beijing
	Beijing Academy of Agriculture and Forestry Sciences	Beijing
Eretmocerus mundus	Quentian ™ Bio-Tech Co., LTD	Beijing
Eremocerus hayati	Institute of Plant Protection, Chinese Academy of Agricultural Sciences	Beijing
Predator mites		
Neoseiulus cucumeris	Quentian ™ Bio-Tech Co., LTD	Beijing
	LandGreen® BioTech Service	Beijing
	Fujian Yanxuan Bio-preventing and Controlling Technology Co., Ltd.	Fuzhou, Fujian Province
Predator ladybird		
Delphastus catalinae	Quentian ™ Bio-Tech Co., LTD	Beijing
	LandGreen® BioTech Service	Beijing
Harmonia axyridis	Quentian ™ Bio-Tech Co., LTD	Beijing
	NewLand™ BioControl Service Co., Ltd	Changchun, Jilin Province
	LandGreen® BioTech Service	Beijing
	Beijing Academy of Agriculture and Forestry Sciences	Beijing
Propylea japonica	Quentian ™ Bio-Tech Co., LTD	Beijing
	LandGreen® BioTech Service	Beijing
Serangium japonicum	Quentian ™ Bio-Tech Co., LTD	Beijing
Predatory stink bug		
Orius sauteri	Quentian ™ Bio-Tech Co., LTD	Beijing
	LandGreen® BioTech Service	Beijing
Orius minutus	Quentian ™ Bio-Tech Co., LTD	Beijing

(continued)

Table 8.1 (continued)

Biocontrol insects	Company/commercial facility	Location
Lacewing		
Chrysopa pallens	Quentian ™ Bio-Tech Co., LTD	Beijing
	LandGreen® BioTech Service	Beijing

Modified from Yang et al. (2014)

laid externally on immature parasitoids inside the whitefly host, either on conspecific species or on heterospecific primary parasitoids, and developed into male progeny (Hunter and Kelly 1998). It is originated from Pakistan, and currently present across northern and southern China (Yang et al. 2012; Li et al. 2011). *Er. hayati* is a primary, solitary parasitoid which oviposits externally under the nymphal host (Yang and Wan 2011). After eclosion, the first instar larva penetrates the host from underneath and develops internally. Both of these two parasitoids attack all nymphal stages (N1-N4) of *B. tabaci*, while *En. sophia* prefer old instar and *Er. hayati* prefer young ones (Yang and Wan 2011). Although *Er. hayati* is capable of high parasitism on *B. tabaci*, its field population often fluctuates because of the severe consumption of hosts by adults (Yang et al. 2012; Xu et al. 2015b). The combined release of the autoparasitoid *En. sophia* will stabilize the effect of *Er. hayati* (Xu et al. 2015b, 2016; Huang et al. 2016).

Predators The lady beetle *Axinoscymnus cardilobus* is a dominant predator of *B. tabaci* in south China, while *Harmonia axyridis* and *Propylea japonica* are dominant predators species in north China (Zhang et al. 2007a, b; Li et al. 2011). *H. axyridis* is a generalist predator extensively employed as a biological control agent in China (Koch 2003). This ladybird species exhibits significantly lower selectivity for *B. tabaci* and a reduced reproduction rate when fed on *B. tabaci* (Tan et al. 2016). However, its generalist character could be helpful in controlling secondary pests. Besides, when it was released in combination with whitefly parasitoids, *En. sophia* and *En. formosa*, *H. axyridis* showed a significant preference for non-parasitized nymphs as prey; and both parasitoid species parasitized more *B. tabaci* as compared to the wasps were released either alone or mixed with the other parasitoid (Tan et al. 2016). This combination release of predator and parasitoids enhanced whitefly control. Li et al. (2014a) also reported an enhanced control effects on *B. tabaci* by combined release of *Orius sauteri*, a generalist predatory stink bug, with *En. formosa*.

Entomopathogenic Fungi Several species of entomopathogenic fungi active against *B. tabaci* are commercially available in China, including *Beauveria bassiana*, *Aschersonia aleyrodis*, *Verticillium lecanii*, and *Isaria fumosorosea*. *B. bassiana* could significantly decrease the survival rate of *B. tabaci* nymphs while increase developmental time and pre-oviposition period (Xia et al. 2013). *A. aleyrodis* could parasitize on 1–3 instar nymphs of *B. tabaci*, and had no contradicted effect on whitefly parasitoid *Eretmocerus* sp. (Qiu et al. 2003b). *I. fumosorosea* infected all immature stages and adults of *B. tabaci* (Tian et al. 2014). *V. lecanii* showed deterrent activity to *B. tabaci* adults and contact toxicity to *B. tabaci* nymphs (Wang et al. 2006). However, *V. lecanii* required high humidity to be effective, this shortage could be made up by using *V. lecanii* toxic (Hong et al. 2011). The deterrent and anti-feeding activity of *V. lecanii* toxic-VIII on *B. tabaci* adults was tested as 41% and 23%, respectively (Hong et al. 2011). Restricted by environmental factors such as temperature and humility, the effect of entomopathogenic fungi in *B. tabaci* management was slow and not very stable. Recent researches revealed that when *I. fumosorosea* combined with thiamethoxam or imidacloprid, and *B. bassiana* combined with non-ionic surfactants, a synergic effect on *B. tabaci* was observed (Mascarin et al. 2014; Zou et al. 2014).

8.4.6 Chemical Control

Even the system of organic crops production inclines to employ increasing biological control measures, chemical control is indispensable when *B. tabaci* occurs in high density and quick eliminating effect is required. However, *B. tabaci* can develop resistance not only to conventional insecticides in China, but also to a number of new insecticides, such as pyriproxyfen, acetamiprid, imidacloprid, and thiamethoxam, especially MED (Crowder et al. 2007; Luo et al. 2010). Besides, thiamethoxam was highly toxic to *Serangium japonicum*, a predator ladybeetle, regardless of exposure routes (e.g., residue contact, egg-dip, and systemic treatment) (Yao et al. 2015). Therefore, thiamethoxam should be used with caution in IPM of *B. tabaci*. Some environmental friendly insecticides, avermectin, spinetoram, sulfozaflor and cyantraniliprole, are recommended to be applied in *B. tabaci* management (Keyimu et al. 2014; Xie et al. 2014).

8.5 Concluding Remarks

Even though many field surveys have been conducted to monitor the spread of *B. tabaci* during the last 15 years, our understanding of the invasion processes of MEAM1 and MED in China remain largely observational and anecdotal. This is mainly due to the lack of efficient methods to easily distinguish different whitefly species in the field and the presence of the *B. tabaci* species complex. Nevertheless, the current surveys have clearly showed that while MEAM1 was dominate near 2000, MED has replaced MEAM1 in many regions of China now. During the last decade, we have identified multiple factors that contribute to the invasions of *B. tabaci* into new regions and habitats, such as mating interactions, host plant adaptation, insecticide resistance, virus-vector-host mutualism and the roles of endosymbionts. Those studies obviously suggest that the successful invasions of MEAM1 and MED in China are largely due to their intrinsic traits, such as high insecticide resistance, wide host range and strong reproduction capacity. More comparative studies on the biological characteristics between invasive and indigenous species may provide better understanding of the mechanism underlying invasions and displacement. It is also crucial to identify the key factors that contribute to the unique features of MEAM1 and MED. The obtained knowledge will help us to develop new strategies to control the spread of invasive *B. tabaci* in China. At this stage, Chinese farmers mainly rely on insecticides as immediate solutions to *B. tabaci* problems. However, the use of insecticides impairs the benefits of consumers and regulators concerned with food safety and environmental protection. This conflict leads to interest and development in alternative control measures. In response, Integrated Pest Management (IPM) has been the main paradigm in modern pest control. In China, IPM has been developed in the last decade with a combination of tactics including cultural control methods, resistant plant varieties, physical control such as mechanical screen and yellow sticky cards, biological control by natural enemies, and the judicious use of insecticides based on sampling and economic thresholds. In the production of some high-value crops, the *B. tabaci* IPM has been tested to be effective and been gradually adopted in organic farms. However, for most farmers, considering the IPM cost and knowledge access, chemical measures are still the prior choice. Thus, the wide demonstration and dissemination of *B. tabaci* IPM in end-users to improve the implement of environmental friendly measures to control *B. tabaci* needs extensive devotion from regulators and researchers.

References

Abuduhani R, Abulimiti M, Jamali, R, Zhang YN, Li J, Ma DY (2013) Population dynamics and spatial distribution of *Bemisia tabaci* and dominat parasitic wasp *Eretmocerus* sp. from the field of jujubi intercropping with cotton. Xinjiang Agric Sci 50(10): 1842–1849 (in Chinese, English abstract)

Alon M, Elbaz M, Ben-Zvi MM, Feldmesser E, Vainstein A, Morin S (2012) Insights into the transcriptomics of polyphagy: *Bemisia tabaci* adaptability to phenylpropanoids involves coordinated expression of defense and metabolic genes. Insect Biochem Mol Biol 42:251–263

Bi JL, Ballmer GR, Hendrix DL, Hennneberry TJ, Toscano NC (2001) Effect of cotton nitrogen fertilization on *Bemisia argentifolii* populations and honeydew production. Entomol Exp Appl 99:25–36

Bi JL, Lin DM, Li KS, Toscano NC (2005) Impact of cotton planting date and nitrogen fertilization on *Bemisia argentifolii* populations. Insect Sci 12:31–36

Bing XL, Ruan YM, Rao Q, Wang XW, Liu SS (2013) Diversity of secondary endosymbionts among different putative species of the whitefly *Bemisia tabaci*. Insect Sci 20:194–206

Boykin LM, De Barro PJ (2014) A practical guide to identifying members of the *Bemisia tabaci* species complex and other morphologically identical species. Front Ecol Evol. doi:10.3389/fevo.2014.00045

Brown JK (1995) The sweet potato or silverleaf whiteflies: biotypes of *Bemisia tabaci* or a species complex. Annu Rev Entomol 40:511–534

Byrne DN, Bellows TS (1991) Whitefly biology. Annu Rev Entomol 36:431–457

Cao ZG, Cao L (2011) The action threshold of *Bemisia tabaci* on protected tomato. Shanghai Agric Technol 6:122 (in Chinese, English abstract)

Chen D, Zhang YH, Reziwanguli, Miriguli, Yang M, Pan WP, Ma DY (2012) Control effects of Yellow Sticky Trap (YST) to *Bemisia tabaci* in the greenhouses in Turpan Area. Xinjiang Agric Sci 49(2):255–260 (in Chinese, English abstract)

Chu D, Zhang YJ, Cong B, Xu BY, Wu QJ (2005) Identification for Yunnan Q-biotype *Bemisia tabaci* population. Chin Bull Entomol 42:54–56

Chu D, Li X, Zhang Y (2012) Microsatellite analyses reveal the sources and genetic diversity of the first-introduced Q-biotype population and the well-established B-biotype populations of *Bemisia tabaci* in China. Acta Entomol Sin 55:1377–1386

Crowder DW, Dennehy TJ, Ellers-Kirk C, Yafuso LC, Ellsworth PC, Tabashnik BE, Carriere Y (2007) Field evaluation of resistance to pyriproxyfen in *Bemisia tabaci* (B biotype). J Econ Entomol 100:1650–1656

Cui XH, Wan FH, Xie M, Liu TX (2008) Effects of heat shock on survival and reproduction of two whitefly species, *Trialeurodes vaporariorum* and *Bemisia tabaci* Biotype B. J Insect Sci 8:1–10

Czosnek H, Ghanim M (2002) The circulative pathway of begomoviruses in the whitefly vector *Bemisia tabaci* – insights from studies with *Tomato yellow leaf curl virus*. Ann Appl Biol 140:215–231

Dalton R (2006) Whitefly infestations: the christmas invasion. Nature 443:898–900

De Barro PJ, Liu SS, Boykin LM, Dinsdale AB (2011) *Bemisia tabaci*: a statement of species status. Annu Rev Entomol 56:1–19

Dennehy TJ, Degain BA, Harpold VS, Zaborac M, Morin S, Fabrick JA, Nichols RL, Brown JK, Byrne FJ, Li X (2010) Extraordinary resistance to insecticides reveals exotic Q biotype of *Bemisia tabaci* in the New World. J Econ Entomol 103:2174–2186

Dik AJ, Albajes R (1999) Principals of epidemiology, population biology, damage relationships, and integrated control of diseases and pests. In: Albajes R, Lodovica-Gullino M, van Lenteren JC, Elad Y (eds) Integrated Pest and Disease Management in Greenhouse Crops. Springer, Amsterdam, pp 69–81

Edwards O, Papanicolaou A (2012) A roadmap for whitefly genomics research: lessons from previous insect genome projects. J Integr Agric 11:269–280

Elbaz M, Halon E, Malka O, Malitsky S, Blum E, Aharoni A, Morin S (2012) Asymmetric adaptation to indolic and aliphatic glucosinolates in the B and Q sibling species of *Bemisia tabaci* (Hemiptera: Aleyrodidae). Mol Ecol 21:4533–4546

Flint HM, Naranjo SE, Leggett JE, Henneberry TJ (1996) Cotton water stress, arthropod dynamics, and management of *Bemisia tabaci* (Homoptera: Aleyrodidae). J Econ Entomol 89:1288–1300

Gao XL, Li JM, Xu HX, Yan GH, Jiu M, Liu SS, Wang XW (2015) Cloning of a putative extracellular Cu/Zn superoxide dismutase and functional differences of superoxide dismutases in invasive and indigenous whiteflies. Insect Sci 22:52–64

Ghanim M, Kontsedalov S (2007) Gene expression in pyriproxyfen-resistant *Bemisia tabaci* Q biotype. Pest Manag Sci 63:776–783

Gibbons JG, Janson EM, Hittinger CT, Johnston M, Abbot P, Rokas A (2009) Benchmarking next-generation transcriptome sequencing for functional and evolutionary genomics. Mol Biol Evol 26:2731–2744

Godfrey LD, Goodell PB, Natwick ET, Haviland DR (2008) Insects and mites. In: UC IPM Pest Management Guidelines: cotton. UC ANR Publication 3444. http://www.ipm.ucdavis.edu/PMG/r114300311.html

Gray S, Cilia M, Ghanim M (2014) Circulative,"nonpropagative" virus transmission: an orchestra of virus-, insect-, and plant-derived instruments. Adv Virus Res 89:141–199

He WB, Li J, Liu SS (2015) Differential profiles of direct and indirect modification of vector feeding behaviour by a plant virus. Sci Rep 5:7682

Henneberry TJ, Jech LJ, De La Torre TM (2002) Effects of cotton plant water stress on *Bemisia tabaci* strain B (Homoptera: Aleyrodidae) honeydew production. Southwest Entomol 27:117–133

Himler AG, Adachi-Hagimori T, Bergen JE, Kozuch A, Kelly SE, Tabashnik BE, Chiel E, Duckworth VE, Dennehy TJ, Zchori-Fein E, Hunter MS (2011) Rapid spread of a bacterial symbiont in an invasive whitefly is driven by fitness benefits and female bias. Science 332:254–256

Hong HJ, Yang YH, Wang LD (2011) Repellence and feeding deterrence of *Verticillium lecanii* toxic-VII against *Bemisia tabaci*. Chin J Appl Entomol 48(1):60–64 (in Chinese, with English abstract)

Horowitz A, Antignus Y, Gerling D (2011) Management of *Bemisia tabaci* Whiteflies. Springer, Dordrecht

Hu J, De Barro P, Zhao H, Wang J, Nardi F, Liu SS (2011) An extensive field survey combined with a phylogenetic analysis reveals rapid and widespread invasion of two alien whiteflies in China. PLoS One 6:e16061

Huang YX, Yang NW, Qin Y, An F, Li ZH, Wan FH (2016) Enhanced stability in host–parasitoid interactions with autoparasitism and parasitoid migration. J Theor Biol 393:43–50

Hunter MS, Kelly SE (1998) Hyperparasitism by an exoticautoparasitoid: secondary host selection and the window of vulnerability of conspecific and native heterospecific hosts. Entomol Exp Appl 89:249–259

Ji XZ (2015) Oviposition and feeding choice of the tabacco whitefly, *Bemisia tabaci* (Gennadius) on different cucumber strains. *Chinese Agricultural University*, Master thesis (in Chinese, English abstract)

Jiu M, Zhou XP, Tong L, Xu J, Yang X, Wan FH, Liu SS (2007) Vector-virus mutualism accelerates population increase of an invasive whitefly. PLoS One 2:e182

Keyimu M, Xu HX, Xian JH, Li J, Wang HQ, Hu AZ, Ma DY (2014) Control effect evaluation of five environmental friendly insecticides against *Bemisia tabaci* MEAM1 cryptic species. Xinjiang Agric Sci 51(6):1137–1142 (in Chinese, with English abstract)

Koch RL (2003) The multicolored Asian lady beetle, *Harmonia axyridis*: a review of its biology, uses in biological control, and non-target impacts. J Insect Sci 3:32

Li SJ, Xue X, Ahmed MZ, Ren SX, Du YZ, Wu JH, Cuthbertson AGS, Qiu BL (2011) Host plants and natural enemies of *Bemisia tabaci* (Hemiptera: Aleyrodidae) in China. Insect Sci 18:101–112

Li S, Lao SB, Wang S, Guo XJ, Zhang F (2014a) Control effect of *Orius sauteri* collaborated with *Encarsia formosa* on *Bemisia tabaci* in the greenhouse. J Environ Entomol 36(6):978–982 (in Chinese, with English abstract)

Li R, Weldegergis BT, Li J, Jung C, Qu J, Sun Y, Qian H, Tee C, van Loon JJ, Dicke M, Chua NH, Liu SS, Ye J (2014b) Virulence factors of geminivirus interact with MYC2 to subvert plant resistance and promote vector performance. Plant Cell 26:4991–5008

Liu SS, De Barro PJ, Xu J, Luan JB, Zang LS, Ruan YM, Wan FH (2007) Asymmetric mating interactions drive widespread invasion and displacement in a whitefly. Science 318:1769–1772

Liu SS, Colvin J, De Barro P (2012) Species concepts as applied to the whitefly *Bemisia tabaci* systematics: how many species are there? J Integr Agric 11:176–186

Lu ZC, Wan FH (2008) Differential gene expression in whitefly (*Bemisia tabaci*) B-biotype females and males under heat-shock condition. Comp Biochem Physiol Part D Genomics Proteomics 3:257–262

Luan JB, Li JM, Varela N, Wang YL, Li FF, Bao YY, Zhang CX, Liu SS, Wang XW (2011) Global analysis of the transcriptional response of whitefly to *Tomato yellow leaf curl China virus* reveals the relationship of coevolved adaptations. J Virol 85:3330–3340

Luan JB, Wang YL, Wang J, Wang XW, Liu SS (2013a) Detoxification activity and energy cost is attenuated in whiteflies feeding on *Tomato yellow leaf curl China virus*-infected tobacco plants. Insect Mol Biol 22:597–607

Luan JB, Yao DM, Zhang T, Walling LL, Yang M, Wang YJ, Liu SS (2013b) Suppression of terpenoid synthesis in plants by a virus promotes its mutualism with vectors. Ecol Lett 16:390–398

Luan JB, Chen W, Hasegawa DK, Simmons AM, Wintermantel WM, Ling KS, Fei Z, Liu SS, Douglas AE (2015) Metabolic coevolution in the bacterial symbiosis of whiteflies and related plant sap-feeding insects. Genome Biol Evol 7:2635–2647

Luo C, Yao Y, Wang R (2002) The use of mitochondrial cytochrome oxidase I (mtCO I) gene sequences for the identification of biotypes of *Bemisia tabaci* (Gennadius) in China. Acta Entomol Sin 45:759–763

Luo C, Jones C, Devine G, Zhang F, Denholm I (2010) Insecticide resistance in *Bemisia tabaci* biotype Q (Hemiptera: Aleyrodidae) from China. Crop Prot 29:429–434

Mahadav A, Kontsedalov S, Czosnek H, Ghanim M (2009) Thermotolerance and gene expression following heat stress in the whitefly *Bemisia tabaci* B and Q biotypes. Insect Biochem Mol Biol 39:668–676

Mascarin GM, Kobori NN, Quintela ED, Arthurs SP, Delalibera I (2014) Toxicity of non-ionic surfactants and interactions with fungal entomopathogens toward *Bemisia tabaci* biotype B. Biocontrol 59(1):111–123

Naranjo SE, Castle SJ, De Barro PJ, Liu SS (2010) Population dynamics, demograpgy, dispersal and spread of *Bemisia tabaci*. In *Bemisia: Bionomics and Management of a Global Pest*, pp 185–226. Springer

Nombela G, Beitia F, Muniz M (2001) A differential interaction study of *Bemisia tabaci* Q-biotype on commercial tomato varieties with or without the *Mi* resistance gene, and comparative host responses with the B-biotype. Entomol Exp Appl 98:339–344

Nuessly GS, Meyerdirk DE, Coudriet DL, Henneberry TJ (1994) The effect of short season cotton production schedules on *Bemisia tabaci* (Gennadius). Southwest Entomol 19:209–217

Pan H, Chu D, Yan W, Su Q, Liu B, Wang S, Wu Q, Xie W, Jiao X, Li R, Yang N, Yang X, Xu B, Brown JK, Zhou X, Zhang Y (2012) Rapid spread of tomato yellow leaf curl virus in China is aided differentially by two invasive whiteflies. PLoS One 7:e34817

Pan H, Preisser E, Chu D, Wang S, Wu Q, Carriére Y, Zhou X, Y Z (2015) Insecticides promote viral outbreaks by altering herbivore competition. Ecol Appl 25: 1585–1595

Qiu BL, Ren SX (2006) Using yellow sticky traps to inspect population dynamics of *Bemisia tabaci* and its parasitoids. Chin Bull Entomol 43(1):53–56. in Chinese, English abstract

Qiu BL, Ren SX, Wen S, Mandour N (2003a) Biotype identification of the populations of *Bemisia tabaci* (Homoptera: Aleyrodidae) in China using RAPD-PCR. Acta Entomol Sin 46:605–608

Qiu BL, Ren SX, Xiao Y, Mandour NS (2003b) Effectiveness of *Eretmocerus* sp. and *Aschersonia aleyrodis* in controlling *Bemisia tabaci* pooulations. Chin J Appl Ecol 14(12):2251–2254 (in Chinese, with English abstract)

Rao Q, Rollat-Farnier PA, Zhu DT, Santos-Garcia D, Silva FJ, Moya A, Latorre A, Klein CC, Vavre F, Sagot MF, Liu SS, Mouton L, Wang XW (2015) Genome reduction and potential metabolic complementation of the dual endosymbionts in the whitefly *Bemisia tabaci*. BMC Genomics 16:226

Ren SX, Wang ZZ, Qiu BL, Xiao Y (2001) The pest status of *Bemisia tabaci* in China and non-chemical control strategies. Entomol Sin 8(3):279–288

Ren SX, Qiu BL, Ge F, Zhang YJ, Du YZ, Chen XX, Guo JY, Lin KJ, Peng ZQ, Yao SL, Hu YH, Wang LD, Zhang WQ (2011) Research progress of the monitoring, forecast and sustainable management of whitefly pests in China. Chin J Appl Entomol 48(1):7–15 (in Chinese, English abstract)

Ren SX, Liu TX, Du YZ, Peng ZQ, Qiu BL, Ge F (2014) Systematic investigation and monitoring of whitefly pests in a vegetable ecosystem. Chin J Appl Entomol 51(3):859–862 (in Chinese, English abstract)

Roberts PA, Thomason IJ (1986) Variability in reproduction of isolates of *Meloidogyne* incognita and *M. javanica* on resistant tomato genotypes. Plant Dis 70:547–551

Rossi M, Goggin FL, Milligan SB, Kaloshian I, Ullman DE, Williamson VM (1998) The nematode resistance gene *Mi* of tomato confers resistance against the potato aphid. Proc Natl Acad Sci U S A 95:9750–9754

Shen BB, Ren SX, Musa PH, Chen C (2005) A study on economic threshold of *Bemisia tabaci*. Acta Agric Univ Jiangxiensis, 27(2):234–237 (in Chinese, English abstract)

Shi X, Pan H, Xie W, Wu Q, Wang S, Liu Y, Fang Y, Chen G, Gao X, Zhang Y (2013) Plant virus differentially alters the plant's defense response to its closely related vectors. PLoS One 8:e83520

Sun D, Liu Y, Qin L, Xu J, Li F, Liu S (2013a) Competitive displacement between two invasive whiteflies: insecticide application and host plant effects. Bull Entomol Res 103:344–353

Sun HL, Zhou FQ, You XY, Hu RL, Lv M, Wu L, Zhu SD (2013b) The selectivity of Q-biotype *Bemisia tabaci* for twenty varities of eggplang, *Solanum melongena*. Plant Prot 39(2):67–71 (in Chinese, English abstract)

Sun D, Li J, Liu Y, Crowder D, Liu S (2014) Effects of reproductive interference on the competitive displacement between two invasive whiteflies. Bull Entomol Res 104:334–346

Tan XL, Hu NN, Zhang F, Ramirez-Romero R, Desneus N, Wang S, Ge F (2016) Mixed release of two parasitoids and a polyphagous ladybird as a potential strategy to control the tobacco whitefly *Bemisia tabaci*. Sci Rep-UK 6:28245

Teng X, Wan F, Chu D (2010) *Bemisia tabaci* biotype Q dominates other biotypes across China. Fla Entomol 93:363–368

Tian J, Hao C, Liang L, Ma RY (2014) Effects of temperature and relative humidity on conidial germination of *Isaria fumosorosea* (Hypocreales: Cordycipitaceae) IF-1106 and pathogenicity of the fungus against *Bemisia tabaci* (Homoptera: Aleyrodidae). Mycosystema 33(3):668–679

Tian XQ, Li N, Liu XC, Xu YJ, Zhang SZ (2015) The occurrence and integrated pest management of *Bemisia tabaci* on vegetables. Shaanxi J Agric Sci 61(1):123–126 (in Chinese, English abstract)

Wang H (2007) Study on the ecology and physiology traits of *Bemisia tabaci* on three brassica vegetables and abutilon and the management strategy. *Yangzhou University*, Master thesis (in Chinese, English abstract)

Wang LD, Huang J, Liu B (2006) Assessment of the control effectiveness of insecticidal toxins from *Verticillium lecanii* on the population of *Bemisia tabaci* (Gennadius) in greenhouse Assessment of the control effectiveness of insecticidal toxins from *Verticillium lecanii* on the population of *Bemisia tabaci* (Gennadius) in greenhouse. Acta Ecol Sin 26(2):391–398

Wang Z, Yao M, Wu Y (2009) Cross-resistance, inheritance and biochemical mechanisms of imidacloprid resistance in B-biotype *Bemisia tabaci*. Pest Manag Sci 65:1189–1194

Wang XW, Luan JB, Li JM, Bao YY, Zhang CX, Liu SS (2010a) *De novo* characterization of a whitefly transcriptome and analysis of its gene expression during development. BMC Genomics 11:400

Wang Z, Yan H, Yang Y, Wu Y (2010b) Biotype and insecticide resistance status of the whitefly *Bemisia tabaci* from China. Pest Manag Sci 66:1360–1366

Wang XW, Luan JB, Li JM, Su YL, Xia J, Liu SS (2011) Transcriptome analysis and comparison reveal divergence between two invasive whitefly cryptic species. BMC Genomics 12:458

Wang XW, Zhao QY, Luan JB, Wang YJ, Yan GH, Liu SS (2012) Analysis of a native whitefly transcriptome and its sequence divergence with two invasive whitefly species. BMC Genomics 13:529

Wang YL, Wang YJ, Luan JB, Yan GH, Liu SS, Wang XW (2013) Analysis of the transcriptional differences between indigenous and invasive whiteflies reveals possible mechanisms of whitefly invasion. PLoS One 8:e62176

Wang LL, Wang XR, Wei XM, Huang H, Wu JX, Chen XX, Liu SS, Wang XW (2016) The autophagy pathway participates in resistance to *Tomato yellow leaf curl virus* infection in whiteflies. Autophagy 12(9):1560–1574

Wei J, Zhao JJ, Zhang T, Li FF, Ghanim M, Zhou XP, Ye GY, Liu SS, Wang XW (2014) Specific cells in the primary salivary glands of the whitefly *Bemisia tabaci* control retention and transmission of begomoviruses. J Virol 88:13460–13468

Wu L (2013) Studies on resistance of different pepper varieties to *Bemisia tabaci* (Gennadius) and its physiological and biochemical mechanism. *Yangzhou University*, Master thesis (in Chinese, English abstract)

Wu S, Wang Z, Wu Y (2010) Competition between the B and Q biotypes of *Bemisia tabaci* and its relevance to insecticide resistance. Chin Bull Entomol 47:1118–1121

Xia J, Zhang CR, Zhang S, Li FF, Feng MG, Wang XW, Liu SS (2013) Analysis of whitefly transcriptional responses to *Beauveria bassiana* infection reveals new insights into insect-fungus interactions. PLoS One 8(7):e68185

Xie W, Liu Y, Wang S, Wu Q, Pan H, Yang X, Guo L, Zhang Y (2014) Sensitivity of *Bemisia tabaci* (Hemiptera: Aleyrodidae) to several new insecticides in china: effects of insecticide type and whitefly species, strain, and stage. J Insect Sci 14:261

Xu J, De Barro PJ, Liu SS (2010a) Reproductive incompatibility among genetic groups of *Bemisia tabaci* supports the proposition that the whitefly is a cryptic species complex. Bull Entomol Res 100:359–366

Xu R, Li W, Zhang LF, Lin YH, Qi B, Xing H (2010b) A study on the inheritance of resistance to whitefly in Soybean. Sci Agric Sin 43(1):80–86 (in Chinese, English abstract)

Xu J, Lin KK, Liu SS (2011) Performance on different host plants of an alien and an indigenous Bemisia tabaci from China. J Appl Entomol 135(10):771–779

Xu HX, Hong Y, Zhang MZ, Wang YL, Liu SS, Wang XW (2015a) Transcriptional responses of invasive and indigenous whiteflies to different host plants reveal their disparate capacity of adaptation. Sci Rep 5:10774

Xu HY, Yang NW, Wan FH (2015b) Field cage evaluation of interspecific interaction of two aphelinid parasitoids and biological control effect on *Bemisia tabaci* (Hemiptera: Aleyrodidae) Middle East-Asia Minor 1. Entomol Sci 18:237–244

Xu HY, Yang NW, Duan M, Wan FH (2016) Functional response, host stage preference and interference of two whitefly parasitoids. Insect Sci 23:134–144

Yang NW, Wan FH (2011) Host suitability of different instars of *Bemisia tabaci* biotype B for the parasitoid *Eretmocerus hayati*. Biol Control 59:313–317

Yang NW, Ji LL, Lövei GL, Wan FH (2012) Shifting preference between oviposition vs. host-feeding under changing host densities in two Aphelinid parasitoids. PLoS One 7:e41189

Yang N, Xie W, Jones CM, Bass C, Jiao X, Yang X, Liu B, Li R, Zhang Y (2013a) Transcriptome profiling of the whitefly *Bemisia tabaci* reveals stage-specific gene expression signatures for thiamethoxam resistance. Insect Mol Biol 22:485–496

Yang N, Xie W, Yang X, Wang S, Wu Q, Li R, Pan H, Liu B, Shi X, Fang Y, Xu B, Zhou X, Zhang Y (2013b) Transcriptomic and proteomic responses of sweetpotato whitefly, *Bemisia tabaci*, to thiamethoxam. PLoS One 8:e61820

Yang NW, Zang LS, Wang S, Guo JY, Xu HX, Zhang F, Wan FH (2014) Biological pest management by predators and parasitoids in the greenhouse vegetables in China. Biol Control 68:92–102

Yao FL, Zheng Y, Zhao JW, Desneux N, He YX, Weng QY (2015) Lethal and sublethal effects of thiamethoxam on the whitefly predator *Serangium japonicum* (Coleoptera: Coccinellidae) through different exposure routes. Chemosphere 128:49–55

Ye XD, Su YL, Zhao QY, Xia WQ, Liu SS, Wang XW (2014) Transcriptomic analyses reveal the adaptive features and biological differences of guts from two invasive whitefly species. BMC Genomics 15:370

Zhang ZL (2000) Some thoughts to the outbreaks of tobacco whitefly. Beijing Agri Sci:1–3

Zhang GF, Lü ZC, Wan FH (2007a) Detection of *Bemisia tabaci* remains in predator guts using a sequencecharacterized amplified region marker. Entomol Exp Appl 123:81–90

Zhang GF, Lü ZC, Wan FH, Lövei GL (2007b) Real-time PCR quantification of *Bemisia tabaci* (Homoptera: Aleyrodidae) B-biotype remains in predator guts. Mol Ecol Notes 6:947–954

Zhang T, Luan JB, Qi JF, Huang CJ, Li M, Zhou XP, Liu SS (2012) Begomovirus- whitefly mutualism is achieved through repression of plant defences by a virus pathogenicity factor. Mol Ecol 21:1294–1304

Zhang YB, Yang NW, Zhang GF, Wang JJ, Wan FH (2015) Identification of naturally occurring *Eretmocerus* parasitoids by CO I sequence analysis and crossing test. J Environ Entomol 37(2):281–292 (in Chinese, English abstract)

Zhu DT, Xia WQ, Rao Q, Liu SS, Ghanim M, Wang XW (2016) Sequencing and comparison of the *Rickettsia* genomes from the whitefly *Bemisia tabaci* Middle East Asia Minor I. Insect Sci 23:531–542

Zou CH, Li L, Dong TY, Zhang BW, Hu QB (2014) Joint action of the entomopathogenic fungus *Isaria fumosorosea* and four chemical insecticides against the whitefly *Bemisia tabaci*. Biocontrol Sci Techn 24(3):315–324

Chapter 9
Rice Water Weevil *Lissorhoptrus oryzophilus* Kuschel

Yunshan Huang, Michael O. Way, and Mingxing Jiang

Abstract The rice water weevil, *Lissorhoptrus oryzophilus* Kuschel (Coleoptera: Curculionidae), native to North America, is one of the most destructive insect pests of rice in the US and eastern Asia. It was introduced into China in the 1980s and so far has spread widely throughout the country. Much work has been done with respect to its ecology and control methods. Here, we review the invasion history, mechanisms underlying its invasions and rapid spread, and major control methods in China. We also discuss factors responsible for its wide spread in China. In addition, we report on options employed to control this weevil. Finally, we propose future directions for research, particularly on the ecology and the development of a more comprehensive control strategy for this weevil in China.

Keywords *Lissorhoptrus oryzophilus* • Invasion • Cultural control • Natural enemies • Integrated management

9.1 Introduction

The rice water weevil, *Lissorhoptrus oryzophilus* Kuschel (Coleoptera, Curculionidae), is one of the most destructive insect pests of rice (*Oryza sativa* L.) in the US and eastern Asia. Adults are mainly terrestrial and can also swim; they feed on tissues of leaf blades leaving typical longitudinal scars along leaf veins. Larvae dwell in soil and feed on host plant roots, causing significant damages to rice. In China, the resulting rice yield losses are normally 10–20%, but in severe cases can reach 40–60% (Reay-Jones et al. 2008).

Y. Huang • M. Jiang (✉)
Ministry of Agriculture Key Laboratory of Agricultural Entomology, Institute of Insect Sciences, Zhejiang University, Hangzhou 310058, China
e-mail: mxjiang@zju.edu.cn

M.O. Way
Texas A&M AgriLife Research and Extension Center,
1509 Aggie Drive, Beaumont, TX 77713, USA

F. Wan et al. (eds.), *Biological Invasions and Its Management in China*,
Invading Nature - Springer Series in Invasion Ecology 11,
DOI 10.1007/978-94-024-0948-2_9

The rice water weevil is native to the marshlands of the Mississippi basin in North America (Isely and Schwardt 1934). It was introduced into Japan in 1976 (maybe several years earlier) and soon became one of most serious rice pests of the country. From 1980 to 1988, the weevil was introduced consecutively to the countries neighboring Japan, including North Korea, South Korea and China (Kisimoto 1993; Jiang and Cheng 2003b; Lupi et al. 2009).

In this chapter, we summarize the invasion, monitoring and management methods of the rice water weevil in China. We also propose perspectives on future research directions. We focus on major factors responsible for the rapid spread of this weevil in China, as well as measures taken to control it.

9.2 Invasions of Rice Water Weevil in China

9.2.1 Invasion History

The risk of *L. oryzophilus* invasions in China was initially discussed in the early 1980s by Gaogan Xu, an entomologist at the Plant Quarantine Institute, Ministry of Agriculture, China (Xu 1981). In the following years, at least three articles were published in morphology, biology, invasion risk and control of this weevil. In 1986, the weevil was listed as a quarantine insect pest of the country (Zhou 1987). Two years later (1988), it was found for the first time in Tanghai, Hebei province, which was suspected to be unintentionally introduced from the Korean peninsula.

Before 1996, rice water weevil occurred only in the provinces along or near the Chinese coast. Since the beginning of the twenty-first century, however, it began to spread inland in mid-east China, and since 2007 it spread to regions in the southwestern, northwestern and northeastern parts of the country. By May 2014, it had spread to ¾ of Chinese provinces (Ministry of Agriculture of the People's Republic of China 2014), becoming one of the most serious and rapidly spreading invasive pests in China.

The dispersal rate of this weevil in China has not been well quantified. In coastal regions of Zhejiang province, it was estimated to be approximately 36 km per year (Wang et al. 2011), which is slightly higher than the rate (20–30 km per year) reported in Japan (Asayama and Nakagome 1993).

9.2.2 Factors Associated with Spread of Rice Water Weevil in China

The rapid spread of rice water weevil in China can be attributed to several biological and life history traits.

Fig. 9.1 Typical habitats in China for overwintering populations of rice water weevil. (**a**) Levee with abundant grassy weeds; (**b**) Fallow field adjacent to rice paddies

First, it can reproduce parthenogenetically. In native regions such as southeastern Texas, this weevil has both female and male adults, and the majority of females must mate for viable eggs to be produced (Jiang et al. 2007). However, all adults so far found in non-native regions are females which reproduce parthenogenetically. Thus, even if only one female adult arrives at a new area, it could theoretically establish (without the need to find mates) and produce a sustainable population in a short period of time through parthenogenesis.

Second, this weevil enters diapause during the summer and winter seasons (Nilakhe 1977; Palrang et al. 1994; Jiang et al. 2004). This allows it to survive harsh environments, in particular high or low temperatures. For instance, in southeastern Zhejiang, most adults emerge in June and July and fly to hills adjacent to rice fields after feeding for a short period of time (Zhai et al. 1997, 1998). In Tongcheng of Anhui province, most weevils enter diapause in the summer and move to levees or fallow fields near rice fields with abundant grassy weeds (Fig. 9.1). Since these habitats (hills, levees and fallow fields) are seldom disturbed, they serve as important sources of adult weevils for the following field season.

Third, rice water weevil has a wide range of host plants. In China, adults were reported to be able to feed on at least 65 plant species (Zhang 1997). Some species of Poaceae and Cyperaceae are much preferred. For instance, barnyardgrass *Echinochloa crusgalli* (L.) P. Beauv. and southern cutgrass *Leersia hexandra* Swartz are widespread and abundant in typical rice agroecosystems (Zhu et al. 2005; Cai et al. 2007; Lu et al. 2013a). This broad host range largely contribute to the population development of the weevil. In particular, it can increase the survival rate of adults that are introduced into a new area.

Moreover, the weevil possesses some life traits that may protect it from natural enemies. These traits include: (i) deposition of eggs inside leaf sheaths, (ii) survival of larvae and pupae in mud, (iii) falling from foliage and feigning death when disturbed, (iv) escaping predators by swimming in water, and (v) small size and cryptic coloration. Additionally, the weevil's life history is relatively plastic. For example, it has been found to have a relatively larger adult body size in northern than southern regions of China (Yunshan Huang, personal observations). The significance of such morphological variance is not yet clear. When environmental conditions are not suitable (no free water and/or no host plants), adults can retain mature eggs in their ovaries until conditions become favorable for successful oviposition (Jiang and Cheng 2003b).

Rice cultural practices, such as water management and use of fertilizer, have certain effects on the weevil. As demonstrated in the US, delaying flooding time can reduce egg deposition and thus mitigate damages; but this finding has not been tested in China. The amount of nitrogen fertilizer applied during the rice seedling stage may significantly affect weevil reproduction (Jiang and Cheng 2003a).

Some rice production or physical factors, such as the distribution pattern of local rice fields, prevailing wind direction and speed, and elevation may also affect the dispersal of rice water weevils. In Zhejiang, connectivity of early-season rice fields can serve as a corridor for among-region dispersal of this weevil, and thus is suspected to have contributed considerably to its dispersal (Wang et al. 2011). The weevil has a low flight ability, thus its dispersal is thought to be little affected by wind (Zhai et al. 1999a, b). However, in Japan, wind might have aided its dispersal (Asayama and Nakagome 1993). High elevation might limit dispersal of this weevil (Saito et al. 2005). Yet, this factor appears not to have deterred its invasion and spread in China, as it has dispersed widely into western China which is separated from the middle and coastal parts of this country by mountainous terrain or desert.

Human activities are believed to contribute greatly to the wide distribution of this weevil in China, as the weevil has been found in inland China despite the existence of geographic barriers. Transporting rice seedlings, rice straw and soil contaminated with weevils may unintentionally introduce the weevil to neighboring villages and counties. After arriving at a new area, the ability of reproducing parthenogenetically, tolerating adverse environments by entering diapause, utilizing large numbers of plant species, and escaping from natural enemies, would increase the probability of this weevil to survive and soon establish a sustainable population.

9.3 Management of Rice Water Weevil in China

Since 1986, the rice water weevil has been listed as a quarantine insect in China. In regions where weevils already occur, the use of chemical insecticides is the most important control strategy. Other alternative control methods may also be employed, such as using light traps to reduce populations, adjusting planting date and reducing depth of standing water in the fields.

9.3.1 Quarantine

During the early years after introductions of the rice water weevil into a new area, an official quarantine was established to curtail its further spread. Rice grains, seedlings, straw and soil were not allowed to be transported from infected to non-infected regions. Rice straw in infected areas should not be used as packing materials in transportation.

9.3.2 Chemical Control

Today in China, the use of chemical insecticides is the first choice for control of rice water weevil. During the years around 1990, some highly toxic chemicals were used, including esfenvalerate, isofenphos-methyl and carbofuran. In addition, cycloprothrin, an environmental friendly chemical widely used in Japan was recommended for the control practices in China.

Since the mid 1990s, some of the chemicals commonly used for control of other rice insect pests, such as stem borers, rice leafrollers and planthoppers, were widely used to control rice water weevil. Such chemicals included triazophos, fipronil, imidacloprid and chlorpyrifos. They are still in use today in China, except fipronil which was prohibited for use in rice paddies in 2009 due to high toxicity to natural enemies, honeybees and aquatic economically important invertebrates (e.g., shrimps and crabs).

In 2009–2010, chlorantraniliprole, as a replacement for fipronil to control lepidopteran insect pests of rice, was recommended for use. At present, chlorantraniliprole, as well as mixtures of this chemical with others (e.g., triazophos and thiamethoxam), have been the major chemicals for the control of rice water weevil in China.

To reduce adverse impacts on non-target organisms, several plant-derived chemicals have been used in the past decade. For example, in southwestern Liaoning, 1% matrine AS was recommended for use in rice fields combined with crab production. In Xinjiang, matrine and pyrethrins can be used in organic rice fields. The major drawback of these chemicals is that they kill weevils slowly and thus require a longer period of time (such as 15 days) to achieve adequate control. Larger time can result in greater damages when compared to traditional chemicals.

The common method of applying chemicals is to spray onto rice plants. In addition, seed treatments have also been recommended in some areas, using the active ingredients such as imidacloprid and thiamethoxam.

9.3.3 Cultural Control

In other countries around the world, such as the US, cultural methods have proved to be highly effective against rice water weevil in some regions. These methods include manipulating planting date, removing grass from levees, and draining fields at intervals (Hesler et al. 1992; Reay-Jones et al. 2008; Stout et al. 2009). Yet, none of them have been widely used in China. Below, we describe their rationale, shortcomings, and potential for use in China.

Manipulating planting date, especially delaying planting, was demonstrated to be very effective in some regions of the US, such as southeastern Texas (Thompson et al. 1994a; Espino et al. 2009). This is because late-planted rice plants are normally more tolerant to weevil damage (Thompson et al. 1994a). However, the use and outcomes of this practice may be dependent on other factors, such as local natural conditions, rice varieties, and the weevil's density and population dynamics. In China, this control tactic is little adopted by rice farmers, particularly in regions where two crops of rice are produced in a single growing season. Planting late would likely preclude production of a second crop. However, late planting of rice in northern China may be beneficial regarding control of the rice water weevil.

Removing grass from levees and adjacent areas, i.e., reducing abundance of host plants available to the weevil, may substantially reduce weevil density and thus perhaps lower the overall density in rice fields (Grigarick 1993; Palrang et al. 1994). This practice can be effective because it can be easily performed by farmers. Yet, it only works when rice plants are not available to weevil adults. If rice plants are available to weevils, there might be a few weevils on levees, or, weevils may move from mowed levees to rice fields. The control efficiency of this practice is probably closely related to the developmental stage of adults. Weevils at an early ovipositional stage may be more affected due to greater requirements for food than those at the preovipositional stage (Jiang and Cheng 2003b).

Drainage can reduce oviposition of rice water weevil since this insect deposits eggs in leaf sheaths that are submerged under water (Hesler et al. 1992). After a period of temporary drainage, oviposition would be delayed/reduced and thus peak larval densities would occur later when rice would be older and less susceptible to damages (Stout et al. 2002). However, this practice has many shortcomings. The most important ones are that weed control may lead to the loss of nitrogen, and thus additional uses of herbicide and nitrogen are required, which increases the cost to farmers. In addition, these drained conditions can foster blast development. Moreover, when the flood is reapplied, weevil adults would return to rice plants and oviposit (Thompson et al. 1994b; Rice et al. 1999). Furthermore, in the southeastern US, rainfall can be abundant and common during the growing season which prevents rice fields from drying, allowing continued rice water weevil attack. For these reasons, in Texas, temporary draining is not an option for weevil control and is not an effective replacement for chemical control (Way and Espino 2014). In China, this practice also has not been recommended to rice farmers.

Rice planting method, i.e., direct seeding versus transplanting, may be closely associated with weevil densities and damages. In some regions of China, such as Xiangtan of Hunan province which is heavily infected by rice water weevil, direct seeding is the major planting method (M. X. Jiang, personal observations). Such fields are not favored by the weevil during the early season, because rice plants are relatively small and thus not suitable for feeding and oviposition. By contrast, fields with transplanted rice are much favored, because such plants normally have developed to the 4–5-leaf stage, which is preferred by the weevil. Research is needed to better elucidate these associations to help manage the rice water weevil.

9.3.4 Host Plant Resistance

The pioneering work in this field was done in 2002 in Shenyang of Liaoning province, one of the regions most seriously infested by the rice water weevil (Tian et al. 2003). In this work, the authors identified a highly resistant rice cultivar (T03), based upon comparisons among different cultivars of adult feeding and oviposition preference, egg hatch rate, and larval survival rate. This cultivar has been suggested as an ideal material for developing resistant rice lines (Li et al. 2011).

9.3.5 Biological Control

There have been several reports regarding natural enemies of rice water weevils. Of the enemies so far identified, *Beauveria bassiana* and *Metarhizium anisopliae* are the most efficient. Yet, they have not been developed into commercial bio-insecticides.

Beauveria bassiana is an entomopathogenic fungus that can parasitize numerous insects. Both in the US and Japan, this fungus was shown to be highly infective to rice water weevil adults (Godfrey et al. 1993; Urtz and Rice 1997). In China, a number of *B. bassiana* strains have been isolated and virulence tested (Jiang et al. 2002; Xu et al. 2013; Tian et al. 2014). Of them, many strains, such as YS03, were shown to have high virulence under laboratory conditions (Xu et al. 2013). Preliminary field assays have also been conducted (Xu et al. 2015). However, *B. bassiana* has several drawbacks that may limit its potential use because adults infected with this fungus can still oviposit before death. Moreover, *B. bassiana* has little efficacy against larvae (Godfrey et al. 1993). In addition, under natural conditions, activity of spores may decrease rapidly due to environmental constraints, such as ultraviolet radiation, and other microorganisms antagonistic to *B. bassiana* (Rice 1996; Urtz and Rice 1997).

Metarhizium anisopliae has received the most attention of any natural enemy of the rice water weevil. In China, a number of strains have been isolated and some of them have been approved as potential biological control agents (Chai et al. 2000).

Their control efficacy against rice water weevil has also been assayed in several locations, such as in Zhejiang (Chen et al. 2000) and Liaoning (Yu et al. 2008). When controlling weevils at high densities, certain chemical insecticides (e.g. triazophid) can be added to *M. anisopliae* spore suspensions to elevate control efficiency (Chen et al. 2000).

Rice water weevil adults can be parasitized by nematodes, as has been reported in native regions (Bunyarat et al. 1977). These parasites may reduce fecundity and cause high mortality in infected adults. In the 1980s, certain nematode species (e.g. *Steinernema* spp.) was suggested as possible biological agents to control the weevil (Carbonell 1983; Nagata 1987). In China, the Otio strain of *Steinernema feltiae* was found to cause high mortality for larvae (Sun et al. 2006). Efficacy of *S. feltiae*, *S. carpocapsae* A24 strain, *S. glaseri* NC34 strain, *Heterorhabditis bacteriophora* and *H. zealandica* have also been detected in adult weevils. Mortality of 82.5% and 97.5% was observed in adults treated with *S. feltiae* and *H. bacteriophora*, respectively, at 10 d after incubation with nematodes (Li et al. 2007). However, none of these strains has been commercially used.

9.4 Conclusion and Future Directions

Since its introduction in the late 1980s, the rice water weevil has become one of the major insect pests of rice in China. This weevil has spread at a faster rate than expected. It now occurs in a great number of far inland regions of China. A number of life history traits, such as the ability to reproduce parthenogenetically and enter diapause under temperature stress, are thought to have greatly contributed to its rapid expansion. Meanwhile, human activities have likely contributed to its dispersal into many far-away regions of China, such as Xinjiang, a province in far northwestern China.

To identify and understand the mechanisms behind the rapid spread of this weevil, its biology and ecology have been extensively studied soon after its introduction. Much work has been done on their diapause, flight and reproductive behaviors. However, many questions remain to be answered. As an example, we still need to learn about the response of the weevil to changes in rice production practices. Will rice water weevil populations and damages be affected in regions where production of two crops of rice in a single season is shifted to one? What changes will be induced in local weevil population dynamics where rice growing methods are changed from direct seeding to transplanting, or *vice versa*, as described above? To what extent does morphological and reproductive plasticity take place in this weevil, and what role does this plasticity play in the weevil's population dynamics?

More recently, symbiotic bacteria in the rice water weevil have attracted much research interest. A number of bacteria have been identified in the weevil's gut and ovary, such as *Wolbachia* spp., *Nardonella* spp., and *Pantoea agglomerans* (Chen et al. 2012; Lu et al. 2013b, 2014; Huang et al. 2016; Zhang et al. 2016). However, we know little about the functions of these bacteria, except for *Wolbachia* spp. which have been shown to be closely linked to host reproduction (Chen et al. 2012).

To limit the spread of this weevil, monitoring should be conducted extensively over a wider range. This work should be given major attention, since it is difficult to detect the presence of rice water weevils when they are at a low density. We suggest training rice farmers to allow them to learn about the major monitoring and detection methods for the rice water weevil.

In terms of control methods, Chinese farmers rely mostly on chemical control to manage the weevil. Other control options, based on cultural practices and natural enemies, have not been developed for farmers, although some options have been tested to be highly effective. Overall, most of the non-chemical control methods need to be further evaluated. In China, it has been the goal to develop a more comprehensive, environmental friendly control strategy for the weevil.

References

Asayama T, Nakagome T (1993) Invasion and spread of the rice water weevil, *Lissorhoptrus oryzophilus* Kuschel, in Japan. In: Hirai K (ed) Establishment, spread, and management of the rice water weevil and migratory rice insect pest in East Asia. NARC, Tsukuba, pp 83–103

Bunyarat M, Tugwell P, Riggs RD (1977) Seasonal incidence and effect of a mermithid nematode parasite on the mortality and egg production of the rice water weevil *Lissorhoptrus oryzophilus*. J Econ Entomol 6:712–714

Cai XT, Zheng YM, Shang HW, Yang Q, Tang QY, Zheng XH, Cheng JA, Zhu ZR (2007) Effect of host plant on relationship between number and total length of feeding scars of the first generation adults of rice water weevil in fields. Chin J Rice Sci 21:309–315

Carbonell RM (1983) Pathogens and nematodes for control of rice water weevil in Cuba. Int Rice Res Newsl 8(6):16–17

Chai YQ, Chen ZA, Feng HY, Pan LC (2000) Bioassay of pathogenicity of *Metarhizium anisopliae* on the rice water weevil *Lissorhoptrus oryzophilus*. Chin J Biol Control 16:22–25

Chen ZA, Feng HY, Shi LC, Liu YG, Pan LC, Wang GE, Fang YJ (2000) Evaluation of *Metarhizium anisopliae* for control of *Lissorhoptrus oryzophilus* in the field. Chin J Biol Control 16:53–55

Chen SJ, Lu F, Cheng JA, Jiang MX, Way MO (2012) Identification and biological role of the endosymbionts *Wolbachia* in rice water weevil (Coleoptera: Curculionidae). Environ Entomol 41:469–477

Espino L, Way MO, Pearson R, Nunez M (2009) Effect of planting date on *Lissorhoptrus oryzophilus* (Coleoptera: Curculionidae) density-yield relationship on rice in southeastern Texas. J Econ Entomol 102:1536–1545

Godfrey LD, Palrang AT, Grigarick AA (1993) Protection of rice from invertebrates. California Rice Experiment Station. http://carrb.com/93rpt/Invertebrate.htm (June 10, 2016)

Grigarick AA (1993) Study of rice water weevil, past, present, and future in US with emphasis on California. In: Hirai K (ed) Establishment, spread, and management of the rice water weevil and migratory rice insect pest in East Asia. NARC, Tsukuba, pp 12–31

Hesler LS, Grigarick AA, Oraze MJ, Palrang AT (1992) Effects of temporary drainage on selected life history stages of the rice water weevil (Coleoptera: Curculionidae) in California. J Econ Entomol 85:950–956

Huang X, Huang YS, Zhang JY, Lu F, Wei J, Jiang MX (2016) The symbiotic bacteria *Nardonella* in rice water weevil (Coleoptera: Curculionidae): diversity, density and associations with host reproduction. Ann Entomol Soc Am 109:415–423

Isely D, Schwardt HH (1934) The rice water weevil problem in Arkansas. J Econ Entomol 25:218–222

Jiang MX, Cheng JA (2003a) Feeding, oviposition and survival of overwintered rice water weevil (Coleoptera: Curculionidae) adults in response to nitrogen fertilization of rice at seedling stage. Appl Entomol Zool 38:543–549

Jiang MX, Cheng JA (2003b) Effects of starvation and absence of free water on the oviposition of overwintered adult rice water weevil, *Lissorhoptrus oryzophilus* Kuschel (Coleoptera: Curculionidae). Int J Pest Manag 49:89–94

Jiang MX, Shang HW, Cheng JA (2002) Bioassay of *Beauveria bassiana* against the rice water weevil, *Lissorhoptrus oryzophilus* Kuschel. Acta Phys Sin 29:287–288

Jiang MX, Zhang WJ, Cheng JA (2004) Reproductive capacity of first-generation adults of the rice water weevil *Lissorhoptrus oryzophilus* Kuschel (Coleoptera: Curculionidae) in Zhejiang, China. J Pest Sci 77:145–150

Jiang MX, Way MO, Zhang WJ, Cheng JA (2007) Rice water weevil females of different elytral color morphs: comparisons of locomotor activity, mating success, and reproductive capacity. Environ Entomol 36:1040–1047

Kisimoto R (1993) Spread and management of the rice water weevil, an imported pest of rice. In: Hirai K (ed) Establishment, spread, and management of the rice water weevil and migratory rice insect pest in East Asia. NARC, Tsukuba, pp 32–41

Li JJ, Qian HT, Dong H, Cong B (2007) Laboratory test on the infectivity of entomopathogenic nematode to *Lissorhoptrus oryzophilus*. Chin J Biol Control 32:188–190

Li ZQ, Yu FQ, Tian CH, Sun FY (2011) Inheritance analysis of the resistance to rice water weevil in rice cultiva T03. Liaoning Agric Sci 5:25–28

Lu F, Zhang WJ, Jiang MX, Way MO (2013a) Southern cutgrass, *Leersia hexandra* Swartz, allows rice water weevils to avoid summer diapause. Southwest Entomol 38:153–161

Lu F, Kang XY, Jiang C, Lou BG, Jiang MX, Way MO (2013b) Isolation and characterization of bacteria from midgut of the rice water weevil (Coleoptera: Curculionidae). Environ Entomol 42:874–881

Lu F, Kang XY, Lorenz G, Espino L, Jiang MX, Way MO (2014) Culture-independent analysis of bacterial communities in the gut of rice water weevil (Coleoptera: Curculionidae). Ann Entomol Soc Am 107:592–600

Lupi DL, Cenghialta CC, Colombo MC (2009) Adult feeding by the rice water weevil *Lissorhoptrus oryzophilus* on different host plants. Bull Insectol 62:229–236

Ministry of Agriculture of the People's Republic of China (2014) Note on the distribution of quarantine pests of plants in China. http://www.moa.gov.cn/govpublic/ZZYGLS/201506/t20150604_4634137.htm (May 27, 2016)

Nagata T (1987) Infectivity of *Steinernema feltiae* on the rice water weevil, *Lissorhoptrus oryzophilus* Kuschel. In: Ishibashi N (ed) Recent advances in biological control of insect pests by entomogenous nematodes in Japan. Ministry of Education, Culture and Science, Japan, pp 65–70

Nilakhe SS (1977) Reproductive status of overwintering rice water weevils. Ann Entomol Soc Am 70:559–601

Palrang AT, Grigarick AA, Oraze MJ, Hesler LS (1994) Association of levee vegetation to rice water weevil (Coleoptera: Curculionidae) infestation in California rice. J Econ Entomol 87:1701–1706

Reay-Jones FPF, Way MO, Tarpley L (2008) Nitrogen fertilization at the rice panicle differentiation stage to compensate for rice water weevil (Coleoptera: Curculionidae) injury. Crop Prot 27:84–89

Rice WC (1996) Evaluation of *Beauveria bassiana* isolates for the control of rice water weevil, *Lissorhoptrus oryzophilus* (Kuschel). In: 26th Rice Technical Working Group. San Antonio, Texas, p 114

Rice WC, Croughan TP, Ring DR, Muegge MA, Stout MJ (1999) Delayed flood for management of rice water weevil (Coleoptera: Curculionidae). Environ Entomol 28:1130–1135

Saito T, Hirai K, Way MO (2005) The rice water weevil, *Lissorhoptrus oryzophilus* Kuschel (Coleoptera: Curculionidae). Appl Entomol Zool 40:31–39

Stout MJ, Riggio MR, Zou L, Roberts R (2002) Flooding influences ovipositional and feeding behavior of the rice water weevil (Coleoptera: Curculionidae). J Econ Entomol 95:715–721

Stout MJ, Harrell D, Tindall KV, Bond J (2009) Impacts of seeding rate on interactions between rice and rice water weevils impacts of seeding rate on interactions between rice and rice water weevils. J Econ Entomol 102:1837–1845

Sun WP, Cong B, Dong H, Qian HT (2006) Study on the efficacy of *S. feltiae Otio* on rice water weevil in laboratory. J Anhui Agric Sci 34:1620–1621

Thompson RA, Quisenberry SS, N'Guessan FK, Heagler AM, Giesler G (1994a) Planting date as a potential cultural method for managing the rice water weevil (Coleoptera: Curculionidae) in water-seeded rice in southwest Louisiana. J Econ Entomol 87:1318–1324

Thompson RA, Quisenberry SS, Trahan GB, Heagler AM, Giesler G (1994b) Water management as a cultural control tactic for the rice water weevil (Coleoptera: Curculionidae) in southwest Louisiana. J Econ Entomol 87:223–230

Tian CH, Yu FQ, Liu PB, Li ZQ, Sun FY (2003) Assays on resistance of cultiva T03 to rice water weevil. Liaoning Agric Sci 5:5–7

Tian ZL, Zhu XM, Luo JY, Lu Y, Zhao Y (2014) Screening of specificity pathogenic strains of *Beauveria bassiana* against *Chilo suppressalis* and *Lissorhoptrus oryzophilus* in Jilin province. Chin J Biol Control 30:665–671

Urtz BE, Rice WC (1997) RAPD-PCR characterization of *Beauveria bassiana* isolates from the rice water weevil *Lissorhoptrus oryzophilus*. Lett Appl Microbiol 25:405–409

Wang ZJ, Wu JG, Shang HW, Cheng JA (2011) Landscape connectivity shapes the spread pattern of the rice water weevil: a case study from Zhejiang, China. Environ Manag 47:254–262

Way MO, Espino L (2014) Insect management. Texas Rice Production Guidelines. University of Texas A&M Agricultural Extension

Xu GG (1981) Rice water weevil: an insect pest to be noted in plant quarantine. Plant Quarant 3:20–21

Xu J, Yang MF, Yang DX, Shi PQ, Zou X (2013) Bioassay of pathogenicity of different *Beauveria bassiana* on rice water weevil. Guizhou Agric Sci 41:69–72

Xu J, Yang MF, Di XY, Shi PQ, Zou X (2015) Field control effects of *Beauveria bassiana* strain YS03 on rice water weevil. Southw China J Agric Sci 28:1630–1633

Yu FQ, Tian CH, Li ZQ, Sun FY, Ji MS (2008) Control efficacy of fungus *Metarhizum anisopliae* against rice water weevil. Liaoning Agric Sci 6:5–8

Zhai BP, Cheng JA, Huang EY, Shang HW, Zheng XH, Wu J, Fang YJ, Xia WQ, Lu XJ (1997) Population dynamics of rice water weevil in double rice cropping area of Zhejiang province. China Scientia Agric Sinica 30(6):23–29

Zhai BP, Shang HW, Cheng JA, Huang EY, Zheng XH (1998) Diapause of first generation adults of rice water weevil in double-cropping rice area. Chin J Appl Ecol 9(4):400–404

Zhai BP, Zheng XH, Shang HW, Cheng JA (1999a) Influence of wind on takeoff of rice water weevils. Chin J Agrometeorol 20(3):24–27

Zhai BP, Zheng XH, Shang HW, Cheng JA, Wang P (1999b) Flight behaviour in the diapause syndrome of the rice water weevil, *Lissorhoptrus oryzophilus* Kuschel. Acta Ecol Sin 19:453–457

Zhang YJ (1997) Occurrence and control of the rice water weevil in Tanghai, China. Plant Quarant 11:40–41

Zhang JY, Huang YS, Huang X, Jiang MX (2016) Infection state of *Pantoea agglomerans* in the rice water weevil, *Lissorhoptrus oryzophilus* (Coleoptera: Curculionidae). Appl Entomol Zool 51:561–569

Zhou YS (1987) Rice water weevil, a novel quarantine insect pest. World Agric 1:35–36

Zhu ZR, Shang HW, Jiang MX, Cheng JA, Yu HZ (2005) Rice water weevil, *Lissorhoptrus oryzophilus* Kuschel. In: Wan FH, Zheng XB, Guo JY (eds) Biology and management of invasive alien species in agriculture and forestry. Science Press, Beijing, pp 129–176

Chapter 10
Colorado Potato Beetle *Leptinotarsa decemlineata* (Say)

Wenchao Guo, Chao Li, Tuerxun Ahemaiti, Weihua Jiang, Guoqing Li, Jiahe Wu, and Kaiyun Fu

Abstract Colorado potato beetle (CPB) was naturally dispersed from Kazakhstan into Xinjiang Autonomous Region of China in 1993. Since then, it has been widely distributed all over the southern region of the Tianshan Mountain. Chinese scientists focused on its monitoring and invasion risk management in China, as well as its invasion biology and ecology related to rapid dispersal, such as developmental threshold and cumulative temperature, diapause condition, and influence factors for flight. In invaded regions of China, several techniques such as improved crop cultivation techniques, friendly environmental chemical control (low or none toxic insecticides), biological control, physical techniques, ecological regulations etc. can be combined into an integrated pest management of CPB. However, some questions still remain in theses fields, for example, the genetic variations, environment (hosts, habitats, climates, soil etc.) adaptabilities and geographical populations of CPB are still unclear, and mechanisms of CPB's rapid resistance development to pesticides have not been well understood. The concerned diapause mechanisms are still unknown. Moreover, the interactions between CPB, hosts, pathogens, predators and parasitoids, environments have not been studied, and the resistance or tolerance of potato plants to CPB and their related mechanisms all need to be understood in order to breed CPB-resistant potato crops.

Keywords *Leptinotarsa decemlineata* • Invasion risk management • Flight • Transgenic CPB-resistant potato • *Beauveria bassina* • RNA interference technology

---

W. Guo (✉) • T. Ahemaiti • K. Fu
Xinjiang Academy of Agricultural Sciences, Ürümqi, China
e-mail: gwc1966@163.com

C. Li
Xinjiang Agricultural University, Ürümqi, China

W. Jiang • G. Li
Nanjing Agricultural University, Nanjing, China

J. Wu
Chinese Academy of Sciences, Beijing, China

F. Wan et al. (eds.), *Biological Invasions and Its Management in China*,
Invading Nature - Springer Series in Invasion Ecology 11,
DOI 10.1007/978-94-024-0948-2_10

10.1 Introduction

The Colorado potato beetle (CPB), *Leptinotarsa decemlineata* (Say) (Coleoptera: Chrysomelidae), is an alien invasive and quarantine species in the world. It is considered to be one of the most destructive insect pests of cultivated potato globally (Alyokhin 2009). Larvae and adults feed on potato or eggplant plants, and their damage can cause a yield loss of 30–50%. Natural migration is the major dispersal strategy of its spreading. In addition, it can also be spread by vehicles loading potato or by products of potato, and even by running water (Boiteau and Heikkilä 2012; Rabitsch 2010). CPB has been spreading continuously in the past 150 years. In the United States, it had been present across the country in 1890s, 40 years after the record of its first serious damage in Colorado (Weber 2003, 2008). In Europe, CPB was multiple-introduced during 1876–1914 into several countries such as Germany, Netherlands and England, and it further spread to other countries from 1930s to 1940s and invaded the European part of the former Soviet Union from 1950s to1970s. In Asia, it was first detected in the Republic of Azerbaijan in 1975, and invaded some Central Asian countries in several years, such as the Kyrgyz Republic. In the past 60 years, CPB had spread westward for about 6000 km, covering a range from France of Europe to Kazakhstan of Central Asia, with a speed of around 80 km per year. In China, CPB was found for the first time in 1993 in Xinjiang Uygur Autonomous Region.

With the rapid development of potato production in the past several decades, China has become the top country since 2010 in both cultivated area and yield of potato. Potato has been ranked as the fourth most widely cultivated crop in this country. Therefore, potato production plays an important role in the economic development of some potato-growing regions of China, such as Gansu, Inner Mongolia, Sichuan, Chongqing, Guangxi and Yunnan (Jansky et al. 2009; Qu et al. 2005). For this reason, CPB has been considered to be a serious threat to the potato production of this country. Since mid 1990s, a lot of research programs have been launched on CPB's invasion biology, invasion risk, population monitoring and control, and a set of integrated sustainable management techniques has been developed and applied for CPB control (Guo et al. 2014).

10.2 Invasive History and Current Distribution of Colorado Potato Beetle in China

Jolivet (1991) speculated that CPB would spread into China in early 1990s and would invade northeastern Asia through China. His speculation coincided with the incident that CPB was found in the boundary region of Kazakhstan and Xinjiang of China at the end of 1980s and spread into Xinjiang in early 1990s. Yet, CPB has not spread into Northeast Asia and so far it distributes only in Xinjiang in Asia.

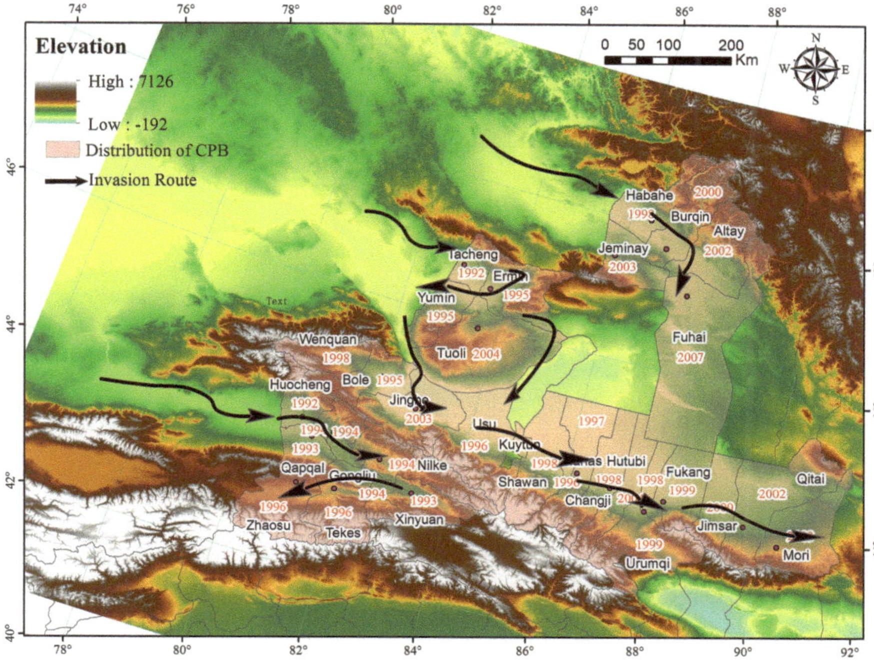

Fig. 10.1 The Invasion route and distribution of Colorado potato beetle in Xinjiang of China, disastrous for China's potato industry

In Xinjiang, CPB was found in 1993 in the Yili River valley and Tacheng, which are near the China-Kazakhstan border (Guo et al. 2011d). Since then, CPB had been constantly spreading from west to east along the oasis in northern slope of the Tianshan Mountain, until it arrived at Mori Kazak autonomous county in 2003, which is about 550 km to the northwest of Gansu province (Fig. 10.1). Its spreading speed was about 80 km on average per year. Since 2003, CPB has been prevented from further spreading and has been intercepted successfully in the west of Sangequanzi of Mori. Such a success was due to the effective control and the existence of naturally isolated deserts between oases, in which there were no host crops and few wild host plants.

In recent years, invasions of CPB from Russia to China have become a risk in northeast China. In 2005, CPB outbreak occurred in the coastal region of Russia which is adjacent to Heilongjiang and Jilin provinces of China (Cheng et al. 2005; Pan et al. 2015; Zhang et al. 2012). Since that time, investigating and monitoring of CPB have been extensively carried out along the border between northeast China and coastal region of Russia. Unfortunately, CPB was found in Chunhua town of Hunchun in eastern Jilin in 2013 (Wang et al. 2014). In 2014, CPB's population in this area was controlled and eliminated completely. In the same year, CPB was also found in Suifenhe, Sanchakou of Dongning and Hutou of Hulin in eastern Heilongjiang province with a distance of <4 km to coastal region of Russia. All CPB population in these regions had been eradicated in time.

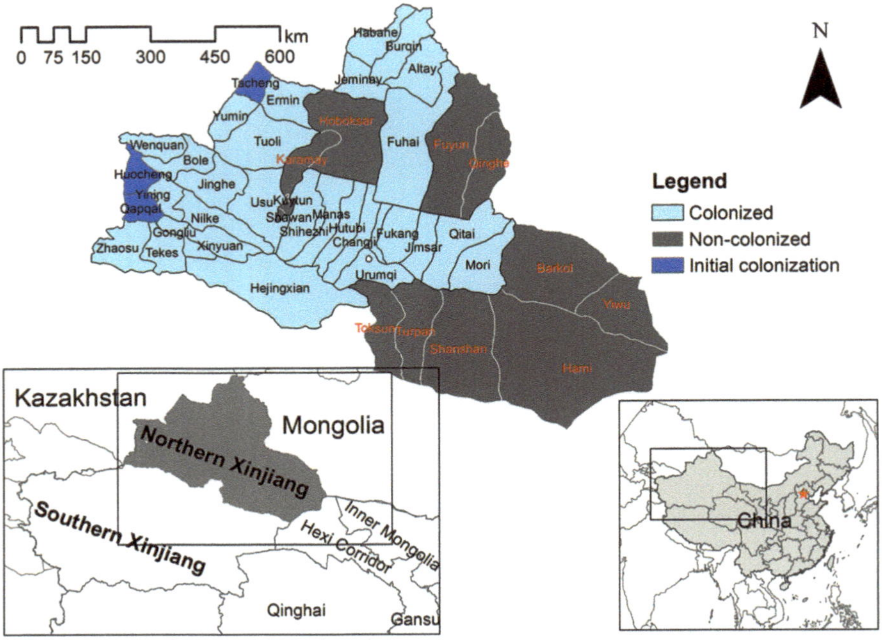

Fig. 10.2 Distribution of Colorado potato beetle in Northern Xinjiang. *Dark blue* regions represent the initial colonization by CPB in 1993. *Light blue* regions stand for the current distribution. *Red names* refer to the non colonization regions in Northern Xinjiang. (Li et al. 2014b)

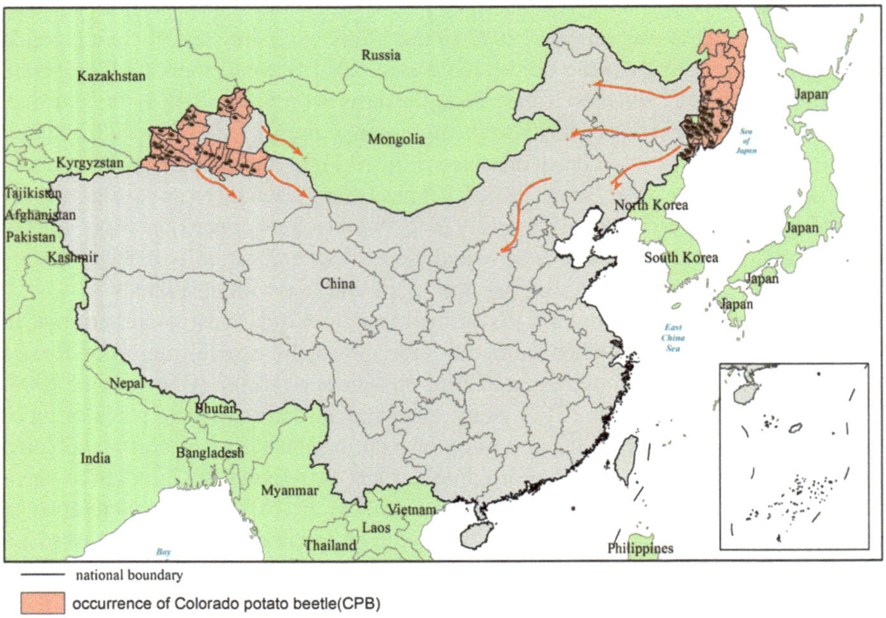

Fig. 10.3 The potential invasion route of Colorado potato beetle in China

So far CPB occurs only in the region of about 30 ha in northern part of the Tianshan Mountain in Xinjiang, covering a total of 38 counties (Fig. 10.2). Despite such a limited distribution, it has the potential of spreading from Xinjiang to neighboring provinces, where potato is grown at a large scale, and from Russia again to northeast China, due to its migrating capacity (Liu et al. 2012; Wang et al. 2014) (Fig. 10.3).

10.3 Invasion Biology of Colorado Potato Beetle in China

CPB possesses several biological and ecological characteristics that might have contributed to its invasions worldwide, such as self-independent migration capacity (Alyokhin et al. 1999; Boiteau 2002; Boiteau et al. 2010), high reproduction rate (Arpaia et al. 2009; Baker et al. 2005; Ferro et al. 1991) and diapauses (Boiteau and Coleman 1996; Boman 2008; Boman et al. 2008). In the past decade, we conducted extensive investigations on the biology of this beetle in Xinjiang, with a particular focus on its invasion and population growth mechanisms. The major findings are presented below.

10.3.1 Life Cycle

CPB has one or two generations every year in Xinjiang, with the two generations to be more common (Ahemaiti et al. 2010). An incomplete third generation may occasionally takes place but it depends on the availability of thermal resource and effective ac*cumulative temperatures* (Guo et al. 2010c) (Fig. 10.4).

10.3.2 Host Plant

More than 20 host plants have been reported in native regions of CPB in the United States. However, the host range is much narrower in China. For example, in Xingjiang there are only 6 host Solanaceae plant species: cultivated potato (*Solanum tuberosum*), eggplant (*S. melongena*), buffalobur (*S. rostratum* Dunal), central Asian henbane (*Hyoscyamus pusillus*), tomato (*Lycopersicon esculentum*), Chinese wolfberry (*Lycium chinense*) and Datura (*Datura suaveolens*). Owing to the lack of proper host plants, CPB may also occasionally feed on Chinese cabbage (*Brassica rapa pekinensis*), which is widely distributed in China. Hosts differ greatly in their suitability for CPB population growth, with the potato being mostly preferred, followed by eggplants and henbanes (Luo et al. 2012). Tomatoes can be fed by larvae of CPB, but it is not suitable because the pupae from tomato-feeding larvae can not develop healthily to adults (Li et al. 2013a) (Fig. 10.5).

Fig. 10.4 Life cycle of Colorado potato beetle

Investigations showed that potato was much more attractive to CPB than other hosts. A number of highly attractive compounds have been extracted from potato leaves, including 2-phenoyl ethanol, (Z)-acetic acaid-3-hexennyl-1-ethenol ester, linalool and methyl salicylate (Li et al. 2010). An acrasin has been synthesized artificially and demonstrated to be highly attractive to CPB adults. This acrasin is formulated with the effective volatiles (linalool, methyl salicylate and hexenyl acetate). The resulted mixture can attract both males and females of CPB and the attraction efficacy is significantly increased (88% for males and 83% for females) (Li et al. 2010).

10.3.3 Reproduction of Colorado Potato Beetle

The overwintered female adults have a very high fecundity, capable of laying 1680 eggs on average (Hare 1990). Female' fecundity is significantly affected by temperature and nutrition conditions. They do not lay egg at temperatures below 20 °C, and their fecundity reduces substantially as the temperature increase beyond 32 °C. First-generation females can lay 310 eggs on average when feeding on young hosts' leaves, whereas only 37 eggs may be laid when feeding on old leaves. In Xinjiang, the reproductive potential of CPB probably varies with regions, because different regions may have different temperature and food conditions.

Fig. 10.5 The damage symptoms of host plants by Colorado potato beetle

Fig. 10.5 (continued)

10.3.4 The Influence Factor to Flight

Flight capacity of CPB has been assayed using the computer-monitored flight-mill and field investigations. The results showed that its fight capacity was related closely to some environmental conditions, such as temperature, food abundance, host species, population density and reproductive status (Guo et al. 2012; Yue 2013; Yue et al. 2013).

1. The CPB flight capacity varies among generations. In general, the overwintered adults are highest and the second-generation adults are lowest. Overwintering adults may be induced to attain a higher flight capacity when population density increases and thus nutrition conditions become moderately deficient (Guo et al. 2011a).
2. Ambient temperature has significant effect on CPB flight capacity (Guo et al. 2011c). They have a highest capacity from 30 to 33 °C. They would not fly under the temperature below 20 °C. When the temperature is 25 °C or lower, food conditions may arise as a major factor affecting flight, that is, availability of sufficient host plants would greatly help the adults in increasing their flight capacity.
3. Flight capacity is closely related to the type of host plants (Guo et al. 2011c). Adult feeding on potato leaves tend to have a highest flight capacity, followed by the adults feeding on henbane or eggplant leaves, while the ones feeding on tomato leaves are lowest (Guo et al. 2011b).
4. For the first-generation CPB adults, neither females and males can fly within 5 days after unearthed. When provided with food, their flight capacity increased steadily. Females and males have a different temporal pattern of flight capacity. That is, females can fly a longer distance than males from 5 to 13 days after unearthed, and females began to reduce their flight capacity and they had a level similar to males at 15 days.

Previous studies by other researchers showed that CPB adults have various flying behaviors. They can fly downwind or upwind for short distance, and fly downwind for long distance under suitable speed and direction of wind flow conditions. Dunn (1949) recorded in the English Channel that CPB adults flied normally for 2.4 km. They flied in crowds under hot and dry weather conditions in the spring season and the speed was usually 8.0 km per hour. The migrating adults could survive for considerably long time even when they dropped into the sea water. Starvation can stimulate flight of CPB and the adults fly more frequently under higher light intensity conditions, while potato crops in spring can reduce CPB flights. CPB adults of the overwintered generation still stored some nutrients in their bodies and the nutrient storage help them survive for 4 weeks and fly over far distance in order to find appropriate host.

From this, the CPB has some autonomous flight capability, its strong population diffusion capacity. To clear the influence factors of CPB adult flight ability, it has important value to make rational planning for the rotation, and decide the timing of the establishment of prevention and treatment to lowering the population density, in order to prevent the proliferation of CPB.

10.3.5 Resistance to Chemical Insecticides

The first instance of CPB resistance to synthetic organic pesticides was noted for DDT in 1952, followed by resistance to other chlorinated hydrocarbons. In subsequent years, most major classes of insecticides in the world lost control efficiency (Alyokhin et al. 2008). Due to the failure of chemical control, the beetle is becoming more difficult to control, and the damage tends to be more severe. The resistance level of CPB to insecticides has been increasing in China. Most populations have developed moderate to high levels of resistance to pyrethroids (cyhalothrin, decamethrin and alphamethrin), which were originally effective for CPB control, and low to medium levels of resistance to neonicotinoids, such as thiamethoxam and acetmiprid (Jiang et al. 2010; Luo et al. 2010; Wang et al. 2010; Yang et al. 2011a).

Studies on the resistance mechanisms of CPB in Xinjiang revealed that point mutations of S291G in the *AChE* and L1014F in the LdVscc1 were responsible for high levels of resistance to beta-cyfluthrin and carbofuran, respectively. A PASA molecular technique has been developed for the detection of CPB resistance to pyrethriod and carbamate pesticides (Jiang et al. 2011; Xiong et al. 2010).

10.4 Risk Analysis and Intercept of CPB in China

10.4.1 Risk Analysis

To identify factors associated with dispersal and spreading of CPB in China, a series of studies have been conducted in heavily infected areas, including Xinjiang, Gansu, Jilin, Heilongjiang of China, Coastal Region of Russia, North Korea and some other neighboring countries of China. Several factors have been identified, such as geographic barriers, types of hosts, soils, climatic factors, wind speed and directions, overwinter conditions and levels of control, etc. (Gui and Boiteau 2010; Guo et al. 2010c; Hiiesaar et al. 2006; Li et al. 2011, 2013a, b, 2014a, b, c; Hansen et al. 2004; Ahemaiti et al. 2010; Zhang et al. 2010).

Geographic barrier is one key factor determining spread routes of CPB (Liu et al. 2012). Xinjiang has very distinctive geographic and ecologic conditions, where oases always mingled with deserts (Jia et al. 2004). For example, the Tianshan Mountain extending from west to east divides Xinjiang into two parts. This Mountain has an attitude higher than 3000 m, so it can serves as a huge geographic barrier against CPB. It has effectively prevented CPB from spreading to southern part of Xinjiang, as well as the spreading from the Yili river valley area to northern or southern part of Xinjiang (Guo et al. 2010c). It had been well proved that CPB in Tacheng was spreading by its natural migration only from west to east, along that oasis at northern slope of the Tianshan Mountain. Owing to absence of host crops and wild host plants, the large Gobi desert is an isolated region that is impossible for CPB to spread through by its natural dispersal (Li et al. 2013b).

Invasion risk analyses showed that most regions in China were suitable for colonization of CPB, except the few regions such as the northern part of Tibet, northwest part of Gansu province and Inner Mongolia (Hong 2009; Wang et al. 2014). There is a high risk for the beetle spreading eastwards from Xinjiang into Gansu and neighboring provinces, and from coastal region of Russia to Heilongjiang province. By contrast, there is a low risk in Taiwan, Hainan, Guangdong and Fujian provinces (Fig. 10.6).

Invasion risk analyses warned that there existed possible invasions into central China. Currently, there are four major potato growing regions in China, which are located in northern, central, southern and southwestern parts of the country. The cropping area in northern and southwestern regions accounted for 90% of the country. Inner Mongolia in the northern region is the largest potato growing province, and some other provinces such as Gansu, Heilongjiang, Guizhou, Yunnan also have a large potato acreage. The Hexi Corridor in Gansu Province had been considered as a critical channel for the CPB invasions to these regions. If central China is invaded by the beetle, no geographical barrier exists to prevent its spread, and its establishment and population expansion would be rapid due to the presence of suitable environments. Therefore, strict inspection and quarantine measures should be taken in and near the Hexi Corridor to prevent CPB invasions. Plants, potato tubers, and packaging materials used for transport are to be given special attention, as they might carry CPB adults and larvae and be brought to other regions by human activities.

10.4.2 Quarantine and Monitoring

China is undertaking strict quarantine and extensive monitoring of CPB in high-risk regions, such as southern part of Xinjiang, Gansu, Helonjiang, Jilin, and Inner Mongolia. In recent years, some effective quarantine measures have been widely

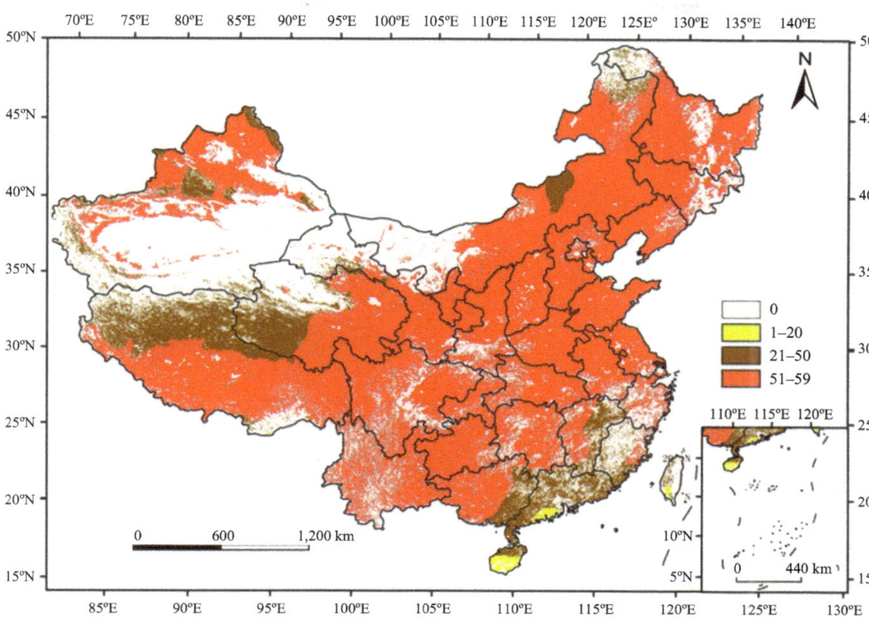

Fig. 10.6 Suitability and colonization risk analysis of Colorado potato beetle

adopted. For instance, a road management station responsible for plant quarantine has been established on Yandun highway toll station located at the junction between Gansu and Xinjiang. This station inspects agricultural products, carries out quarantine treatments on vehicles loading agricultural products from Xinjiang, and inspects vehicles to prohibit transporting infested potatoes to regions outsides of Xinjiang. In addition, in Colorado potato beetle colonized regions (Fig. 10.7) and in the high-risk regions of Gansu (Fig. 10.8), Heilongjiang, Jilin and Inner Mongolia, monitoring and investigations have been conducted at large scales in potato growing seasons each year.

To increase the nationwide awareness of CPB invasion in China, a guideline for the monitoring and eradicating of this pest has been formulated. Furthermore, based on lots of research work, an internet platform has been constructed. This platform contains information of CPB monitoring, warning, forecast, control technology, as well as CPB identification, biology and ecology.

10.4.3 Establishment of "The Controlling and Intercepting Belt"

The controlling and intercepting belt for CPB has established in the frontier of its occurrence regions, which located from the eastern of Mori Kazak autonomous county to the western of Barkol county. This belt exceeds more than 250 km with

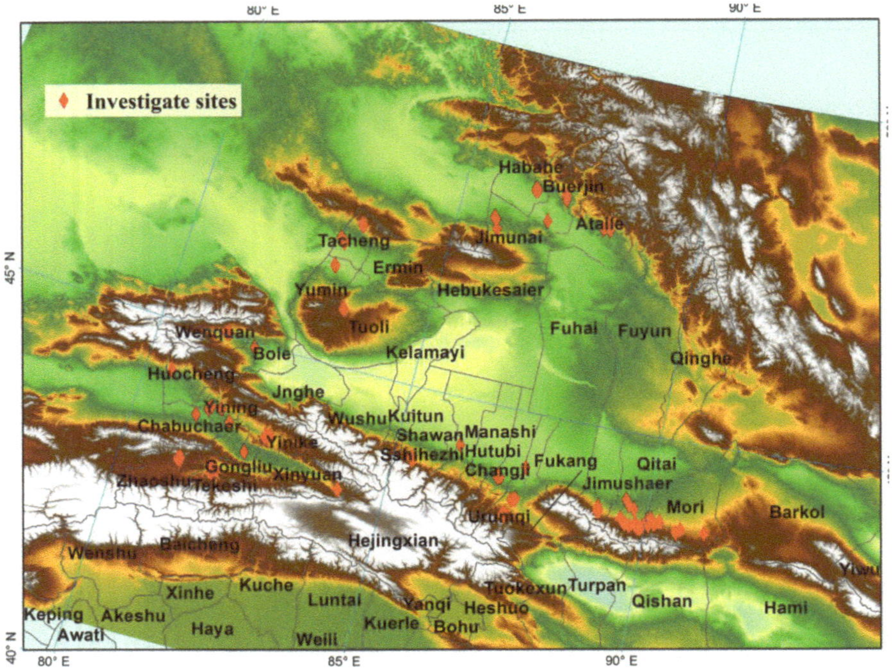

Fig. 10.7 Population investigation in Colorado potato beetle colonized regions

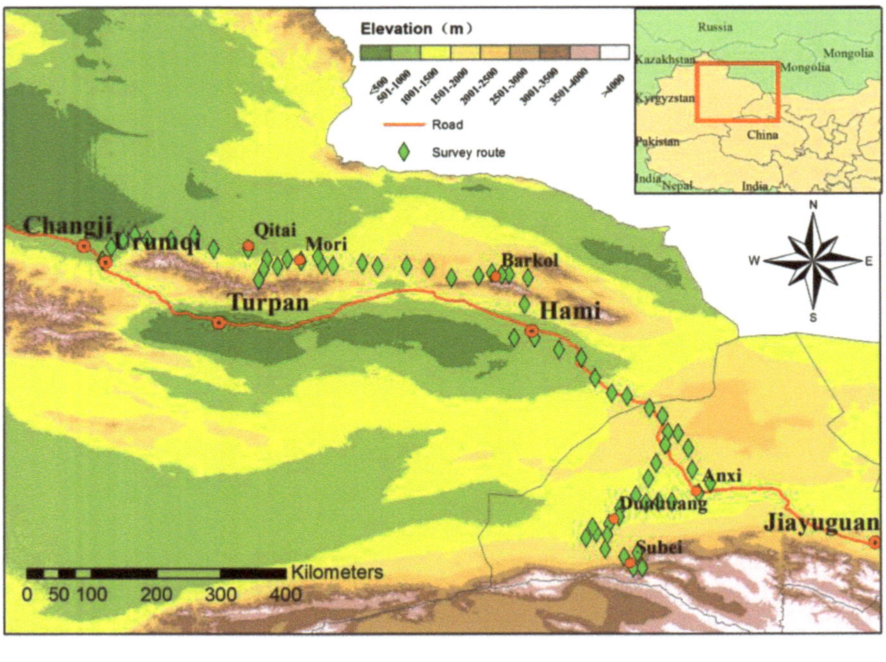

Fig. 10.8 Population investigation in high risk invasion regions

desert. In this belt, we have taken a series of effective measures (Fig. 10.9). Firstly, we reduce the density of CPB in frontier region by chemical control. Secondly, we thoroughly eliminate henbane (*Hyoscyamus niger*), one wild host plant which is only distributed in the desert in northwest of China. Thirdly, we monitor the occurrence of CPB on host crops which are planted in the desert at regular intervals, and seven monitoring sites have been established in Mori County (Fig. 10.10). In case that CPB is found in these plants, we eradicate it immediately.

By widely and continuously applying effective measures nationwide, especially in the regions of high invasion risk, CPB has been effectively prevented from spreading to other potato growing region of China.

10.5 Sustainable Management of Colorado Potato Beetle in China

10.5.1 Cultural Control

Some cultural practices have been used widely in Xinjiang for CPB control. By ploughing potato fields in autumn, the death rate of overwintering CPB adults reached 33–76%, which was twice higher than that in the fields without ploughing. Rotating potato fields with non-host crops can delay occurrence of overwintered adults for 2–7 days, and the delay could be over seven days if the distance of rotating and non-rotating fields was beyond 400 m. Moreover, cropping rotation can reduce CPB population density significantly. For example, in the Yili River Valley, the density of first-generation larvae in the rotated potato fields was only 29–84% (65% on average) of that in continuous cropping potato fields (Guo et al. 2011d).

Plastic mulching after seeding can increase soil temperature and promote seedling emergence and growth, and thus may increase plants' tolerance to CPB damage. Moreover, plastic membrane can stop CPB adults from getting out of the soil or stop the transfer of CPB larvae among different plants. In the Ili River Valley area, such a technique, together with usual irrigation, has been demonstrated to be able to reduce CPB density by 50–74% and potato yield was increased by 57% (Guo et al. 2011d, 2014).

In addition, increasing the fertilization of field prior to potato growing may also reduce CPB damages, due to the increased tolerance of plants (Xu et al. 2011). Applications of micronutrient fertilizers such as zinc sulfate, manganese sulfate, boron and ferrous sulfate, have also been reported to be potential effective control measures (Guo et al. 2011d).

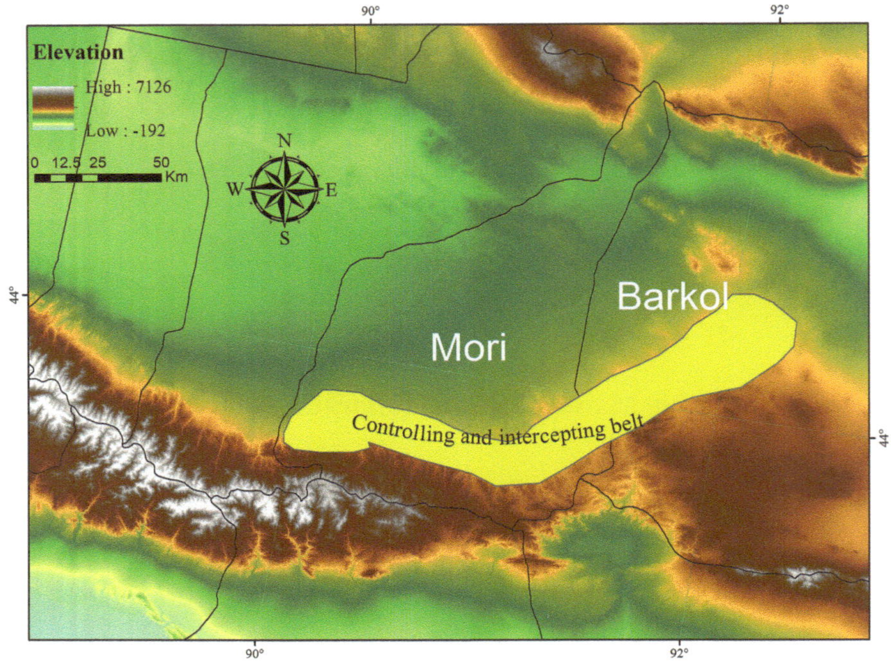

Fig. 10.9 Controlling and intercepting belt in risk regions

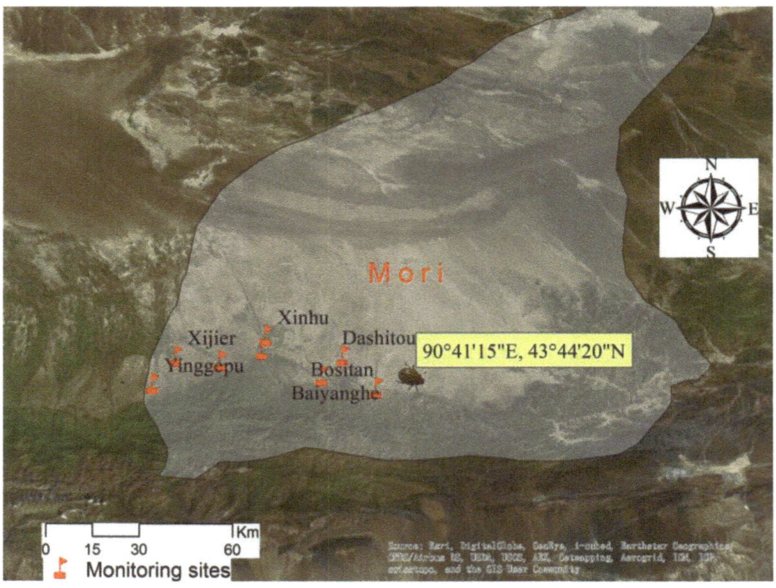

Fig. 10.10 Seven monitoring sites in Mori County

10.5.2 Transgenic CPB-Resistant Potato Breeding

A single gene expression vector was constructed with the *Bacillus thuringiensis* (Bt) toxic protein gene *cry1B3*, and a gene stacking vector was constructed containing *cry3A* gene and the *Vitreoscilla* hemoglobin gene (*vhb*) (Wu et al. 2011). These genes have been introduced into the local potato cultivars "Zihuabai" and "Longshu", respectively, through Agrobaterium-mediated transformation method and they were expressed successfully (Lu et al. 2015). Greenhouse tests showed that both the transgenic plants are highly effective in killing CPB larvae and adults. Particularly, the CPB death rate reached 100% four days after feeding the insects with leaves of the *crylB3* transgenic potato plants under greenhouse conditions. A number of 13 potato lines of the gene stacking (*cry3A+ vhb*) were obtained, and greenhouse tests indicated that all of the 1st- and 2nd-instar CPB larvae feeding on these potato lines died (Zhou et al. 2012). Limited releases of these transgenic potato lines in field experiments also resulted in high death rates and effective control efficacy of CPB with significantly higher yields obtained than the conventional cultivars. Additionally, none of these Bt-transgenic potato lines had any reverse effects on natural enemies and other arthropods. And the tube shape, tube color and other product natures of these transgenic potato were not changed compared to mother cultivars.

10.5.3 Biological Control of Colorado Potato Beetle

10.5.3.1 Survey and Evaluation of Predatory Natural Enemies

In recent years, lots of surveys have been conducted for natural enemies of CPB in Xinjiang. Forty-six predatory species of CPB have been identified, which include 25 insects and 21 spiders (Feng et al. 2013). The frequently found predatory species include *Chrysopa sinica*, *Zicrona caerula*, *Adonia variegate*, *Hippodamia tredecimpunctata libialis*, *Coccinella septempunctae*, *Arma chinensis*, *Adetphocoris lineotatus*, *Lyygus pratensis*, *Nabis sinoferus*, *Nabis ferus*, *Erigonidium graminicolum*, *Philodromus cespitum* and *Oenopia conglobata* (Shu et al. 2011) (Fig. 10.11). The richness and abundance of predatory species in China were quite different from those in the US and European countries (Greenstone et al. 2010; Li et al. 1997; Mallampalli et al. 2002; O'Neil et al. 2005). A few species have been evaluated in the laboratory for their predatory capacity on CPB, such as *C. sinica*, *Z. caerula* and *A. lineotatus* (Feng et al. 2013; Shu et al. 2011, 2012; Zhang 1994a, b).

Fig. 10.11 Natural enemies of Colorado potato beetle in Xinjiang

10.5.3.2 Entomogenous Pathogens and Development of an Effective Bio-insecticide

Several entomogenous pathogen isolates have been obtained from infected CPB adults collected from Xinjiang. The most often isolated pathogens include *Beauveria bassina* (Deng et al. 2012; Ma et al. 2014), *Veriticillium* sp., *Metarhizium anisopliae*, *Fusaroim* sp., *Trichoderma* sp. and *Bacillus thuringiensis* (Lu et al. 2015). Some *B. bassina* and *B. thuringiensis* isolates are highly virulent and pathogenic to CPB larvae and thus are very effective in controlling CPB in the field. An isolate of *B. bassiana* has been well studied and mass produced by fermentation (Luo et al. 2011). A bio-pesticide was formulated into a wettable powder containing 3×10^{10} spores/g and an oil suspension containing 1×10^{10} spores/ml. Field trials showed that these bio-pesticide formulations were very effective for CPB control (He et al. 2015). Additionally, an engineered Bt bacterium has been obtained by transforming the modified *cry*3A toxic protein gene and pre-experiment was made from the engineered bacterium (Yang et al. 2011b). The CPB control efficacy of this preparation can reach 99% when applied at a dosage of 1200 g per hectare. Both the myco- and the bacterio-insecticide are being registered for commercial uses and they are promising to be widely applied in sustainable management of CPB.

10.5.4 RNA Interference Technology

Complete sequence of the proline dehydrogenase gene has been cloned from CPB cells and the difference of the gene expressions before and after overwintering has been compared (Guo et al. 2015b; Wan et al. 2013a; Wu et al. 2011). This gene has been modified and transformed into *E. coli* to produce an engineered bacterium (Li et al. 2014d). The engineered bacterium was then fermented to produce significant amount of double stranded RNA (Li et al. 2015; Shi et al. 2012). Based on this technique, a CPB-lethal gene was synthesized to inactivate the juvenile hormone synthesis in CPB (Guo et al. 2015a). The juvenile hormone synthesis related genes were also cloned and these were the juvenile hormone esterase (*JHE*), juvenile hormone epoxide hydrolase (*JHEH*) and juvenile hormone diol kinase (*JHDK*) genes. A dsRNA expression vector (pET-2P-X) was constructed with 3-hydroxyl-3-methyl glutarie usions-CoA reductase, mevalonate kinase, S-adeno-sylhom ocysteine hydrolase and juvenile hormone esterase genes, and the expected products could be expressed successfully in *E. coli* (Fua et al. 2014; Liu et al. 2014; Zhou et al. 2013). Tests showed that the transformed bacterium with S-adeno-sylhom ocysteine hydrolase gene expressed could exert significant lethal effect on CPB (Wan et al. 2013b).

10.5.5 Chemical Control

The economic threshold of different chemicals for CPB control have been determined for the populations in Xinjiang (Guo et al. 2014). Several high efficiency and low toxic insecticides had been screened, such as acetamiprid and imidacloprid that belong to neonicotinoid insecticides. These insecticides can be used as seed-dressing for controlling overwintered CPB adults, or sprayed to kill first-generation larvae (Fu et al. 2013; Guo et al. 2010a; b; Lu et al. 2010; Lu 2011). In addition, pesticides have been developed for seed or potato tuber dressing, such as 24% thiamethoxam and 3.2% methyclothiazide. Greenhouse experiments showed that more than 65% of CPB larvae were killed when fed with the leaves of plants obtained at 63–68 days after dressing treatment of potato tubers with the pesticides. Similarly, the field investigations showed that 75.6 and 68.5% of CPB larvae died when feeding on the plants produced for 60 days after the treatments of thiamethoxam and methyclothiazide, respectively. All these results suggest that both pesticides can maintain their control efficacy for at least 2 months. These two pesticides have been used in Urumqi and Qitai.

A number of strategies and techniques have been adopted to reduce CPB resistance to chemical insecticides (Alyokhin et al. 2008; Jiang et al. 2010, 2011; Xiong et al. 2010). For example, potato rotation with wheat and other crops can reduce CPB density, and consequently can reduce pesticide use. Concentrated plant potato as trap crops which can attract CPB adults to lay eggs. The appropriate adjustment of sowing time can avoid the harm to potato by unearthed CPB and its reproduction peak. Plastic-lined trenches (20–30 cm width, 30–40 cm depth) placed between overwintering sites and commercial potato fields reduced recruitment of adults by 85%. A vacuum CPB-sucker and a propane flame deinsetizator can be used at seedling stages to kill CPB adults and the control efficiency can reach 80–95%.

10.5.6 Integrated Pest Management

In the past 5 years, a series of techniques for controlling CPB were investigated systematically in Xinjiang. Some of these techniques have been selected and integrated into an effective management strategy. Compared with traditional control measures, the integrated CPB management emphasizes more on the adoption of environments-friendly measures, such as cultural, physical and biological techniques as described above. In addition, several healthy crop cultivation techniques, which can help in improving potato plants' resistance to CPB damage, have also been used in the integrated control system. Two typical options of such cultivation techniques are the film and drip irrigation production model, which was referenced from cotton production in Xinjiang, and the fertilization optimization control technology. The aims of undertaking integrated management are to reduce the risk of resistance to pesticides, and ensure favorable economic, ecological and sociological consequences.

The integrated management is formulated with following principles: (*a*) it is simple, easy, economical and practical to use by local farmers; (*b*) the control must be relatively quick, effective and sustainable; and (*c*) it is safe and friendly to environments. So far, different integrated programs have been established in various potato growing regions. Of them, some have been commercially used in Xinjiang in the past decades. The use of some programs have produced large amount of economic profits, which was estimated as around 200 million CN *yuan* each year, and have benefited the ecosystem evidently.

10.6 Conclusions and Future Perspectives

After CPB has successfully invaded China, great efforts have been made on its monitoring and control. Several national projects have been sponsored to study CPB's basic biology and to develop control methods. So far a number of biological aspects that are closely associated with CPB's invasions have been extensively investigated. Moreover, a set of integrated sustainable management techniques have been developed and applied successfully for CPB control in Xinjiang (Guo et al. 2014). This pest has been prevented from spreading to other potato growing regions in China and has been intercepted in the west of Mori county successfully for more than 20 years.

In addition to the progress described above, several other achievements have been made. For example, some useful experiences and advanced techniques on the eradicating, intercepting and monitoring of CPB have been accumulated and are ready to be applied nationwide. Some of the experience is of important reference for the management of other invasive species in China. Moreover, a guideline for monitoring and eradicating CPB has been formulated. Also, an internet platform for sharing information of CPB monitoring, warning and forecast nationwide has been preliminarily constructed.

The import of agricultural products from other countries is increasing, and thus there are new risks for the country to be invaded by CPB that might hide in imported agricultural products. For the regions that have been invaded by CPB, it is still necessary to reduce CPB damages, prevent it from spreading into other potato cropping regions. Some more advanced control techniques have not been investigated thoroughly and thus have not been cooperated into the integrated management programs. For instance, the breeding of transgenic CPB-resistant potatoes are still at its infant stage. Little work has been done on the attractants for control purpose. The methods for conserving, mass rearing and releasing natural enemies have not been studied. New effective and least-contaminative chemical and biological pesticides need to be developed. Plant quarantine needs to be more strengthened and the integrated management programs be more widely used.

References

Ahemaiti T, Xu J, Guo W, Liu J, He J, Xia Z, Fu W, Zhang D (2010) Study on major biological characteristics and occurrence regulation of Colorado potato beetle. J Xinjiang Agric Sci 47:1147–1151

Alyokhin A (2009) Colorado potato beetle management on potatoes: current challenges and future prospects. Fruit Veg Cereal Sci Biotech 3:10–19

Alyokhin A, Ferro DN, Hoy CW, Head G (1999) Laboratory assessment of flight activity displayed by Colorado potato beetles (Coleoptera: Chrysomelidae) fed on transgenic and Cry3a toxin-treated potato foliage. J Econ Entomol 92:115–120

Alyokhin A, Baker M, Mota-Sanchez D, Dively G, Grafius E (2008) Colorado potato beetle resistance to insecticides. Am J Potato Res 85:395–413

Arpaia S, Schmidt JEU, Di Leo GM, Fiore MC (2009) Oviposition of the Colorado potato beetle (*Leptinotarsa decemlineata*) and natural predation on its egg masses in Bt-expressing fields. Biocontrol Sci Tech 19:971–984

Baker MB, Alyokhin A, Dastur SR, Porter AH, Ferro DN (2005) Sperm precedence in overwintered Colorado potato beetle (Coleoptera: Chrysomelidae) and its implications for insecticide resistance management. Ann Entomol Soc Am 98:989–995

Boiteau G (2002) Flight takeoff behavior of Colorado potato beetle. Can Entomol 134:229–240

Boiteau G, Coleman W (1996) Cold tolerance in the Colorado potato beetle, *Leptinotarsa decemlineata* (Say)(Coleoptera: Chrysomelidae). Can Entomol 128:1087–1099

Boiteau G, Heikkilä J (2012) Successional and invasive colonization of the potato crop by the Colorado potato beetle: managing spread. Insect Pests Potato 339

Boiteau G, McCarthy PC, MacKinley PD (2010) Wind as an abiotic factor of Colorado potato beetle (Coleoptera: Chrysomelidae) flight take-off activity under field conditions. J Econ Entomol 103:1613–1620

Boman S (2008) Ecological and genetic factors contributing to invasion success: the northern spread of the Colorado potato beetle (*Leptinotarsa decemlineata*). University of Jyväskylä
Boman S, Grapputo A, Lindström L, Lyytinen A, Mappes J (2008) Quantitative genetic approach for assessing invasiveness: geographic and genetic variation in life-history traits. Biol Invasions 10:1135–1145
Cheng Y, Qu L, Chen Y, Shi D, Wang W (2005) An analysis of possibility of occurrence of Colorado potato beetlel in Jilin province. J Jilin Agric Sci 30:34–35
Deng C, Zhang Y, Zhang M, Geng B, Wu W, Fu W, Zhang D (2012) Effect of *Beauveria bassiana* wettable powder against *Leptinotarsa decemlineata* (Say). Chin J Biol Control 28:62–66
Dunn E (1949) Colorado beetle in the channel Islands. Ann Appl Biol 36:525–534
Feng L, Wang P, Shu M, Guo W, Wang P (2013) Preliminary Investigation on species and predatory capacity of predatory enemies of *Leptinotarsa decemlineata* Say in Xinjiang. J Environ Entomol 35:334–342
Ferro DN, Tuttle AF, Weber DC (1991) Ovipositional and flight behavior of overwintered Colorado potato beetle (Coleoptera: Chrysomelidae). Environ Entomol 20:1309–1314
Fu W, He J, Hui M, Ahemaiti T, Guo W (2013) Preliminary exploration into the control eddects of six new kinds of medicaments to prevent and treat Colorado potato beetles in the fields. Xinjiang Agric Sci 50:1482–1487
Fu K, Guo W, Lü F, Liu X, Li G (2014) Response of the vacuolar ATPase subunit E to RNA interference and four chemical pesticides in *Leptinotarsa decemlineata* (Say). Pestic Biochem Physiol 114:16–23
Greenstone MH, Szendrei Z, Payton ME, Rowley DL, Coudron TC, Weber DC (2010) Choosing natural enemies for conservation biological control: use of the prey detectability half-life to rank key predators of Colorado potato beetle. Entomol Exp Appl 136:97–107
Gui L, Boiteau G (2010) Effect of food deprivation on the ambulatory movement of the Colorado potato beetle, *Leptinotarsa decemlineata*. Entomol Exp Appl 134:138–145
Guo J, Liu Y, Zhang H, Lu H, Guo W (2010a) The effects of dressing potato with 70% thiaethoxa WS ans 10% imidacloprid WP on control efficiency to potato beetle. Plant Prot 36:151–154
Guo J, Zhang H, Liu Y, Lu H, Guo W (2010b) Effect of seed dressing with neomnicotinoid insecticides on food utilization and development of *Leptinotarsa decemlineata* (Coleptera: Chrysomelidae) larvae. Acta Entomol Sin 53:748–753
Guo W, Ahemaiti T, Xu J, Liu J, He J, Li J, Ma D, Wang J (2010c) Research on the identification of Colorado potato beetle & its distribution dispersal and damage in Xinjiang. Xinjiang Agric Sci 47:906–909
Guo L, Guo W, Liu M, Ahemaiti T, He J, Xu J (2011a) Effects of density on fight capacity of Colorado potato beetle. Xinjiang Agric Sci 48:320–327
Guo L, Guo W, Ahemaiti T, He J, Xu J (2011b) Effects of different hosts on flight capcaity of Colorado potato beetle. Xinjiang Agric Sci 48:853–858
Guo L, Guo W, Ahemaiti T, He J, Xu J (2011c) Effects of temperature and feeding status on flight capacity of overwintering Colorado potato beetles. Plant Prot 37:56–61
Guo W, Ahemaiti T, Xu Y, Liu J, Xu J, Wang P, He J, Xia Z, Fu W, Jing X, Zhang D (2011d) Study and application on sustained and integrated control techniques of Colorado potato beetle. Xinjiang Agric Sci 48:197–203
Guo W, Ahemaiti T, Guo L, He J, Xu J (2012) Effects of different nutrients on fight capacity in Colorado potato beetles. Xinjiang Agric Sci 49:461–469
Guo W, Ahemaiti T, Cheng D, Tan W, Zhang Z, Li G, Jiang W, Deng J, Wu J, Deng C, Li J, Liu X, Lu H (2014) Main progress on biology & ecology of Colorado potato beetle and countermeasures of its monitoring and controlling in China. Plant Prot 40:1–11
Guo W, Fu K, Yang S, Li X, Li G (2015a) Instar-dependent systemic RNA interference response in *Leptinotarsa decemlineata* larvae. Pestic Biochem Physiol 123:64–73
Guo W, Liu X, Fu K, Shi J, Lü F, Li G (2015b) Functions of nuclear receptor HR3 during larval-pupal molting in *Leptinotarsa decemlineata* (Say) revealed by in vivo RNA interference. Insect Biochem Mol Biol 63:23–33

Hare JD (1990) Ecology and management of the Colorado potato beetle. Annu Rev Entomol 35:81–100

Hansen LM, Nielsen GC, Philipsen H, Moller L, Secher B (2004) Colorado beetles: biology and control Gron Viden, Markbrug, Tjele, Denmark

He J, Guo W, Ahemaiti T, Wang Y (2015) Evaluation of the field control efficacy of biogenic pesticides to Colorado potato beetles. Xinjiang Agric Sci 52:258–262

Hiiesaar K, Metspalu L, Judu J, Jgar K (2006) Over-wintering of the Colorado potato beetle (*Leptinotarsa decemlineata* Say) in field conditions and factors affecting its population density in Estonia. Agron Res 4:21–30

Hong B (2009) The research on the GIS-based pest distribution prediction system. Master degree. North West Agriculture and Forestry University, Yanglin

Jansky S, Jin L, Xie K, Xie C, Spooner D (2009) Potato production and breeding in China. Potato Res 52:57–65

Jia B, Zhang Z, Ci L, Ren Y, Pan B, Zhang Z (2004) Oasis land-use dynamics and its influence on the oasis environment in Xinjiang, China. J Arid Environ 56:11–26

Jiang W, Xiong M, Wang Z, Lu W, Liu P, Guo W, Li G (2010) Incidence and synergism of resistance to conventional insecticides in larvae of the Colorado potato beetle, *Leptinotarsa decemlineata* (Say) (Coleoptera: Chrysomelidae), in northern Xinjiang, China. Acta Entomol Sin 53:1352–1359

Jiang W, Guo W, Lu W, Shi X, Xiong M, Wang Z, Li G (2011) Target site insensitivity mutations in the AChE and LdVssc1 confer resistance to pyrethroids and carbamates in *Leptinotarsa decemlineata* in northern Xinjiang Uygur autonomous region. Pestic Biochem Physiol 100:74–81

Jolivet P (1991) The Colorado beetle menaces Asia (*Leptinotarsa decemlineata* Say 1824)(Col. Chrysomelidae). L'Entomologiste 47:29–48

Li Q, Liu A, Gu J, Qiu Z (1997) Investigation and analysis on species of *Leptinotarsa decemlineata* say in Xinjiang: Proceedings of The Sixth Annual Meeting of Entomological Society of China

Li Y, Zhao P, Yin C, Liu X, Zhang Q (2010) Attraction of *Leptinotarsa decemlineata*(Say) (Coleoptera:Chrysomelidae) by several plant volatiles and aggregation pheromone. Acta Entomol Sin 53:734–740

Li C, Zhang Z, Guo W, Zhang Y, Sun J, Cheng D, Liu H (2011) The relationship between the occurrence of Colorado potato beetle, *Leptinotarsa decemlineata*, and rivers based on GIS: a case study of Shawan County. Acta Ecol Sin 31:6488–6494

Li C, Cheng D, Guo W, Liu H, Zhang Y, Sun J (2013a) Attraction effect of different host plant to Colorado potato beetle *Leptinotarsa decemlineata*. Acta Ecol Sin 33:2410–2415

Li C, Cheng D, Liu H, Zhang Y, Sun J (2013b) Effects of temperature on the distribution of the Colorado potato beetle (*Leptinotarsa decemlineata*)—effect of high temperature on Its emergence in Turpan, Xinjiang. Sci Agric Sin 46:737–744

Li C, Cheng D, Guo W, Liu H, Zhang Y, Sun J (2014a) Emergence mechanism and the factors influencing overwintering Colorado potato beetle in Xinjiang, China. Acta Phys Sin 41:1–6

Li C, Liu H, Huang F, Cheng D, Wang J, Zhang Y, Sun J, Guo W (2014b) Effect of temperature on the occurrence and distribution of Colorado potato beetle (Coleoptera: Chrysomelidae) in China. Environ Entomol 43:511–519

Li C, Peng H, Cheng D, Guo W, Liu H, Zhang Y, Sun J (2014c) Diffusion of colorado potato beetle, *Leptinotarsa decemlineata*, adults in field. Acta Ecol Sin 34:359–366

Li C, Qu Y, Han Z, Jiang W (2014d) Cloning and expression analysis of three genes encoding nicotinic acetylcholine receptpor a subunit in *Leptinotarsa decemlineata*. J Nanjing Agric Univ 37:65–71

Li X, Guo W, Fu K, Li G, Ahemaiti T, He J (2015) The toxicity measurement of ds SAHase on different age Colorado potato beetles. Xinjiang Agric Sci 52:56–64

Liu N, Li Y, Zhang R (2012) Invasion of Colorado potato beetle, *Leptinotarsa decemlineata*, in China: dispersal, occurrence, and economic impact. Entomol Exp Appl 143:207–217

Liu X, Fu K, Lü F, Meng Q, Guo W, Li G (2014) Involvement of FTZ-F1 in the regulation of pupation in *Leptinotarsa decemlineata* (Say). Insect Biochem Mol Biol 55:51–60
Lu W (2011) Toxicitues of several newly-typed insecticides to Colorado potato beetle, *Leptinotarsa decemlineata* (Say) and molecular cloning of two kinds of targets. Master degree, Nanjing Agricultural University, Nanjing
Lu W, Shi X, Guo W, Jiang W, Xia Z, Fu W, Li G (2010) Susceptibilities of *Leptinotarsa decemlineata* (Say) in the North Xinjiang Uygur Autonomous region in China to two biopesticides and three conventional insecticides. J Agric Urban Entomol 27:61–73
Lu H, Lin Z, Tan W, Luo H, Xian F, Bi C, Yu Y, Yang Y (2015) Insecticidal protein genes of *Bacillus thuringiensis* strain CPB012 and its effects in controlling different insect pests. Sci Agric Sin 48:1112–1121
Luo X, Yang X, Guo W, Tian Y, Fu W, Wu J (2010) The analysis of expression of cry3A gene and resistance to Colorado potato beetle in transgenic potato lines. J Shanxi Agric Sci 38(3–5):12
Luo H, Wang L, Tan W, Zhou J, Du X (2011) Fungal isolates from Colorado potato beetles and their pathogenicity to silkworm larvae. J Nanjing Agric Univ 34:61–67
Luo J, Liu C, Zhou Z (2012) A comparative study of development and reproduction of the Colorado potato beetle, *Leptinotarsa decemlineata* (Say) (Coleoptera: Chrysomelidae), on different host plants. Acta Entomol Sin 55:84–90
Ma J, Deng C, Zhang Y, Wang N, Geng B (2014) Identification of four *beauveria bassiana* strains from Colorado potato beetle. Xinjiang Agric Sci 51:876–884
Mallampalli N, Castellanos I, Barbosa P (2002) Evidence for intraguild predation by *Podisus maculiventris* on a ladybeetle, *Coleomegilla maculata*: implications for biological control of Colorado potato beetle, *Leptinotarsa decemlineata*. Biocontrol 47:387–398
O'Neil RJ, Cañas LA, Obrycki JJ (2005) Foreign exploration for natural enemies of the Colorado potato beetle in central and South America. Biol Control 33:1–8
Pan X, Zhang J, Xu H, Zhang X, Zhang W, Song H, Zhu S (2015) Spatial similarity in the distribution of invasive alien plants and animals in China. Nat Hazards:1–14
Qu D, Xie K, Jin L, Pang W, Bian C, Duan S (2005) Development of potato industry and food security in China. Sci Agric Sin 38:358–362
Rabitsch W (2010) Pathways and vectors of alien arthropods in Europe. Chapter 3. *Alien terrestrial arthropods of Europe*. BioRisk 4:27–43
Shi X, Xiong M, Jiang W, Wang Z, Guo W, Xia Z, Fu W, Li G (2012) Efficacy of endosulfan and fipronil and joint toxic action of endosulfan mixtures against *Leptinotarsa decemlineata* (Say). J Pest Sci 519-526
Shu M, Liu W, Fei L, Mou L, Ma N, Wang P (2011) Study on the functional response of *Chrysopa sinica Tjeder* to predation at the eggs of *Leptinotarsa decemlineata* Say. Xinjiang Agric Sci 48:328–333
Shu M, Eyou M, Luo Q, Liu W, Mou L, Feng L, Ma N, Wang P (2012) Predation potential of *Zicrona caerula* (Linnaeus) to the *Leptinotarsa decemlineata* (Say) low instar larvae. J Environ Entomol 34:38–44
Wan P, Lu D, Guo W, Ahemaiti T, Yang L, Mu L, Li G (2013a) Molecular cloning and characterization of a putative proline dehydrogenase gene in the Colorado potato beetle, *Leptinotarsa decemlineata*. Insect Sci 21:147–158
Wan P, Shi X, Kong Y, Zhou L, Guo W, Ahemaiti T, Li G (2013b) Identification of cytochrome P450 monooxygenase genes and their expression profiles in cyhalothrin-treated Colorado potato beetle, *Leptinotarsa decemlineata*. Pestic Biochem Physiol 107:360–368
Wang Z, Jiang W, Li G (2010) Insecticide resistance in adult of the Colorado potato beetle among North Xinjiang Uygur Autonomous region. Agrochemicals 49:206–208
Wang D, Chen L, Zhang Y, Luo G, Zhang Y, Li C (2014) Research on the model of quantitative evaluation technology in northeast China invasion risk assessment of Colorado potato beetle as an example. Plant Quarantine 28:39–45
Weber D (2003) Colorado beetle: pest on the move. Royal Soc Chem 14:256–259
Weber D (2008) Colorado potato beetle. Encycl Entomol 2:324–328

Wu J, Luo X, Zhang X, Shi Y, Tian Y (2011) Development of insect-resistant transgenic cotton with chimeric TVip3A* accumulating in chloroplasts. Transgenic Res 958:247–258

Xiong M, Jiang W, Li G (2010) Monitoring of insecticide resistance in larvae of *Leptinotarsa decemlineata* in Xinjiang Uygur autonomous region by diagnostic dose. Chin Bull Entomol 47:763–766

Xu Y, Guo W, Xie X, Xia Z, Xu J, Zhang D, He J, Fu W, Ban X (2011) Response of Colorado potato beetle quantity in Xinjiang under different variety and fertilizer. Acta Agric Boreali-occidentalis Sin 20:179–185

Yang X, Liang A, Luo X, Wu J (2011a) Cloning and characterization of mannose-6-phosphate somerase gene (*manA/pmi*) from *Escherichia coli*. Acta Agric Boreali-Sin 26:107–110

Yang X, Zhang X, Zhang M, Guo W, Tian Y, Xia Q, Wu J (2011b) Transgenic potato overexpressing the *Amaranthus caudatus* agglutinin gene to confer Aphid resistance. Crop Sci 51:2119

Yue R (2013) The study on the relation between the development and flight capacity of Colorado potato beetle (*Leptinotarsa decemlineata*). Shihezi University, Shihezi

Yue R, Guo W, Ahemaiti T, He J (2013) Relations between the fecundity and fight capacity of Colorado potato beetles. Xinjiang Agric Sci 50:829–838

Zhang S (1994a) Insectivorous natural enemies of *Leptinotarsa decemlineata* in Europe. Plant Quarantine 8:276–278

Zhang S (1994b) Natural enemies of *Leptinotarsa decemlineata* introduced from America to Europe. Plant Quarantine 8:342–344

Zhang R, Liu N, Li Y (2010) Colorado potato beetle, a disastrous invasive species in China infesting Solanaceae vegetables. China Bull Life Sci 11:10

Zhang Z, Lu X, Li J, Zhang Y, Ding H, Xu W, Du Q, Li Q, Guo W (2012) Monitoring of *Leptinotarsa decemlineata* in northeast China and analysis of its incasion risk. J Jilin Agric Sci 37:42–44

Zhou Z, Pang J, Guo W, Zhong N, Tian Y, Xia G, Wu J (2012) Evaluation of the resistance of transgenic potato plants expressing various levels of Cry3A against the Colorado potato beetle (*Leptinotarsa decemlineata* Say) in laboratory and field. Pest Manag Sci 68:1595–1604

Zhou L, Jia S, Wan P, Kong Y, Guo W, Ahemaiti T, Li G (2013) RNA interference of a putative S-adenosyl-L-homocysteine hydrolase gene affects larval performance in *Leptinotarsa decemlineata* (Say). J Insect Physiol 59:1049–1056

Chapter 11
Red Turpentine Beetle *Dendroctonus valens* LeConte

Min Lu and Jianghua Sun

Abstract The red turpentine bettle (RTB), *Dendroctonus valens* LeConte (Coleoptera:Curculionidae: Scolytinae), is a secondary pest of pines in its native ranges in North and Central America. Outbreaks and tree mortality caysed by RTB alone are rare in its native ranges. RTB was detected in China in the early 1980s and spread rapidly from Shanxi province to the adjacent provinces, and has caused extensive tree mortality since 1999. In this chapter, we discuss factors likely contributing to *D. valens* invasion success and control strategies in China. Genetic variation of RTB, interactions between RTB and its associated fungi, behavioral differences in Chinese RTB, and other factors favoring RTB outbreaks are considered in an effort to explain the invasiveness of RTB in China. Promising management options for controlling RTB, as with any bark beetle pest, including regulatory, silvicultural, insecticidal, and semiochemical tactics, are also discussed.

Keywords *Dendroctonus valens* • Invasiveness • *Leptographium*

11.1 Introduction

The red turpentine beetle (RTB) , *Dendroctonus valens* LeConte (Coleoptera: Curculionidae: Scolytinae), native to North America, is the most widely distributed pine bark beetle in the American continent (Owen 1985). It has been recorded on at least 40 species of domestic and foreign conifers. Despite its abundance and wide distribution, outbreaks of RTB have generally not been extensive or severe in its native ranges, and most cases in North America and parts of Central America (Smith 1971; Sun et al. 2013).

M. Lu • J. Sun (✉)
State Key Laboratory of Integrated Management of Pest Insects and Rodents,
Institute of Zoology, Chinese Academy of Sciences, 1 Beichen West Road,
Chaoyang District, Beijing 100101, China
e-mail: sunjh@ioz.ac.cn

F. Wan et al. (eds.), *Biological Invasions and Its Management in China*,
Invading Nature - Springer Series in Invasion Ecology 11,
DOI 10.1007/978-94-024-0948-2_11

This beetle was introduced into China in the early 1980s when logs without being quarantined were imported from the Pacific Coast of the United States (Zhang et al. 2006). In China, its distribution expands rapidly since 1999 when it began to spread from Shanxi province to three other adjacent provinces (Hebei, Henan, and Shaanxi). One of the major possible problems is the loss of large areas of reforestation forests. In such forests, many pine species are widely planted, such as *Pinus tabuliformis* Carrière, yet these pines are very vulnerable to RTB's invasions. So far more than ten million *P. tabuliformis* trees have been killed, as well as other pines such as *Pinus bungeana* Buee (Yan et al. 2005). In addition, the invasions of RTB into China can cause a series of negative effects, for example extensive loss of forest cover can result in dramatic changes to ecosystems, with concurrent losses of biodiversity and carbon sequestration capacity (Kurz et al. 2008; Yan et al. 2005).

Several factors may have contributed to the invasion success of RTB in China, including the abundance of native hosts, lack of natural enemies, and mutualistic symbioses with microorganisms. Increasing global temperature might also have favored the RTB invasions by broadening its latitudinal ranges. In addition, RTB has changed some of its behaviors as it spreads to non-native regions, which might also be a reason for its successful invasions in China. To control this pest, several management methods have been developed in China. Meanwhile, extensive researches have been conducted to investigate invasion mechanisms of RTB.

11.2 Invasion Mechanisms

11.2.1 Genetic Diversity and Variation of D. valens

The genetic structure of native and non-native RTB populations has been compared based on partial sequences of the mitochondrial cytochrome c oxidase subunit I gene (COI) in populations collected from multiple locations in North America and China. Overall, high haplotype diversity was found in this beetle, and Chinese populations shared some of the haplotypes with beetles from Pacific northwestern North America. This supports the hypothesis that the *D. valens* was recently introduced to China and originated from the PNW of the USA (Cognato et al. 2005). To learn more information about haplotype diversity within populations and genetic similarity among populations, Cai et al. (2008) conducted a more extensive survey on population genetic structure. They also used a partial sequence of COI. Moreover, they sampled from more sites, took larger sample sizes, and collected more paleoclimate and geological data. The result showed that marked genetic differences exited between eastern and western North American populations as well as between Mexican/Guatemalan and U.S. populations.

11.2.2 Distinct Characteristics of Chinese **D. valens**

RTB has an excellent dispersal ability. Its flight distance exceeded 16 km in North America (Smith 1971). Amazingly, its flight distance was up to 35 km in China (Zhang et al. 2002). The long-distance flight capability enables the beetle to migrate over large geographical areas, including beyond barriers such as the Luliang and Taihang mountains (Zhang et al. 2002).

This beetle is able to detect volatiles of host trees. It is one of the important biological traits of this pest and was thought to have contributed to its successful establishment in China (Shi and Sun 2010). Volatiles from frass expelled from female nuptial chambers are used by males to locate these chambers (Liu et al. 2006a, b). In China, RTB has been shown to harbor *trans*-verbenol, *cis*-verbenol, myrtenol, myrtenal, and verbenone in its hindgut (Shi and Sun 2010). These volatiles are known pheromone components in other species of *Dendroctonus* (Symonds and Elgar 2004).

Of them, *trans*-verbenol, myrtenol, and myrtenal can be detected by antennal receptors and elicit attraction in RTB (Zhang and Sun 2006). In field tests, addition of both *trans*-verbenol and myrtenol in baits significantly increased catches of beetles in traps baited with host kairomones (Zhang and Sun 2006). Newly emerged females and males have only minute amounts of *trans*-verbenol and myrtenol, but the quantity of volatiles from female adults increase after feeding (Shi and Sun 2010). Male adults produce larger quantities of trans-verbenol and myrtenol after they join females in galleries, which could be a means for RTB males to accelerate mass colonization on host trees (Shi and Sun 2010). Verbenone functions as a multipurpose pheromone for RTB-attractive at very low concentrations but repellent at high concentrations (Zhang et al. 2006). This phenomenon has been reported in other species of *Dendroctonus*. For practical purposes, verbenone dominantly serves as an interruptant to host or mate location (Rappaport et al. 2001).

Shi and Sun (2010) found that all of the volatiles described above were not synthesized de novo by the beetle, instead they were all synthesized through oxidation of the host monoterpene, α-pinene. Oxidation of α-pinene is a relatively simple chemical conversion and hence may involve lower metabolic costs and provide faster mobilization and release of attractant semiochemicals.

11.2.3 Multispecies Interactions

Intriguing questions concerning the RTB in China is about the fungi associated with the beetle in its new environments and whether such an association might, more or less, account for its unusual behaviors there. Various studies have been undertaken to compare the fungal associates of RTB in China and North America (Lu et al. 2008a, b, 2009a, b). Interestingly, of all the fungi isolated from North American and Chinese beetles, only two shared species, *Leptographium procerum* and *Ophiostoma ips*, have been found (Lu et al. 2009a). Of all the fungi, *L. procerum* is the most consistently isolated from RTB in China (Lu et al. 2009a). This fungus is not a primary pathogen in North America, although it has been associated with tree decline syndromes and with other Scolytinae that feed on living conifers (Wingfield 1983). On the other hand, strains of *L. procerum* collected from RTB of China had higher virulence to *P. tabuliformis* than a strain from the beetle of North America (Lu et al. 2010). Intriguingly, these Chinese strains may also increase production of 3-carene, the most attractive host volatile for RTB, in the inoculated pine seedlings. Although these results are of a preliminary nature, Lu et al. (2010, 2011) suggested that this intrinsic multitrophic interaction could be an important factor involved in the RTB invasions in China.

In China, an important semiochemical relationship exists between RTB and the root-feeding Scolytinae *Hylastes parallelus* (Lu et al. 2007). The two species can infest trees simultaneously with RTB infesting the lower trunks and upper roots while *H. parallelus* infesting the lower roots (Wu et al. 2002). This synergistic relationship appears to be mediated by cross-attraction via chemical cues (Lu et al. 2007).

11.3 Risk Prediction and Assessment

The risk of spreading to a wide geographic range has been assessed for RTB. This was done by modeling analysis using the data from 305 weather stations with software @ RISK (Tang et al. 2008; Wang et al. 2007). The results showed that there is an extremely high risk of this beetle continuing to outbreak as well as to spread in China. Some regions would have to face a high risk. In northern China, for example, these assessments were used regionally to regulate transport of potentially infested wood materials in order to limit human-assisted RTB dispersal. According to the model, the north China has a higher likelihood of drought, so if RTB were to spread to this area, the damage would likely increase largely. Analyses of host volatiles that serve as beetle attractants (monoterpenes) showed that *Pinus bungeana* and *Picea asperata,* which grow in this region, might be more susceptible to RTB attack than other hosts in the infested area (Wang et al. 2007), and global warming may increase their susceptibility.

11.4 Monitoring and Management

Before 1999, RTB was not considered a forest pest in China. However, its pest status escalated as the beetle continued to spread and levels of damage significantly increased. The Chinese State Forestry Administration now ranks RTB as the second most harmful forest pest in this country, and a National Management Project was initiated for RTB in 2000 (Sun et al. 2013). Options for controlling RTB include regulatory, silvicultural, insecticidal, and semiochemical tactics, all of which have been implemented in China to mitigate the damages. The control in some regions has been very successful. About 30% of the 85,300 ha of pine forest in eastern Shanxi province has been infested, causing death in 7% of *P. tabuliformis* trees in 2001. Data provided by the Shanxi Forestry Bureau indicated that the RTB-infested area within the province had decreased from 256,700 ha in 1999 to 29,900 ha in 2010. Over a 12-year period (1999–2010), the control projects covered nearly 39,000 ha on average per year, with a peak of 79,300 ha in 2001. RTB infestations extended into Henan province, but the extension end due to drought and various control methods such as physical, chemical, and biological control by the State Forestry Administration in 2002 (Yan et al. 2005).

11.4.1 Semiochemical-Based Control

Semiochemical tactics such as application of antiattractants, trap-out of beetles with attractants, or combinations thereof, are much environmental friendly and not labor-intensive (Gillette and Munson 2009; Zhang et al. 2007). Promising semiochemicals include attractant kairomones produced by hosts and conspecific beetles, sex attractant pheromones, aggregation pheromones, and antiattractants.

(−)-α-pinene, (+)-α-pinene, (+)-3-carene were produced by pines in natural environments. The antipode (−)-α-pinene interrupted response to (+)-α-pinene is the first reported example of chiral specificity of kairomones (Hobson et al. 1993). The studies in China showed that (+)-3-carene, was the best RTB attractant in invaded regions (Sun et al. 2004a). This suggests that founder effects had yielded an invasive RTB population that differed from native populations in semiochemical response (Cognato et al. 2005; Yan et al. 2005). A large follow-up study, however, showed that the populations in North America also responded strongly to (+)-3-carene (Erbilgin et al. 2007). Release rates of (+)-3-carene differed vastly in the two studies, which may explain the differences. Because of its superior performance, (+)-3-carene has been used successfully in a series of trapping projects in China (Li et al. 2006a; Liu et al. 2006a, b; Sun et al. 2004b; Yuan 2008).

In addition, RTB can also show kairomonal responses to some volatiles. Lu et al. (2007) showed that RTB had both antennal and behavioral responses to the volatiles produced by a native Chinese root beetle, *Hylastes parallelus*. This is the first report on kairomonal interactions between native and introduced insects. Joseph et al. (2001) reported increased attraction of RTB by kairomones with the addition of ethanol, but Fettig et al. (2004) found such an enhancement to be not significant, suggesting that ethanol may not function consistently to be used in operational trapping programs. The potential for combining effective semiochemicals in binary and ternary blends for enhanced trapping merits need further exploration (Table 11.1).

11.4.2 Regulatory Enforcement Strategies

Any pine material with intact bark could potentially harbor RTB, and harvesting of dying, infested trees might facilitate RTB spread through the movement of infested logs. Thus, human-mediated movement appears to be principle means of spread in

Table 11.1 Response of *D. valens* to semiochemicals

Semiochemicals	Response	References
(−)-β-Pinene	Attractive	Hobson et al. (1993) and Rappaport et al. (2001)
	Non-attractive	Guo et al. (2003) and Sun et al. (2003)
(+)-3-Carene[a]	Attractive	Hobson et al. (1993), Erbilgin et al. (2007), and Rappaport et al. (2001)
	Attractive	Sun et al. (2004a, b)
Myrcene	Attractive	Hobson et al. (1993), and Rappaport et al. (2001)
(−)-Limonene	Non-attractive	Hobson et al. (1993), Rappaport et al. (2001)
(+)-Limonene	Non-attractive	Hobson et al. (1993), and Rappaport et al. (2001)
(−)-α-Pinene	Non-attractive	Hobson et al. (1993), and Rappaport et al. (2001)
	Attractive	Erbilgin et al. (2007)
(+)-α-Pinene	Attractive	Hobson et al. (1993), Rappaport et al. (2001), and Erbilgin et al. (2007)
	Non-attractive	Guo et al. (2003), and Sun et al. (2003)
Ethanol	Attractive	Gladwin et al. (2001)
Verbenone	Disruption	Hobson et al. (1993), and Rappaport et al. (2001)
	Disruption	Guo et al. (2003), and Sun et al. (2003)
Mixture of (−)-α-Pinene, (−)-β-Pinene, (+)-3-Caren	Attractive	Guo et al. (2003), and Sun et al. (2003)
Frontalin	Attractive	Luxova et al. (2007)
Trans-verbenol	Attractive	Zhang et al. (2006)
Myrtenol	Attractive	Zhang et al. (2006)
Mytenal	Attractive	Zhang et al. (2006)
1-Octen-3-ol	Disruption	Zhang et al. (2006)
(Z)-3-Hexen-1-ol	Disruption	Zhang et al. (2006)
(E)-2-Hexen-1-ol	Disruption	Zhang et al. (2006)

[a]Used in trapping project.

areas where pine stands are widely separated. For this reason, it is essential to prevent or slow down the spread of RTB by monitoring and regulatory enforcement in China (Yan et al. 2005). Unauthorized tree harvesting and the movement of infested material (e.g. logs, wood blocks, and wood boxes with bark) should be strictly restricted. Moreover, the Forest Pest Control Station system was established in China, with the aim of facilitating enforcement of quarantine regulations from the central government to the provincial, city, and county levels. So far, strict quarantine regulations are enforced at ports and along highways and railways.

11.4.3 Silvicultural Control

In the stands that have not been invaded by RTB, several silvicultural methods have been advised to minimize the risk, including (a) preventing tree wounds, (b) eliminating chip piles or other sources of attractive host volatiles, (c) carefully timing thinning, pruning, and soil ripping to avoid periods when trees might be drought-stressed and dispersing beetles in high number (Fetting et al. 2004), and (d) reducing stand density to minimize stress caused by competition (Li et al. 2006b). Avoiding monocultures is also frequently recommended to minimize resource concentration that might favor the increase of pest populations (Li et al. 2006a; Liu et al. 2004). In China, RTB outbreak status has been monitored annually since 1999, primarily using baited traps combined with summer and fall plot inspections for signs of attack, such as pitch tubes and frass on stems. In the infected stands, and four forest management measures are applied to infested stands: ceasing turpentine collection from living trees, reducing stand density, removing infested trees, and increasing tree species diversity (Li et al. 2006b).

11.4.4 Insecticidal Control

In North America, several chemicals are effective in protecting individual tree from RTB attack, such as fenitrothion, carbaryl, and permethrin. They are normally used before beetles' attack. In China, the effective methods include fumigating holes with aluminum phosphide under plastic cover, injecting dichlorvos or omethoate into newly initiated galleries, and spraying insecticides (e.g., phorate, monocrotophos, cypermethrin, phoxim; Sun et al. 2013). Of them, fumigation and injection with insecticides can be conducted in all reasons; early June to October is the optimal time, during which pupation and development of new larvae occur (Yan et al. 2005).

11.5 Conclusions and Perspectives

In China, RTB has rapidly expanded its geographical ranges and caused tree extensive mortality since 1999. Some factors may explain RTB's successful colonization and establishment in China, including more aggressive attack behavior, high dispersal capability, an abundance of native hosts, lack of predators and pathogens, positive interactions with native bark beetles, an effective symbiosis with new fungal associates, and favorable climate patterns. Major differences in RTB's behavior between native and invaded ranges likely reflect adaptations to new environments. Thus, numerous researches have been conducted to explain the benefits of microbial associates on RTB fitness improvement. But multitrophic interactions between RTB and pines, fungi, bacteria, mites, and other beetles remains unclear. A comprehensive understanding of these ecological interactions is crucial to understanding the biology of RTB and other invasive beetle-fungus mutualisms.

Promising management options for controlling RTB include regulatory, silvicultural, insecticidal, and semiochemical tactics, and all these approaches have been implemented in China as integrated pest management programs to mitigate RTB damage. Fumigation during the flight period is a direct control method that is effective at killing beetles at large scales. In addition, trapping beetles with semiochemicals lures (host volatiles) has been tested as a promising labor-saving and environmental friendly method for RTB management. In addition, planting strategies (mixed-species stands and mixed-age stands) should be designed to reduce outbreak potentials. More improved lures, including host volatiles and pheromones, and improved trapping techniques will enhance monitoring, detection, interruption, and trap-out. Similarly, epidemiological models for RTB in urban, managed, and natural forests will benefit from refinement.

References

Cai YW, Cheng XW, Xu RM, Duan DH, Kirkendall LR (2008) Genetic diversity and biogeography of red turpentine beetle *Dendroctonus valens* in its native and invasive regions. Instr Sci 15:291–301

Cognato AI, Sun JH, Anducho-Reyes MA, Owen DR (2005) Genetic variation and origin of red turpentine beetle (*Dendroctonus valens* LeConte) introduced to the People's Republic of China. Agric For Entomol 7:87–94

Erbilgin N, Mori S, Sun J, Stein JD, Owen DR, Merrill LD, Campos Bolaños R, Raffa KF, Méndez Montiel T, Wood DL, Gillette NE (2007) Response to host volatiles by native and introduced populations of *Dendroctonus valens* (Coleoptera: Scolytidae) in North America and China. J Chem Ecol 33:147–156

Fettig CJ, Borys RR, Cluck DR, Smith SL (2004) Field response of *Dendroctonus valens* (Coleoptera: Scolytidae) and a major predator, *Temnochila chlorodia* (Coleoptera: Trogositidae), to host kairomones and a *Dendroctonus* spp. pheromone component. J Entomol Sci 39:490–499

Gillette NE, Munson AS (2009) Semiochemical sabotage: behavioral chemicals for protection of western conifers from bark beetles. pp 85–109. In: Hayes JL, Lundquist JE, comps (2009) The Western Bark Beetle Research Group: a unique collaboration with Forest Health Protection-proceedings of a symposium at the 2007 Society of American Foresters conference. Gen. Tech. Rep. PNW-GTR-784. Portland, OR: U.S. Department of Agriculture, Forest Service, Pacific Northwest Research Station, 134 p

Gladwin J, Kelsey RG, Peck RW, Niwa CG (2001) Response of some scolytids to ethanol and 4-allylanisole in pine forest of central Oregon. J Chem Ecol 27:697–715

Guo YY, Wu GL, Miao ZW. 2003. Trapping effect and sex pheromone prescription of *Dendroctonus valens* LeConte. J Shanxi Agric Univ 23:216–219 (in Chinese).

Hobson KR, Wood DL, Cool LG, White PR, Ohtsuka T, Kubo I, Zavarin E (1993) Chiral specificity in responses by the bark beetle *Dendroctonus valens* to host kairomones. J Chem Ecol 19:1837–1847

Joseph G, Kelsey RG, Peck RW, Niwa CG (2001) Response of some scolytids and their predators to ethanol and 4-allylanisole in pine forests of central Oregon. J Chem Ecol 27:697–715

Kurz WA, Dymond CC, Stinson G, Rampley GJ, Neilson ET, Carroll AL, Ebata T, Safranyik L (2008) Mountain pine beetle and forest carbon feedback to climate change. Nature 452:987–990

Li YZ, Li JK, Li WA, Wang Y, Guo XR, Li ML (2006a) A study on the control of *Dendroctonus valens* by sex attractant. J North For Univ 21:109–112

Li YZ, Wang FH, Wang PX, He H, Li ML (2006b) The study on control effect of silvicultural treatment to *Dendroctonus valens*. J North For Univ 21:113–116

Liu GM, Li MC, Wen XJ, Wang ZL, Shi GH (2004) Using ecosystem stability to manage *Dendroctonus valens*. Hebei For Sci Technol 6:43–44

Liu GM, Xu ZH, Zhao YF, Liu YW, Gao GR (2006a) Analysis of affecting factors on trapping *Dendroctonus valens*. For Sci Tech 20:28–31

Liu Z, Zhang L, Sun J (2006b) Attacking behaviors and behavioral responses to boring dust volatiles in the red turpentine beetle, *Dendroctonus valens* (Coleoptera: Scolytidae), in China. Environ Entomol 35:1037–1048

Lu M, Miller DR, Sun J (2007) Cross-attraction between an exotic and a native pine bark beetle: a novel invasion mechanism? PLoS One 2:e1302. doi:10.1371/journal.pone.0001302

Lu Q, Decock C, Zhang X, Maraite H (2008a) *Leptographium sinoprocerum* sp. nov., an undescribed species associated with *Pinus tabuliformis-Dendroctonus valens* in northern China. Mycologia 100:275–290

Lu Q, Zhang X, Yang Z, Maraite H, Yin D, Ren X (2008b) Progress in study on the fungi associated with *Dendroctonus valens*. Sci Silvae Sin 44:134–142

Lu M, Zhou XD, De Beer ZW, Wingfield MJ, Sun J (2009a) Ophiostomatoid fungi associated with the invasive pine-infesting bark beetle, *Dendroctonus valens*, in China. Fungal Divers 38:133–145

Lu Q, Decock C, Zhang X, Maraite H (2009b) Ophiostomatoid fungi (Ascomycota) associated with *Pinus tabuliformis* infested by *Dendroctonus valens* (Coleoptera) in northern China and an assessment of their pathogenicity on mature trees. Anton Leeuw Int J G 96:275–293

Lu M, Wingfield MJ, Gillette NE, Mori SR, Sun J (2010) Complex interactions among host pines and fungi vectored by an invasive bark beetle. New Phytol 187:859–866

Lu M, Wingfield MJ, Gillette NE, Sun J (2011) Do novel genotypes drive the success of an invasive bark beetle/fungus complex? Implications for potential reinvasion. Ecology 92(11):2013–2019

Luxova A, Graves AD, Gries R, Wakarchuk D, Seybold SJ, Hamud SM (2007) An attractive pheromone for the red turpentine beetle, *Dendroctonus valens* Le Conte (Coleoptera: Scolytidae). Poster D0339. In: 55th annual meeting of entomology society of America, San Diego

Owen DR (1985) The role of *Dendroctonus valens* and its vectored fungi in the mortality of Ponderosa pine. PhD dissertation, University of California, Berkeley, p 64

Rappaport NG, Owen DR, Stein JD (2001) Interruption of semiochemical-mediated attraction of *Dendroctonus valens* (Coleoptera: Scolytidae) and selected nontarget insects by verbenone. Environ Entomol 30:837–841

Shi Z, Sun J (2010) Quantitative variation and biosynthesis of hindgut volatiles associated with the red turpentine beetle, *Dendroctonus valens* LeConte, at different attack phases. Bull Entomol Res 100:271–277

Smith RH (1971) Red turpentine beetle USDA. For Pest Leaflet 55, 9 pp

Sun JH, Gillette NE, Miao ZW, Kang K, Zhang ZN, Owen DR, Stein JD (2003) Verbenone interrupts attraction to host volatiles and reduces attack on *Pinus tabulaeformis* (Pinaceae) by *Dendroctonus valens* (Coleoptera: Scolytidae) in the People's Republic of China. Can Entomol 135:721–732

Sun J, Miao Z, Zhang Z, Zhang Z, Gillette NE (2004a) Red turpentine beetle, *Dendroctonus valens* LeConte (Coleoptera: Scolytidae), response to host semiochemicals in China. Environ Entomol 33:206–212

Sun YM, Fan HP, Miao JL (2004b) Study on attractant effect on different traps on *Dendroctonus valens*. Shanxi For Sci Tech 4:21–22

Sun J, Lu M, Gillette NE, Wingfield MJ (2013) Red turpentine beetle: innocuous native becomes invasive tree killer in China. Annu Rev Entomol 58:293–311
Symonds MRE, Elgar MA (2004) The mode of pheromone evolution: evidence from bark beetles. Proc R Soc B-Biol Sci 271:839–846
Tang WD, Shi J, Luo YQ (2008) Evaluation of the potential risk of *Dendroctonus valens* with @ Risk software. For Pest Dis 27:7–14
Wang HB, Zhang Z, Kong XB, Liu SC, Shen ZR (2007) Preliminary deduction of potential distribution and alternative hosts of invasive pest, *Dendroctonus valens* (Coleoptera: Scolytidae). Sci Silvae Sin 43:71–76
Wingfield MJ (1983) Association of *Verticicladiella procera* and *Leptographium terebrantis* with insects in the Lake States. Can J For Res 13:1238–1245
Wu JG, Zhao MM, Zhang CM, Guo BP, Li JZ, Niu X (2002) Damage of *Dendroctonus valens* on *Pinus tabuliformis* and its distribution on trunk and root before and after overwintering period. For Pest Dis 21:38–41
Yan ZL, Sun J, Owen DR, Zhang Z (2005) The red turpentine beetle, *Dendroctonus valens* LeConte (Scolytidae): an exotic invasive pest of pine in China. Biodivers Conserv 14:1735–1760
Yuan YQ (2008) Further study on optimizing test of different RTB (Red Turpentine Beetle) vegetable Lure agent formula. Shanxi For Sci Tech 1:7–9
Zhang L, Sun J (2006) Electrophysiological and behavioral response of *Dendroctonus valens* (Coleoptera: Scolytidae) to candidate pheromone components identified in hindgut extracts. Environ Entomol 35:1232–1237
Zhang LY, Chen QC, Zhang XB (2002) Studies on the morphological characters and bionomics of *Dendroctonus valens* LeConte. Sci Silvae Sin 28:95–99
Zhang LW, Sun JH, Clarke S (2006) Effects of verbenone dose and enantiomer on the interruption of response of the red turpentine beetle, *Dendroctonus valens* LeConte (Coleoptera: Scolytidae), to its kariomones. Environ Entomol 35:655–660
Zhang L, Lu M, Liu Z, Sun J (2007) Progress on invasion biology and chemical ecology of red turpentine beetle, *Dendroctonus valens*. Chin Bull Entomol 44:171–178

Chapter 12
Coconut Leaf Beetle *Brontispa longissima* Gestro

Baoqian Lu and Zhengqiang Peng

Abstract *Brontispa longissima* Gestro, known as the Coconut leaf beetle, is a leaf beetle that feeds on young leaves and damages seedlings and mature coconut palms. *B. longissima* has become an increasingly serious pest of coconuts in the Asian-Pacific region, especially over the last three decades. *B. longissima* broke out in Hainan Province, China in 2002. Quarantine and surveillance are the best prevention methods for introductions and further spread. Chemical spraying has been recommended, but foliar sprays pollute environments and are not suitable for certain areas and tall trees. To reduce pollution caused by chemical sprays, chemical sachets, Yejiaqing which is eluviation powder, were widely used to control *B. longissima* in scenic spots and cities in China. Great efforts have been put on biological controls. *Metarhizium anisopliae* which was isolated from *B. longissima* and formulated as a bioinsecticide was applied in China, as well as in Vietnam and Samoa. The larval parasite *Asecodes hispinarum* Bouček and pupal parasite *Tetrastichus brontispae* Ferrière were introduced into China in 2004 and have approved to be very functional. The parasitisation of *T. brontispae* and *A. hispinarum* was ~90% and ~100%, respectively. Field recoveries were made in some regions after parasitoids releasing however the parasitisation is not always so high. It is necessary to release substantial numbers of the parasite into suitable populations of *B. longissima* to ensure establishment. Environment and climate might effect on the survival and effectiveness of the parasitoid. Such information on control measure is necessary to study.

Keywords *Brontispa longissima* Gestro • Invasion biology • Chemical control • Biological control • *Metarhizium anisopliae* • *Asecodes hispinarum* Bouček • *Tetrastichus brontispae* Ferrière

---

B. Lu • Z. Peng (✉)
Environment and Plant Protection Research Institute, Chinese Academy of Tropical Agricultural Sciences, Haikou Shi, Hainan Sheng, China
e-mail: lypzhq@163.com

F. Wan et al. (eds.), *Biological Invasions and Its Management in China*,
Invading Nature - Springer Series in Invasion Ecology 11,
DOI 10.1007/978-94-024-0948-2_12

Fig. 12.1 *B. longissima* (from *left* to *right*: adult; pupa; 5th, 4th, 3rd, 2nd, 1st instar larvae; eggs)

12.1 Introduction

Palms, most of which are restricted to tropical, subtropical, and warm temperate zones, are among the best studied and most extensively cultivated plant families. They have been important to humans throughout much of the human being's history. Many common products and foods are made from palms, and palms are also widely used in landscaping for their unique appearance, making them one of the most economically important plants. For example, coconut palm (*Cocos nucifera* L.) is an important crop in the tropics supporting the livelihood of millions of people. Global production of coconut is around 61.2 billion nuts from an area of 12.1 million hectares. Approximately 78% of global production is contributed by Indonesia, Philippines, India and Sri Lanka (Rethinam and Singh 2007). The production value of coconut is about 0.5 billion RMB from an area of 40,000 ha in China, most in the Hainan Island.

The coconut leaf beetle, *Brontispa longissima* Gestro (Fig. 12.1) is one of the most serious pests of palms. It is native to Indonesia and possibly also to Papua New Guinea. The pest has spread to most regions of coconut production areas. Both adults and larvae damage the leaflets of young unopened fronds. They graze away the surface in streaks, which are typically parallel to the midrib. Destruction of young leaf spike tissues restricts growth for a long time and heavy attack may cause death (Fig. 12.2). Coconut production losses due to the damage of *B. longissima* have been recorded to be as high as 30–50% in Vietnam and 50–70% in Samoa (Tran 2004; Voegele 1989). The coconut hispid beetle has the potential to have drastic affects on the livelihoods of whole villages; in some localities 90% of the people rely on the coconut for their livelihood (Quirante 2007). Where the coconut hispid beetle has been left unchecked and coconut processing factories have been shut

Fig. 12.2 Symptoms of coconut tree damaged by *B. longissima*

down, thousands of workers have been let off and farmers have been left without work (Bernama 2007). To control this pest, extensive research has been conducted worldwide, including in China, to learn its biology and ecology. Moreover, great efforts have been exerted to the development of various control methods, in particular biological control. In this chapter, we reviewed pest status, distribution, biology, ecology and control measures to *B. longissima*.

12.2 Distribution and Impact

The beetle is capable of flight. Laboratory experiment showed that females were able to fly ~400 m (Zhou et al. 2004a). They might fly further far with strong wind. It is believed that human activities have largely contributed to its dispersal. This pest has probably been introduced frequently along with global trade. This is supported by the documented detection of *B. longissima* in products imported into southern China. For example, *B. longissima* was detected in ornamental palms imported from Vietnam by Guangdong Entry-Exit Inspection and Quarantine Bureau in 1999 (Huang et al. 2000). The pests were also found in Fanyu, Guangdong in same year (Huang and Liang 2000). *B. longissima* was detected in coconut trees imported from Vietnam by local authority in Guangxi in 2000 (Gong and Bai Zh 2001). *B. longissima* was discovered in Haikou, Hainan Province in June 2002 and more infested

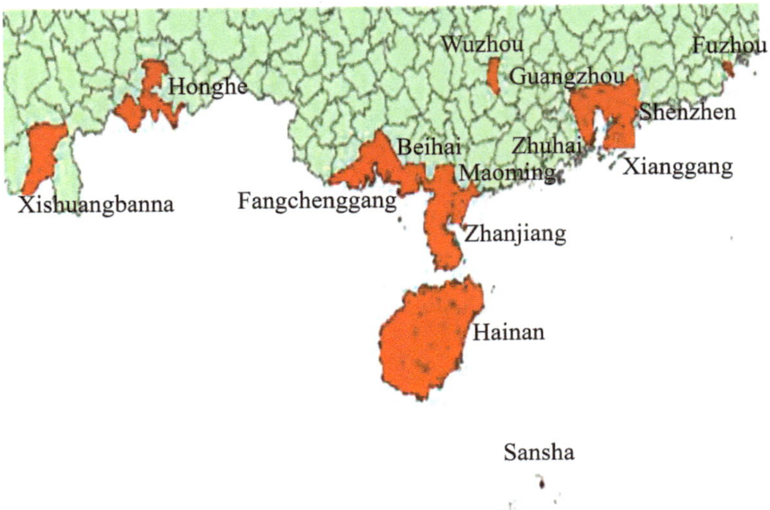

Fig. 12.3 Current geographical distributions of *B. longissima* in China

palms in and around Haikou and Sanya cities were confirmed by field surveys. In the year later, the pests spread throughout the Hainan Island (Lu et al. 2005a). The pests also dispersed northwards and reached a number of regions, including Zhanjiang, Maoming and Shenzhen of Guangdong Province, Beihai and Wuzhou of Guagnxi Province, and Honghe of Yunnan province (Fig. 12.3). Modeling analysis indicated that the beetle could be established in most provinces in south China ranging from 16.53°N to 25.73°N and from 97.85°E to 118.91° (Lu et al. 2004; Li et al. 2005; Peng et al. 2006). China's coconut industry consists of smallholders with 40,000 ha. Survey in 2003 showed 100,000 palm plants were infested and 90% of them were coconuts. Infestation of *B. longissima* seriously reduces the income of coconut farmers, but also the rapidly developing tourism. In support of tourism industry, Hainan has commenced a "beautification campaign" which includes the planting of coconuts and many ornamental species in cities and along major roads. Yet, most of these palms have been affected by the beetle since 2002.

12.3 Biology and Ecology

Understanding the biological and ecological traits provides a basis for managing invasiveness. Researchers in China conducted a series of studies on biology and ecology of *B. longissima*, e.g., fecundity, generation time, and cold tolerance.

There are 4–5 overlapping generations per year in Hainan Province and about three overlapping generations per year in Guangdong Province (Zhong et al. 2005; Zhou et al. 2004a). Larval stages last for 30–40 days, and adults of *B. longissima* live up to 220 days. A female can lay more that 100 eggs. Both adults and larvae seem to avoid light and prefer to feed in young unopened leaflets (Zhou et al. 2004b). When facing harsh environmental conditions, adults can adjust their breeding efforts so as to increases the survival of their progeny. The beetle was able to live up to 6 days without food and starvation can enhance the beetle mature (Xu et al. 2007). Supercooling points of the beetle for all stage were below minus 5 °C indicating the pest had reasonable cold hardness (Xiao et al. 2006).

Up to 34 plant species in 25 genus have been recorded as hosts for *B. longissima*. However, various hosts may differ significantly in their suitability (Li et al. 2009). In China, *C. nucifera* is one of the most favorable hosts, on which the beetle has a higher population growth capacity as compared on other hosts (Zeng et al. 2003; Li et al. 2007). For this reason, *C. nucifera* is often very seriously damaged.

Palm plants are attacked severely in dry seasons, however, plants would recover in rainy seasons, as found in Hainan Province, China (Lu unpublished). Such a phenomenon had also been demonstrated in previous studies (Tjoa 1953; Kalshoven, 1981). There are some possible explanations. Firstly, rain might depress *B. longissima* density and dry periods favour the growth of its population. Secondly, palm plants can grow more rapidly in rain season than dry season. Thus, in rainy season hosts may compensate for the damage at a given pest density.

The response of this beetle to *C. nucifera* leaf volatiles was investigated in laboratory bioassays (Fang et al. 2011). Both sexes are attracted to a mixture of beta -myrcene, (−)-limonene and E-2-hexen-1-ol (1: 6: 1), which are key components of coconut palm leaf volatiles. A blend of beta -myrcene and (−)-limonene (0.7: 1–1: 0.7) in low amounts (100 ng) elicits aggregation and oviposition in females. Chemical analyses of food-deprived, gravid female *B. longissima* show high concentrations of beta -myrcene and (−)-limonene in their accessory glands, suggesting that female beetles sequester both compounds and release them during oviposition.

12.4 Control

12.4.1 Quarantine and Cultural Control

Quarantine is the first step in blocking the pest, and keeping the community aware that the control program is essential to ensure substantial control of pest. International collaboration is important as the pest is relating to more than one nation responsibility.

A surgical method of control was ever attempted in the Solomon Islands which involved cutting out and destroying the central unopened frond which harbours the pest (Brown and Green 1958). This procedure must be conducted over a large area at same time to reduce re-infestation from neighbouring palms. It also has to be repeated fairly often to be effective. Palms which were 3–6 years old could stand the loss of one leaf every 6 months, but younger palms could not as this caused too much reduction in growth rate. However, this method is expensive and will not greatly affect the Brontispa population as a whole unless mature palms (more than 5 years) are also treated. Mechanical control of the pest by removing affected heart leaves is laborious and has very little effect (Kalshoven, 1981). In China, this measure was taken when the isolated infested spot were detected.

Coconut varieties in the Solomon Islands considerably varied in their susceptibility to Brontispa attack. Some varieties from the Ivory Coast and Fiji also show high degrees of resistance (Stapley 1973, 1980). Six local coconut cultivars were tested in Western Samoa, five of which were highly susceptible, and green dwarf was fairly resistant (FAO 1983). The preference of adult *B. longissima* to six local varieties in China was investigated by Yu et al. (2009). The development of *B. longissima* was the longest on the Red Dwarf and the shortest on the Hainan Tall; Female *B. longissima* laid the lowest eggs on Mawa and the highest on Hainan Tall; survival were the highest on Hainan Tall and the lowest on Aromatica Green Dwarf. The trend index of experimental population of *B. longissima* fed on the 6 coconut cultivars was 96.07 (78F1), 82.34 (Hainan Tall), 73.93 (Yellow Dwarf), 60.16 (Red Dwarf), 60.07 (Mawa) and 55.26 (Aromatica Green Dwarf). The resistant variation needs to be further tested in the filed.

12.4.2 Chemical Control

For emergency method, chemical control is first choice at early stages of *B.longissima* invasions. Particularly for young palms which are vulnerable to damage, chemical control has to be applied in time. Chemicals used to control the pest must reach the insect in the narrow crevices between leaflets and chemical treatment must be maintained throughout the year because *B. longissima* breeds continuously, with several generations a year. If beetles start breeding as soon as they reach young palms, larvae will probably be present after about 1 week; if there has been any residual effect from insecticide treatment, it will be considerably longer before the larval population presents a serious risk (Brown and Green 1958). Insecticide resistance in *B. longissima* has been documented by Georghiou and Lagunes-Tejada (1991).

Chemicals originally recommended for control include nicotine sulphate, lead arsenate, dichlorodiphenyltrichloroethane (DDT) and dieldrin, but carbaryl, trichlorfon and lindane are now advocated (Maddison 1983; Stapley 1973, 1980; Wu and Tao 1976). In Australia, carbary was recommended to be sprayed on unopened fronds thoroughly (Waterhouse and Norris 1987). In China, Cypermethrin, ethiofencarb, Acetamiprid, Fipronil and imidacloprid were recommended as they are highly efficient and much less toxic to untargeted organisms. Trichlorfon eliminated Brontispa from isolated areas of young palms in Western Samoa (Bourke 1981). There were attempts to develop botanical pesticide, however the application of botanical pesticide was at the laboratory stage (Qin et al. 2007; Feng et al. 2010; Zhang and Feng 2010).

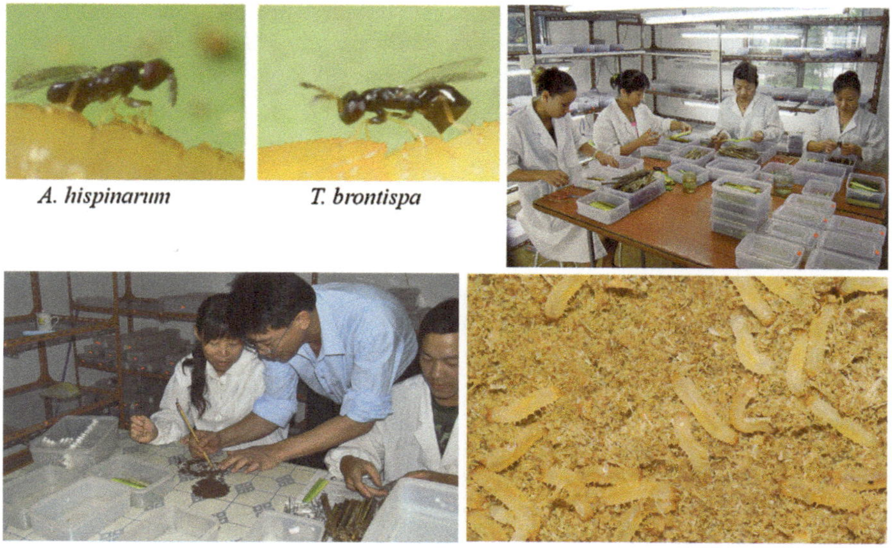

Fig. 12.4 Parasitoids breeding (*Top left*: two parasitoids; *Top right*: *B. longissima* breeding; *Bottom left*: parasitoids breeding; *Bottom right*: artificial diets

The pesticide is applied to the central spike of the palm. Satisfactory control can be achieved at low cost using a fine, low-volume spray applied from above to the central spike of each individual palm (Brown and Green 1958). The application of chemicals at 10-day intervals was more effective than 3-weekly applications (Hollingsworth et al. 1986; Peters et al. 1984).

Chemicals applied through foliar spray may pollute environments through drifts and kill both harmful and beneficial insects. Foliar sprays were also limited for high palms or plant parts where were inaccessible. To overcome these disadvantages, several specific chemical-applying methods have been developed. For example, crown application with carbofuran sachets produced the most effective results, even for tall palms (Choo-Toh 1999). A new pesticide powder mixture (Thiosultap sodium and acetamipri), developed by South China Agriculture University, was mainly used in China (Lu et al. 2012a). The eluviation powder put into bags and then hung on the palm shoots can work with the rainfall (Fig. 12.4). Treatment with this pesticide powder could not only effectively control the beetle with long duration and little side-effect to environment, but also effectively prevent the beetle from spreading. Satisfactory control could be achieved if insecticide bags were used in tourist areas, on sides of street and isolated areas. There are some shortcomings for insecticide sachets. Firstly, eluviation powder depends on rainfall. Therefore it does not work well in dry seasons or dry areas. Secondly, placing the sachets onto palm shoots is labour intensive with high cost.

Researchers have attempted to develop other delivery methods and optimum dosages to control the beetle. Experiments of soil drenching of systemic insecticides and trunk injection were conducted (He et al. 2005; Zhao et al. 2003; Zheng et al. 2010). All methods were effective in delivering the chemicals for beetle control, however soil drench and trunk injection were not applied in field widely.

12.4.3 Biological Control

The history of biological control of this pest was reviewed by Waterhouse and Norris (1987). The prospects for control are also discussed. In China, studies have been focused on natural enemies and entomopathogenic fungus of this pest and their use for biological control.

12.4.3.1 *Metarhizium anisopliae*

A suspension of *Metarhizium anisopliae* was used in the 1980s in Samoa for a pest on young coconut seedlings in fields and in nurseries, but its application to tall trees was difficult and impracticable. In Taiwan, a domestic strain of *M. anisopliae* var. *anisopliae* (MA-1) was isolated from infected coconut leaf beetles (Liu et al. 1989). Two field trials of microbial control of *B. longissima* were conducted in the Pingtung area (southern part of Taiwan) in 1986 and 1987. *B. longissima* could not be detected after three applications of MA-1 formulated as a homogeneous biomass, in granules or in a conidial suspension. In Vietnam, *M. anisopliae* was isolated from naturally infected *B. longissima*, which exhibited the highest virulence to pest, particularly to larvae (Nguyen et al. 2004). The field test showed *M. anisopliae* had good control to the beetle and long persistence, even up to 21 days.

In mainland China, high efficient strains of *M. anisopliae* were selected after strain screening (Qin et al. 2006; Song et al. 2006; Ding et al. 2006). A fermentation technology was also developed and *M. anisopliae* was produced 10 tones per year. To be easily applied, powder, solutions and powder sachet of *M. anisopliae* were developed. In the field, *M. anisopliae* could kill 80% *B. longissima* after 1 week. Laboratory tests showed *M. anisopliae* mixed with Monosultap improved efficacy (Qin et al. 2008). Similarly, *M. anisopliae* mixed with *Bacillus thuringiensis* showed synergic control (Bian et al. 2009).

12.4.3.2 Natural Enemies

Despite the habit to hide in leaflets of unopended fronds, *B. longissima* is attacked by a number of natural enemies, including many species of ants, earwigs and parasitic hymenopterans. There are two parasitoids of coconut leaf beetle *Tetrastichus brontispae* and *Asecodes hispinarum,* which have been successfully used in several countries to control the beetle (Table 12.1).

Table 12.1 Introdutions for the biological control of *B. longissima*

Country	Species	Liberated	From	Result
Cambodia	*A. hispinarum*	2005	Vietnam	?
China	*T. brontispae*	2004	Taiwan	+
China	*A. hispinarum*	2004	Vietnam	+
Thailand	*A. hispinarum*	2004	Vietnam	+
Maldives	*A. hispinarum*	2004	Vietnam	+
Lao PDR	*A. hispinarum*	2004	Vietnam	+
Vietnam	*A. hispinarum*	2003	Western Samoa	+
Nauru	*A. hispinarum*	2004	?	?
American Samoa	*T. brontispae*	1985	Western Samoa	+
American Samoa	*T. brontispae*	1984	Guam	?
Australia	*T. brontispae*	1984	New Caledonia	+
Taiwan	*T. brontispae*	1983	Guam	+
Western Samoa	*T. brontispae*	1982	Tahiti	
Western Samoa	*T. brontispae*	1982	Papua New Guinea	+
Western Samoa	*T. brontispae*	1981–1982	New Caledonia	+
Tahiti	*T. brontispae*	1979	New Caledonia	?
American Samoa	*T. brontispae*	1978	New Caledonia	?
American Samoa	*T. brontispae*	1978	Vanuatu	?
American Samoa	*T. brontispae*	1975	Vanuatu	–
American Samoa	*T. brontispae*	1974	Vanuatu	–
American Samoa	*T. brontispae*	1973	New Caledonia	–
American Samoa	*T. brontispae*	1973	Vanuatu	–
Vanuatu	*T. brontispae*	1969–1970	New Caledonia	+
Solomon Island	*T. brontispae*	1968	Java	+
Solomon Island	*T. brontispae*	1968	Tahiti	+
Tahiti	*T. brontispae*	1962–1963	Sulawesi	+
New Caledonia	*T. brontispae*	1963	Saipan	+
New Caledonia	*T. brontispae*	1947	Java	–
Papua New Guinea	*T. brontispae*	1939	Solomon Island	+
Solomon Island	*T. brontispae*	1935	Java	+
Sulawesi	*T. brontispae*	1932	Java	+

+ represent positive results; – represent negative resutls; ? represent unknown results

However, in most of the invaded regions in China, only a few natural enemy species were detected so far. For example, in Hainan Island where *B. longissima* were found in 2002, only a few predators (earwigs, ants and mantis) have been found. As local natural enemies were not efficient, the pest broke out and spread throughout the Hainan Island quickly. To reduce outbreak of *B. longissima*, China mainland has introduced several natural enemies from overseas. Both *T. brontispae* and *A. hispinarum* were introduced to the Hainan Island in 2004, from Taiwan and Vietnam, respectively. To breed at a large scale and release in the field, Lu et al. (2005a, b, c, 2006) conducted a series of experiments on the two parasitoids, including the comparison of their biology. *A. hispinarum* and *T. brontispae* were found to have some similar

Table 12.2 Comparison of two parasitoids biology

Biological characteristics	*A. hispinarum*	*T. brontispae*
Stage attacked	Larval stage	Pupal stage
Adult body length (mm)	Female, 0.5–0.85;male, 0.35–0.65	Female,0.85–1.4;male, 0.98–1.25
Adult longevity (hr)	47.4	84.0
Fecundity per female	43.0	20.6
Adult emergence per host	60.3	21.5
Female percentage	74.8%	77.4%
Development in Hainan (d)	19.5	21.0
Initial development temperature (°C)	10.7	10.9
Effective accumulated temperature (day degree)	261.2	298.8
Generations a year in Hainan	19	17
Phototaxis	Attracted to light	Attracted to light
Parthenogenesis	Arrhenotoky	Arrhenotoky

biological characteristics such as phototaxis and parthenogenesis. Meanwhile, they do have their own biological traits. Generally, *T. brontispae* has a lower fecundity, bigger body size and a longer life cycle than *A. hispinarum* (Table 12.2).

To breed and produce parasitoid wasps at large scale to control *B. longissima*, *techniques of mass rearing of B. longissima* using coconut leaf were developed. Leaflets of coconut spears were cut into small 5–7 cm pieces and kept inside a plastic container (20 cm length, 13 cm width, 7 cm high) covered with a screened and ventilated lid. To reduce the cost, coconut new leaves were mixed with old leaves. Adult *B. longissima* were introduced into the container and continuously reared at rearing rooms. The rearing rooms are maintained at temperature from 22 to 26 °C, 70–80% RH (Relative Humidity) with natural lighting. Adults of *A. hispinarum* or *T. brontispae* were introduced into the container at the stages of 4th-instar larvae or pupae (host : parasitoid = 1 : 1). A piece of tissue paper saturated with honey solution 10% (w/v) was fixed to the wall of the box to provide food for adult wasp. Parasitized hosts were then collected and stored at 14 °C for less than 10 days. To get large numbers of hosts required for parasitoid production, the artificial diet had been developed for mass rearing of *B. longissima* (Ichiki et al. 2009; Lu et al. 2005c, 2012b). When reared on this diet, the beetle has a survival rate of approximately 40% from hatching to adult stage. *B. longissima* larvae produced from this diet performs well as hosts for both *A. hispinarum* and *Tetrastichus brontispae*. So far five beneficial insectaries have been established in Danzhou, Haikou and Wenchang for mass culture of *A. hispinarum* and *T. brontispae*. These insectaries can produce a total of 300,000 *T. brontispae* and 1,200,000 *A. hispinarum* per day.

To release the two parasitoids in the fields, a container was design (Fig. 12.5). The container can protect parasitoids from rain and predators. One container was hanged for every 15 palm plants. Before releasing, *B. longissima* population density was investigated, and the number of parasitoids to be released was double of the

Before releasing After releasing

Container for parasitoids releasing Palm trees recovery

Fig. 12.5 Parasitoid releasing

pest. *A. hispinarum* and *T. brontispae* can be released at same time with 3:1 ratio. The release is normally performed 4–6 times at 1 month intervals. From 2004 to now, parasitoids have been released at more than 50 spots in Hainan, Guangdong, Yunnan, Guangxi Province. Almost all of the parasited larvae died within 3–4 days after hatching. To integrate chemical control and biological control, parasitoids should be released after chemical residue is below sub-lethal. It was suggested in China that 1 month was safe to release parasitoids after chemical spraying and 3 months were safe after pesticide sachet applying. Field recoveries made in some regions in following year after parasitoids releasing showed substantially reduced injury to coconut plantations by the beetle (Fig. 12.5). The parasitisation of *T. brontispae* was up to 100% and the parasitisation of *A. hispinarum* was up to 90%. However, the level of simultaneous parasitisation by two parasitoids is not always so high.

Earwigs are common predators of *B. longissima* as reported in many countries. In China, *Chelisoches morio* was the dominant species (Fig. 12.6). This predator takes 100 days for development from egg to adult in Hainan. One female lays 138 eggs on average, and one adult can consume about seven 2nd-instar *B. longissima* larvae per day. There was a new species, *Paralabella* sp. which is much smaller than *C. morio* in China.

Fig. 12.6 *Chelisoches morio*

12.5 Conclusion and Perspectives

The breakout of the coconut leaf beetle in Southeast Asia is a typical example of pest invasions into a new area where no or few natural enemies exist. Quarantine technologies need to be further developed, such as fumigation for import or export palm plants.

Based on monitoring, IPM is key to control the pest substantially. There are advantages and disadvantages for each control strategy. For example, chemical control might cause resistance, resurgenee and residue of pest and biological control is unstable as natural enemies are affected by environmental and biotic factors. In the future, culture control, chemical control and biological control should be integrated.

Parasitoids introduction and releasing achieved positive results in many countries. However, parasitisation is not always high and it is hard to establish population of parasitoids in some regions. In Hainan, winter season is relative cold and dry which might have negative effects on the survival and effectiveness of the parasitoids. The landscape might affect parasitoids establishment. Scatted palm plants might limit the parasitoids dispersal and establishment. Therefore, more information on control measures is necessary.

Local enemies is suitable to local whether and environment. Great efforts should be put to find and test local natural enemies. Although *A. hispinarum* and *T. brontispae* were widely introduced and released, we could find local natural enemies to suppress the pest. In Hainan, earwigs are very common on damaged palm crown and its consumption of *B. longissima* is high. Collection, assessment, rearing and releasing of local natural enemies need to be conducted in the future.

References

Bernama (2007) Vietnam to help Philippines to solve coconut beetle problem. Available from: http://www.bernama.com/bernama/v3/news_lite.php?id=278707. Accessed 12 Dec 2008

Bian Q, Wang GJ, Zhang ZH, Zhang J, Li H, Qin WQ (2009) Synergism between Bacillus thuringiensis and *Metarhizium anisopliae* against *Brontispa longissima*. Plant Prot 35(3):130–132

Bourke TV (1981) Quarterly report for quarter ending 31 October 1981. FAO Plant Protection Advisor, Apia

Brown ES, Green AH (1958) The control by insecticides of *Brontispa longissima* (Gestro) (Coleopt., Chrysomelidae-Hispinae) on young coconut palms in the British Solomon Islands. Bull Entomol Res 49:239–272

Choo-Toh GT (1999) An outbreak of *Plesispa reichei* Chapuis on palms in Singapore. In: Sivapragasam et al. (eds) Proceedings of the 5th International Conference on Plant Protection in the Tropics, 15–18, Kuala Lumpur, Malaysia, pp 390–393

Ding FZ, Zhang ZH, Zhang LS, Ding W (2006) Biological control of *Metarhizium anisopliae* against *Brontispa longissima*. J Southwest Agric Univ (Natural Science) 28(3):454–456

Fang Y, Sun J, Zhang Z (2011) Response of *Brontispa longissima* to coconut palm (*Cocos nucifera*) leaf volatiles. Physiol Entomol 36:321–326

FAO (1983) Control of the coconut hispid beetle, Samoa. Terminal Statement Prepared of the Government of Samoa by FAO, Rome, 1983. Rome, Italy: FAO, AG: TCP/SAM/0101 and 0102,1983

Feng G, Zhang J, Jin QA, Wen HB, Peng ZQ (2010) Bioactivity of rotenone against *Brontispa longissima*. Chin J Tropic Crops 31(4):636–639

Georghiou GP, Lagunes-Tejada A (1991) The occurrence of resistance to pesticides in arthropods. An index of cases reported through 1989. FAO, Rome

Gong XZ, Bai Zh L (2001) *Brontispa longissima* was detected in palm plants inported from Vietnam. Guangxi Plant Protection 14(4):29–30

He LS, Ong KH, Yik CP, Fong YK, Chan HJA (2005) Chemical control of hispid beetles (*Brontispa longissima*) on palms. Singap J Prim Ind 32:80–92

Hollingsworth RG, Meleisea S, Iosefa T (1986) Life history notes on *Brotispa longissima* (Gestro) in Western Samoa. Alafua Agric Bull 11

Huang FY, Liang GQ (2000) Quarantine and control of *Brontispa longissima*. Plant quarantine 14(3):158–160

Huang FY, Liang QC, Lai TZ (2000) *Brontispa longissima* and *Rhynchophorus ferrugineus* were detected in South Sea port of China. Plant quarantine 14(2):69

Ichiki RT, Dung DT, Takano SI, Nakamura S (2009) Artificial diets for rearing the coconut hispine beetle *Brontispa longissima* and its parasitoid *Asecodes hispinarum*. J Appl Entomol 133(7):539–545

Kalshoven LGE, van der Laan PA (Reviser and translator) (1981) Pests of crops in Indonesia (revised). Jakarta, Indonesia, Ichtiar Baru, p 701

Li HM, Sun JH, Han HX, Xiao H, Xue DY (2005) Prediction of potential distribution of the coconut leaf beetle in China. For Pest Dis 24(6):8–11

Li Y, Cheng LS, Peng ZQ, Ju RT, Wan FH (2007) Effects of host plants on development and fecundity of *Brontispa longissima*(Gestro). Chin J Appl Ecol 18(9):2050–2054

Li CX, Qin WQ, Huang SC, Ma ZL, Wang DH (2009) Host Preference of *Brontispa longissima* to seven varieties of coconut palms. China For Sci Technol 23(6):61–63

Liu SD, Lin SC, Shiau JF (1989) Microbial control of coconut leaf beetle *Brontispa longissima* with greent muscardine fungus *Metarhizium anisopliae* var. *anisopliae*. J Invertebr Pathol 53(3):307–314

Lu YY, Zeng L, Wang L, Zhou R (2004) Risk analysis of palm leaf beetle *Brontispa longissima*(Gestro). Entomol J East Chin 13(2):17–20

Lu BQ, Peng ZQ, Jin QA, Wen HB, Fu YG (2005a) Toxicity of 15 insecticides to *Brontispa longissima* (gestro) and *asecodes hispinarum* boucek, a parasitoid of *B.longissima*. Chin J Tropic Crops 26(1):47–51
Lu BQ, Peng ZQ, Tang C, Wen HB, Jin QA, Fu YG, Du YZ (2005b) Biological characteristics of *Asecodes hispinarum* Boucek (Hymenoptera:Eulophidae),a parasitoid of *Brontispa longissima* Gestro (Coleoptera:Hispidae). Acta Entomol Sin 48(6):943–948
Lu BQ, Chen YQ, Bao Y, Han RD, Peng ZQ (2005c) The feasibility of the controlling coconut leaf beetle (*Brontispa longissima*)with introducing natural enemies Asecodes hispinarum. Chin Bull Entomol 42(3):254–258
Lu BQ, Zheng-Qiang P, Xu C-A, Chao T, Yue-Guan F, Yu-Zhou D, Wan F-H (2006) Biological characteristics of *Tetrastichus brontispae* Ferriere (Hymenoptera : Eulophidae), a parasitoid of *Brontispa longissima* (Gestro) (Coleoptera: Hispidae). Acta Entomol Sin 49(4):643–649
Lu BQ, Jin Q-A, Wen H-B, Du Y-Z, Peng Z-Q (2012a) Current status of *Brontispa longissima* outbreaks and control. Chin J Appl Entomol 6:1708–1715
Lu BQ, Jin Q A, Wen H B, Tang C, Peng Zh Q, Du Y Z (2012b) Development of the artificial diet for *Brontispa longissima* (Gestro) and the efficacy for mass rearing. Acta Phytophyl Sin 39(4): 347–351
Maddison PA (1983) Coconut hispine beetle. Advisory Leaflet, South Pacific Commission, No. 17, pp. 4
Nguyen TL, Vo TBC, Nguyen TN, Nguyen DT, Tran TBH, Pham QH (2004) Biocontrol potential of *Metarhizium anisopliae* against coconut beetle, *Brontispa longissima*. Omonrice 12:85–91
Peng ZQ, Cheng LS, Ju RT, Li YZ, Chi XZ, Yan W, Xu Y, Du YZ, Wan FH (2006) Adaptive distribution of *Brontispa longissima* (Gestro) in China. Chin J Tropic Crops 27(1):80–83
Peters A, Skatulla U, Aukuso Q, Meleisea S and Hammans H (1984) The coconut hispid beetle-*Brontispa longissima*. Crop Protection Leaflet 2.3 PP. Department of agriculture, forestry and fisheries and samoan crop protection project. Apia, Western Samoa
Qin CS, Xu JZ, Liao FY (2006) Separation and biological character of metarhizium strains obtained from *Brontispa longissima* corpse. Guangdong For Sci Technol 22(4):23–25. 35
Qin WQ, Yu FY, Huang SC, Li CX, Ma ZL (2007) Effects of plant alcohol extracts on bioactivities of B*rontispa longissima* population in laboratory. Chin J Tropic Crops 28(4):84–88
Qin CS, Xu JZ, Xie PH, Xu ZB, Liao FY (2008) Screening of chemical pesticides compatible with *Metarhizium anisopliae* for Control of *Brontispa longissima*. J South China Agric Univ 29(2):44–46
Quirante NB (2007) *Brontispa* infests 52 cocotrees in Paranas, Samar. Philippine Information Agency. Available from: http://www.pia.gov.ph/?m=12&sec=reader&rp=2&fi=p070925.htm&no=11&date=. Accessed 12 Dec 2008
Rethinam P, Singh SP (2007) Current status of the coconut beetle outbreak in the Asia-Pacific Region. In: Developing an asian-pacific strategy for forest invasive species: the coconut beetle problem – bridging agriculture and forestry. [Report of the Asia-Pacific Forest Invasive Species Network Workshop 22–25 February 2005, Ho Chi Minh City, Viet Nam (RAP PUBLICATION 2007/02)]. FAO: Regional Office for Asia and the Pacific, Bangkok
Song Y, Zhan RL, Zhang SQ, Guo LJ, Xu TW, Huang JS (2006) The virulence selection of Metarhizium spp.against the Brontispa longissium(GastrO). Plant Quarantine 20(6):333–336
Stapley JH (1973) Insect pests of coconuts in the Pacific region. Outlook Agric 7(5):211–217
Stapley JH (1980) Coconut leaf beetle (*Brontispa*) in the Solomons. Alafua Agric Bull 5(4):17–22
Tjoa TM (1953) Memberantas hama-hama kelapa dan kopra. Noorhoff, Jakarta, p 270
Tran TV (2004) Classical biological control of coconut hispine beetle with the parasitoid *Asecodes hispinarum* Boucek (Hymenoptera: Eulophidae) in Viet Nam. Available from: https://www.ippc.int/cds_upload/1115175069697_AD522E02.pdf. Accessed 12 Dec 2008
Voegele JM (1989) Biological control of *Brontispa longissima* in Western Samoa: an Ecological and economic evaluation. Agric Ecosyst Environ 27:315–329

Waterhouse DF, Norris KR (1987) *Brontispa longissima* (Gestro). Biological control pacific prospects. Inkata Press, Melbourne, pp 134–141

Wu KC, Tao CCC (1976) Natural enemies of the transparent scale and control of the leaf bud beetle attacking coconut palm in Taiwan. J Agricl Res China 25:141–156

Xiao GJ, Zeng L, Li Q, Lu YY (2006) Cold hardiness of palm leaf beetle, *Brontispa longissima*. Chin Bull Entomol 4:527–530

Xu CA, Peng ZQ, Tang C, Jin QA, Wan FH (2007) Study on the survival time of *Brontispa longissima*. Plant quarantine 21(4):205–207

Yu FY, Qin WQ, Li CX, Han CW, Peng ZQ (2009) Effects of coconut cultivars on development and fecundity of *Brontispa longissima* (Gestro). Chin J Trop Crops 30(6):846–850

Zeng L, Zhou R, Cui ZX, Lu YY, Liang GW (2003) Effect of host plants on development and survival of *Brontispa longissima* (Gestro). J South Chin Agric Univ (Natural Science Edition) 24(4):37–39

Zhang J, Feng G (2010) Insecticidal activity of extract from *Allamanda cathartica* Linn against *Brontispa longissima*. Chin J Trop Crops 31(7):1152–1156

Zhao ZY, Lin J, Yin YZ (2003) Experimented on the control of *Brontispa longissima*(Gestro)on tall Coconut with 5 kinds of pesticides. Trop For 31(3):8–10

Zheng CG, Chen HJ, Jiang DX, Tang L, Xu HH (2010) Efficacy of comparative test on preventing *Brontispa longissima* (Gestto)by two methods. Guangdong Agric Sci 9:137–138

Zhong YH, Li H, Liu K, Wen HB, Jin QA, Peng ZQ (2005) Effects of temperature on *Brontispa longissima* population growth. Chin J Appl Ecol 16(12):2369–2372

Zhou R, Zeng L, Liang GW (2004a) Life history of the laboratory population of the palm leaf beetle *Brontispa longissima*. Entomolo Knowl 41(4):336–339

Zhou R, Zeng L, Lu YY, Liang GW, Cui ZX (2004b) Feeding behavior of *Brontispa longissima*. J South Chin Agric Univ 25(4):50–52

Chapter 13
Red Palm Weevil *Rhynchophorus ferrugineus* (Olivier)

Lu Peng and Youming Hou

Abstract The red palm weevil (RPW), *Rhynchophorus ferrugineus* (Coleoptera, Curculionidae) is the most destructive pest of palm trees worldwide. RPW can potentially spread to all palm-growing areas in China, mainly ranging across much of South, East, and Southwest areas and causing severe economic loss. Here, we introduce the current distribution of RPW in China, discuss the invasive mechanisms, including thermal adaptability, host adaptability, immune response, and population genetic differentiation, and summarize the frequently-used prevention and control strategies in China, with chemical controls accompanied by trapping are the dominated measures. We aim to summarize research progress in the past several decades in RPW and to provide sound prospects for improving the strategies and tactics currently employed in ecologically-based pest management in China. In addition, further studies will carry out from two aspects of prevention and control, mainly focus on building the early observing and monitoring system, better elucidating invasive mechanisms, as well as developing the new techniques for pest management, such as genetic and behavior regulation, biological pesticides screening, and natural enemies release.

Keywords *Rhynchophorus ferrugineus* • Distribution • Damage • Invasive mechanisms • Pest management

13.1 Introduction

The red palm weevil (RPW), *Rhynchophorus ferrugineus* (Olivier) (Coleoptera: Curculionidae), also known as Asiatic palm weevil, coconut weevil and Indian palm weevil (Faghih 1996), is one of major pests for palms. It can attack more than 20

L. Peng • Y. Hou (✉)
State Key Laboratory of Ecological Pest Control for Fujian and Taiwan Crops and Fujian Provincial Key Laboratory of Insect Ecology, Fujian Agriculture and Forestry University, Fuzhou 350002, China
e-mail: ymhou@fafu.edu.cn

F. Wan et al. (eds.), *Biological Invasions and Its Management in China*,
Invading Nature - Springer Series in Invasion Ecology 11,
DOI 10.1007/978-94-024-0948-2_13

palm species in 16 genera worldwide (Giblin-Davis et al. 2013). It feeds within palm trunks and creates criss-cross tunnels in trunks. The damaged trees tend to die and collapse, and eventually lose application values (Ju et al. 2011; Montagna et al. 2015). This weevil is native to southeastern Asia and Melanesia, and was described as a deadly coconut pest first in northern India (Lefroy 1906). In the last 30 years, RPW has spread to 34 countries and regions in Europe, Asia and Oceania.

Early detection of this weevil is difficult, because its presence can be confirmed only when the damage symptom has become visible (Dembilio and Jacas 2011). Since it resides inside palm trunks until the emergence of adults, it can be rarely reached by chemical insecticides, natural enemies, and pathogens, and thus the control efficiency is normally low.

13.2 Distribution and Damage of RPW in China

In China, RPW was first discovered in Guangdong province in 1997 (Wu et al. 1998). The pest has spread to many regions of southern and eastern China such as Hainan, Fujian, Taiwan, Guangxi, Yunnan, Hong Kong, Jiangxi, Shanghai, Zhejiang, Sichuan, and Chongqing (Ju et al. 2006; Wang 2007; Zhang et al. 2008; Hou et al. 2011). In addition to these regions, more provinces would be potentially invaded by this weevil (Fig. 13.1; Ge et al. 2015). RPW has been listed as one of the 19 quarantine forestry pests since 2005 and one of 233 dangerous forestry pests since 2003 by the State Forestry Administration of China.

13.3 Invasion Mechanisms of RPW in China

13.3.1 Thermal Adaptability

Insects have an optimal temperature range for growth and development, and their population increase would be significantly limited under the temperature beyond this range (Zhou et al. 2010; Olson et al. 2013). Laboratory experiments revealed that the most suitable temperature for population growth of RPW in China was from 26 to 32 °C. However, the situation may slightly vary with the geography. For example, the population in Shanghai (31.14°N, 121.29°E) prefers the temperature from 26 to 30 °C (Zhao and Ju 2010), while in Fuzhou, Fujian (26.08°N, 119.28°E), 27 °C was favorable (Peng et al., unpublished). In addition, the temperature threshold required for development is also different among geographical populations (Li et al. 2010; Zhao and Ju 2010). These results suggest that genetic differentiation and adaptive evolution might have occurred in RPW, which as a result has possibly assisted it to adapt to new environments quickly and spread successfully (De Barro et al. 2011; Fand et al. 2015).

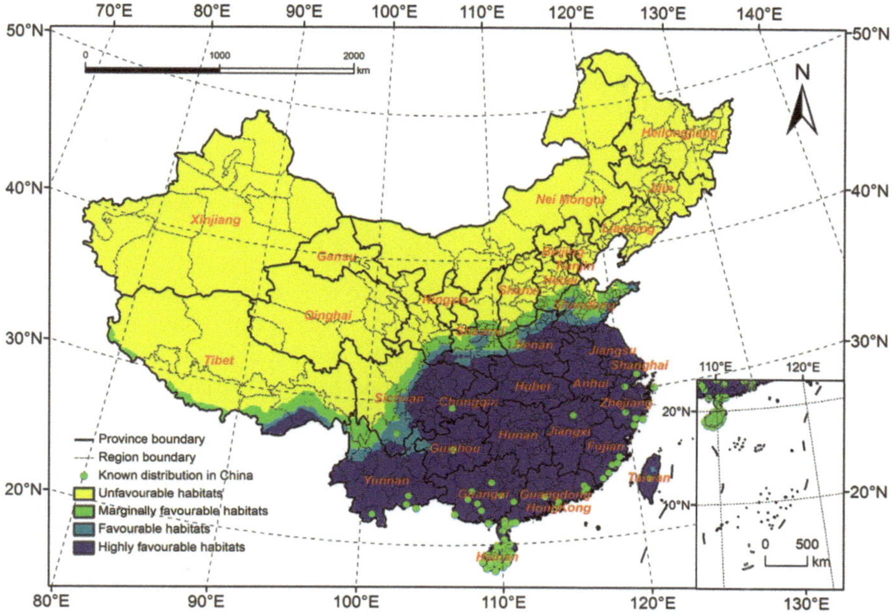

Fig. 13.1 Potential distribution for *Rhynchophorus ferrugineus* under B2 scenario (2020s) (Ge et al. 2015)

13.3.2 Adaptability to Host Plants

The weevil has a broad host rang, which covers most of primary palms (Arecaceae). A total of 32 plant species belonging to three families (Agavaceae, Arecaceae and Poaceae) have been reported as suitable hosts (Faleiro 2006; EPPO 2008; Dembilio et al. 2012). In China, the main host plants are *Cocos nucifera*, *Phoenix dactylifera*, *Phoenix canariensis*, *Phoenix hamceana* var. *formosana*, *Phoenix sylvestris*, *Melroxylon sagu*, *Arenga pinnata*, *Elaeis guineensis*, *Corypha gebanga*, *Borassus flabelliformis*, *Caryota maxomat*, *Caryota cumingii*, *Areca catechu*, *Archontophoehix alexandra*, *Hyophorbe lagenlcaulise*, *Neodypsis decaryi*, *Saccharum officinarum*, *Oreodox aregia*, *Corypha umbraculifer*, *Livistona cochinchinensis*, *Livistona chinensis*, *Livistona saribus*, *Caryota ochlandra*, *Phoenix roebel*, *Chrysalidocarpus lutescens*, *Washingtonia robusta*, *Trachycarpus fortunei*, *Butia capitata*, *Washingtonia filifera* (Qin et al. 2009; Feng and Liu 2010; Ju et al. 2011). Most of these plants are widespread in Hainan, Yunnan, Guangxi, Guangdong, Fujian, and Taiwan. They can also be found in some regions of Guizhou, Hunan, Sichuan, Zhejiang, Jiangxi, Tibet, Hongkong, and Macao.

In recent years, the region of palms planting has been expanding along with the adjustment in agricultural industry. In some regions such as Hainan province, *C. nucifera* and *A. catechu* have become the second and third most important tropical crops, respectively, following the natural rubber. *S. officinarum* is one of major sugar and biomass-energy crops in China. Planting of ornamental palms is becoming a new booming industry in coastal regions of southern China. Thus, it is suggested that increased planting and transportation of palm plants have probably contributed to the rapid spread of RPW in China (Ju et al. 2006; Qin et al. 2009).

13.3.3 Immune Response

Previous studies indicated that many insects have the ability to increase immune protection when exposed to bacteria, and they can recover from its infection (Parmakelis et al. 2008; Shi and Sun 2010; Bali and Kaur 2013; Urbański et al. 2014). This response is often defined as immune priming (Roth et al. 2009; Yue et al. 2013). Shi et al. (2014) induced the immune response of RPW by injecting the bacteria pathogen (*Escherichia coli* DH5α), which showed that previous challenges of bacteria pathogen enhanced the magnitude of phenoloxidase activity and antibacterial activity in RPW larvae against the secondary infection. Furthermore, transgenerational immune priming (TGIP) was also determined in this pest, and only challenged RPW females transferred the immune protection to their offspring. These results suggest that males and females of this pest might have evolved different strategies on the investment of delivering immune protection to their offspring. The results obtained from RPW are highly consistent with those in insects (Roth et al. 2010; Zanchi et al. 2011; Trauer and Hilker 2013). As an invasive pest, successful invasions of RPW largely rely on its colonization ability in new environments, and pathogens represents one of the most important factors for rapid adaptation. Therefore, immune priming and transgenerational immune priming may make RPW itself and offspring more resistant to pathogens, thus facilitate its adaptation and dispersal in new regions.

13.3.4 Population Genetic Differentiation

Information of population structures and invasion pathways are important for understanding invasion processes and developing control strategies against invasive pests. Loss of allelic diversity is a typical consequence of the introduction of species into new environments (Dlugosch and Parker 2008; Ficetola et al. 2008). RPW populations have a low level of genetic diversity and a similar genetic structure in Fujian, China. Such genetic patterns are highly related to the limited flight capacity of RPW and human-aided recent introductions (Wang et al. 2015). Further studies showed genetic variation was generally low among the RPW populations in the southern China, including those collected from Hainan, Guangdong, Guangxi, Fujian,

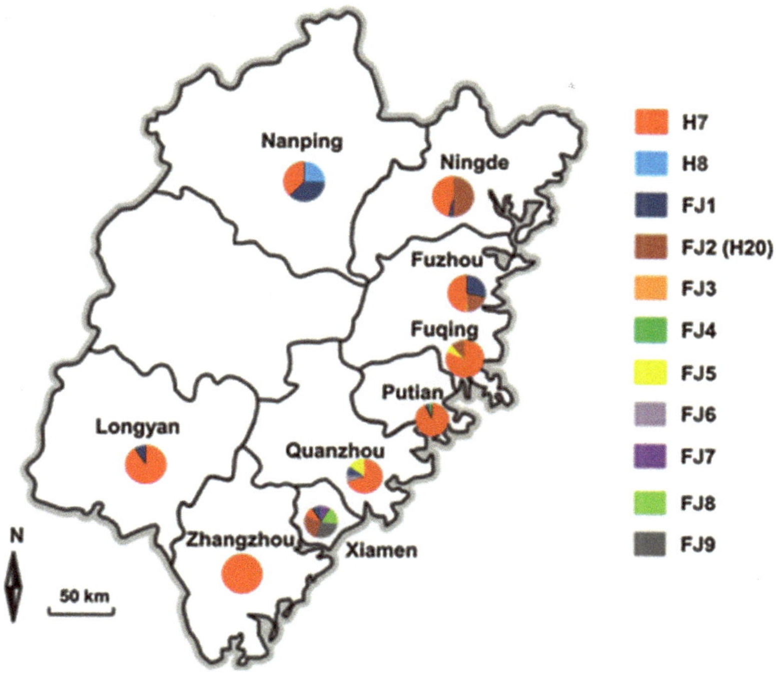

Fig. 13.2 Distribution of haplotypes among the surveyed RPW populations in Fujian. The pie charts indicate the frequencies of COI haplotypes (corresponding to the various colors) at each collection site (Wang et al. 2015)

Taiwan, and Sichuan, especially in populations from Taiwan province, which had the lowest genetic diversity (Y. Hou, unpublished data). In addition, the obvious haplotype diversity of RPW occurred both in southern China and in Fujian Province, which indicated the RPW populations in China might face great invasion pressure, likely promoted by anthropogenic transport (Wang et al. 2015; Y. Hou, unpublished data). For example, eleven haplotypes of RPW populations were identified in Fujian, China. Of them, H7, H8, and FJ2 (H20) were also found in Japan, the Mediterranean Basin, and Aruba, respectively (El-Mergawy et al. 2011; Rugman-Jones et al. 2013). FJ1, FJ3 ~ 9 were the new haplotypes detected for the first time. FJ1 was the dominant haplotype, and distributed in most places of Fujian (Wang et al. 2015, Fig. 13.2). The diversity of haplotypes suggests that the RPW populations in China might originate from various introduction events or from the introduction of more than one haplotype in a single event (Wang et al. 2015).

13.4 Prevention and Control Strategies

13.4.1 Predicting Population Dynamics

There is a close relationship between population dynamics and a number of stand factors, such as site conditions, climate conditions, vegetation compositions, natural enemies, and human factors. Multiple factors do not work alone, and they always interact with each other. Using a linear stepwise regression analysis method, Yan et al. (2011) developed an efficient model for predicting RPW population dynamics in medium and short terms. In this model, four factors were used as parameters, i.e., soil moisture, crown density, infertile extent of soil and coconut age, all of which were closely related to population growth of RPW. Simultaneously, field test suggests that the model can be used in production practice.

13.4.2 Chemical Control

Spraying or injecting of chemical insecticides into palm trees is widely used to control RPW. The commonly used chemical pesticides in China include emamectin benzoate, malathion, phoxim, omethoate, carbofuran, dichlorvos, acephate, chlorpyrifos, acetamiprid, imidacloprid and carbaryl. Ou et al. (2009) found that applying 500 times liquid of 3% emamectin benzoate and 45% malathion in the field produced a control efficacy of 93.4% and 84.5%, respectively. Both of these two chemicals were safe to palmaceous plants. Liu et al. (2011) compared the control efficacy of four chemical-applying methods for controlling larvae, i.e., fumigation, hole injections, topical application and root-irrigation. The fumigation was demonstrated to be more efficient than the other methods, and was recommended for field use. The fumigation can be performed in the following steps. First, drill on the stem at ca. 20 cm above the ground to make a tunnel inside. The tunnel is required to be 8–10 cm in length and in a direction of ca. 45° to the ground. Then, put the chemical (e.g., calcium phosphate at a concentration 9.6 g per tree) inside the hole. Last, fill the left space of the tunnel with mud and cover the entrance with a plastic film to prevent the chemical from diffusing outwards. A field testing in Wenchang of Hainan using such a fumigation method produced a control efficacy of 83% on RPW larvae at 5 days after application (Liu et al. 2011).

13.4.3 Bait Trapping

In the past decade, some countries have exerted much efforts on the development of mass trap systems based on pheromone or synthetic chemicals (El-Shafie et al. 2011; Abuagla and Al-Deeb 2012; Guarino et al. 2011; Vacas et al. 2013;

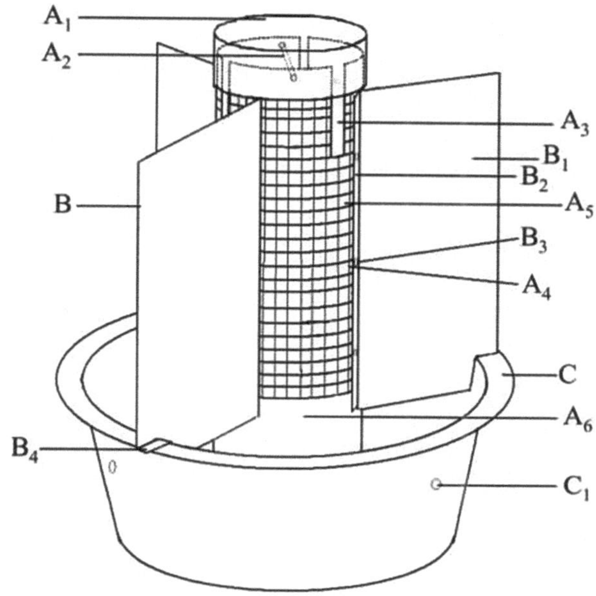

Fig. 13.3 The structure diagram of new- type trapper for *Rhynchophorus ferrugineus* (Huang et al. 2011). A_1: pipe closer, A_2: pheromones hanger, A_3: wormhole, A_4: fixing holes of stents, A_5: cloth cover for crawling, A_6: stents, *B*: blocking pest board, B_1: blocking pest surface, B_2: conjugation surface, B_3: fixing holes for conjugation surface, B_4: legs, *C*: basin for collecting insects, C_1: overflow hole

Chakravarthy et al. 2014). So far, trapping has been considered as the most effective method for controlling palm weevils (Hajjar et al. 2015). In China, Huang et al. (2011) developed a new-type trapper for RPW, which consists of three parts, including trapping tube, blocking pest board and basin for collecting insects. Trapping tube includes six parts: tube closer, pheromones hanger, wormhole, fixing holes of stents, cloth cover for crawling, and stents. Moreover, there are four parts for the blocking pest board, including blocking pest surface, conjugation surface, fixing holes for conjugation surface and Legs (Fig. 13.3). The application results indicated that it was easy and convenient to prepare and use, and the adults of RPW in the field could be well trapped.

13.4.4 Biological Control

At least 14 strains of *Metarhizium anisopliae* var. *anisopliae*, a well known entomopathogenic fungus, have been isolated by Chinese researchers (Zhu et al. 2010; Zhang et al. 2012). Of them, some strains, such as Red2, had high pathogenicity in the laboratory. Field tests are needed to clarify practical effect of these strains against RPW. Some *M. anisopliae* strains isolated from the RPW of other countries have been demonstrated in laboratory bioassays. They were highly efficient against RPW larvae and adults (Francardi et al. 2012, 2013). In addition, a *Serratia marcescens* subsp. strain (HN-1) capable of producing red pigment was isolated from RPW. This bacterium was demonstrated to have certain pathogenecity again eggs and larvae (Zhang et al. 2011; Pu and Hou 2016). In addition, Shi et al. (2014) found that the magnitude of immune response upon the exposure to *Staphyloccocus aureus* in the offspring from Ec-challenged parents was significantly greater than that of the offspring from PBS-challenged and control parents. This suggests that the different kinds of biological control agents might be used alternatively or in combination to fight against RPW because of the existence of immune priming with a low species-specific level.

13.5 The Large-Scale Molecular Dataset for RPW

Since the RPW genome has not been sequenced, a large-scale gene discovery effort is of essence for promoting molecular studies and future genome annotation. Wang et al. (2013) reported a large-scale de novo complementary DNA (cDNA) sequencing of RPW. They identified more than 60,000 single nucleotide polymorphisms and 1200 microsatellite markers. In addition, the systematic transcriptomic libraries of RPW at five embryonic developmental stages were also obtained using RNA-seq, which included 22,532 genes (Yin et al. 2015). Meanwhile, the expression dynamics of several conserved and related signaling pathways (such as Hedgehog, JAK-STAT, Notch, TGF- β, Ras/MAPK and Wnt), as well as key developmental genes, including those related to apoptosis, axis formation, Hox complex, neurogenesis and segmentation, were analysis (Yin et al. 2015).

To better understand the mechanisms of host adaptation of RPW, the intestinal metagenomic library was constructed. The diverse bacterial community in the RPW gut consisted of 50 common species, accounting for >80% of the total bacterial species present. The composition of gut bacterial community varied greatly among seasons. For example, in November and March the most abundant specie was *Klebsiella pneumoniae*, whereas in July it was *Lactococcus lactis*. Consequently, the high-temperature stress was thought to have significant influence on RPW gut microbiota (Jia et al. 2013).

13.6 Conclusions and Prospectives

R. ferrugineus is an invasive and extremely devastating pest for many palm species in many countries around the world (Ju et al. 2006; EPPO 2008). In China, this weevil was first reported in 1997 and has been listed as a forestry quarantine pest since 2005. In recent 20 years, some of *R. ferrugineus*'s host plants have expanded in their growing regions and/or have arisen as major resources for industrial production. For example, coconut and areca have become the second and third tropical crop industry in Hainan province, respectively. Palm nursery stock and flower industry is gradually developing into a new industry in southern coastal areas of

China. This suggests that more serious economic loss might be resulted from RPW damage and should be given more concerns.

Comprehensive researches have been carried out in recent years, mainly focusing on the biological and ecological characteristics of RPW. In the future, an early detection and monitoring system is to be built to prevent the introduction of RPW. Moreover, its invasive mechanisms are to be elucidated which include such aspects: interactions between RPW and host plants, molecular mechanisms of immunity, insecticide resistance, chemical communication in RPW, and population genetic evolution of RPW. In term of control strategies, use of chemical pesticides is still serving as a major control method in China, which, however, may cause serious ecological problems. Thus new control techniques will have to be developed, which may include genetic and behavioral regulation of RPW, and screening efficient biological pesticides. Based these work, we will hopefully develop a novel strategy for safe, cost-effective and sustainable management of this pest.

References

Abuagla AM, Al-Deeb MA (2012) Effect of bait quantity and trap color on the trapping efficacy of the pheromone trap for the red palm weevil, *Rhynchophorus ferrugineus*. J Insect Sci 12:120

Bali GK, Kaur S (2013) Phenoloxidase activity in haemolymph of *Spodoptera litura* (Fabricius) mediating immune responses challenge with entomopathogenic fungus, *Beauveria bassiana* (Balsamo) Vuillmin. J Entomol Zool Stud 1(6):118–123

Chakravarthy AK, Chandrashekharaiah M, Kandakoor SB, Nagaraj DN (2014) Efficacy of aggregation pheromone in trapping red palm weevil (*Rhynchophorus ferrugineus* Olivier) and rhinoceros beetle (*Oryctes rhinoceros* Linn.) from infested coconut palms. J Environ Biol 35:479–484

De Barro PJ, Liu SS, Boykin LM, Dinsdale AB (2011) Bemisia tabaci: a statement of species status. Annu Rev Entomol 56:1–19

Dembilio Ó, Jacas JA (2011) Basic bio-ecological parameters of the invasive red palm weevil, *Rhynchophorus ferrugineus* (Coleoptera: Curculionidae), in *Phoenix canariensis* under Mediterranean climate. Bull Entomol Res 101:153–163

Dembilio Ó, Tapia GV, Téllez MM, Jacas JA (2012) Lower temperature thresholds for oviposition and egg hatching of the red palm weevil, *Rhynchophorus ferrugineus* (Coleoptera: Curculionidae), in a Mediterranean climate. Bull Entomol Res 102:97–102

Dlugosch KM, Parker IM (2008) Founding events in species invasions: genetic variation, adaptive evolution, and the role of multiple introductions. Mol Ecol 17:431–449

El-Mergawy RAAM, Nasr MI, Abdallah N, Silvain JF (2011) Mitochondrial genetic variation and invasion history of red palm weevil, *Rhynchophorus ferrugineus* (Coleoptera: Curculionidae), in Middle-East and Mediterranean Basin. Int J Agric Biol 13:631–637

El-Shafie HAF, Faleiro JR, Al-Abbad AH, Stoltman L, Mafra-Neto A (2011) Bait-Free attract and kill technology (Hook™ RPW) to suppress red palm weevil, *Rhynchophorus ferrugineus* (Coleoptera: Curculionidae) in date palm. Fla Entomol 94:774–778

EPPO (2008) Data sheet on quarantine pests: *Rhynchophorus ferrugineus*. EPPO Bull 38:55–59

Faghih AA (1996) The biology of red palm weevil, *Rhynchophorus ferrugineus* Oliv (Coleopter, Curculionidae) in Savaran region (Sistan province, Iran). Appl Entomol Phytopathol 63:16–18

Faleiro JR (2006) A review of the issues and management of the red palm weevil *Rhynchophorus ferrugineus* (Coleoptera: Rhynchophoridae) in coconut and date palm during the last one hundred years. Int J Trop Insect Sci 26(3):135–154

Fand BB, Sul NT, Bal SK, Minhas PS (2015) Temperature impacts the development and survival of common cutworm (*Spodoptera litura*): simulation and visualization of potential population growth in India under warmer temperatures through life cycle modelling and spatial mapping. PLoS ONE 10:e0124682

Feng Y, Liu H (2010) Potential suitability analysis of *Rhychophorus ferrugineus* (Olvier) in China based on Maxtent and GIS. J Huazhong Agric Univ 29(5):552–556. (in Chinese)

Ficetola GF, Bonin A, Miaud C (2008) Population genetics reveals origin and number of founders in a biological invasion. Mol Ecol 17:773–782

Francardi V, Benvenuti C, Roversi PF, Rumine P, Barzanti G (2012) Entomopathogenicity of *Beauveria bassiana* (Bals.) Vuill. and *Metarhizium anisopliae* (Metsch.) Sorokin isolated from different sources in the control of *Rhynchophorus ferrugineus* (Olivier) (Coleoptera Curculionidae). Redia 95:49–55

Francardi V, Benvenuti C, Barzanti GP, Roversi PF (2013) Autocontamination trap with entomopathogenic fungi: a possible strategy in the control of *Rhynchophorus ferrugineus* (Olivier) (Coleoptera Curculionidae). Redia 96:57–67

Ge X, He S, Wang T, Yan W, Zong S (2015) Potential distribution predicted for *Rhynchophorus ferrugineus* in China under different climate warming scenarios. PLoS ONE 10:e0141111

Giblin-Davis RM, Faleiro JR, Jacas JA, Peña JE, Vidyasagar PSPV (2013) Biology and management of the red palm weevil, *Rhynchophorus ferrugineus*. In: Peña JE (ed) Potential invasive pests of agricultural crops. CAB International, Wallingford, pp 1–34

Guarino S, Lo Bue P, Peri E, Colazza S (2011) Responses of *Rhynchophorus ferrugineus* adults to selected synthetic palm esters: electroantennographic studies and trap catches in an urban environment. Pest Manag Sci 67:77–81

Hajjar MJ, Ajlan AM, Al-Ahmad MH (2015) New approach of *Beauveria bassiana* to control the red palm weevil (Coleoptera: Curculionidae) by trapping technique. J Econ Entomol 108(2):425–432

Hou YM, Wu ZJ, Wang CF (2011) The status and harm of invasive insects in Fujian, China. In: Xie LH, You MS, Hou YM (eds) Biological invasions: problems and countermeasures. Sci Press, Beijing, pp 121–122. (in Chinese)

Huang SC, Li CX, Yan W, Liu L, Qin WQ (2011) Preparation and application of new-type trapper for *Rhynchophorus ferrugineus*. Acta Agric Jiangxi 23(9):86–87. (in Chinese)

Jia SG, Zhang XW, Zhang GY, Yin A, Zhang S, Li FS, Wang L, Zhao DJ, Yun QZ, Tala, Wang JX, Sun GY, Baabdullah M, Yu XG, Hu SN, Al-Mssallem IS, Yu J (2013) Seasonally variable intestinal metagenomes of the red palm weevil (*Rhynchophorus ferrugineus*). Environ Microbiol 15(11):3020–3029

Ju RT, Li YZ, Du YZ, Chi XZ, Yan W, Xu Y (2006) Alert to the spread of alien invasive pest, red palm weevil, *Rhynchophorus ferrugineus* (Oliver). Chin Bull Entomol 43:159–163. (in Chinese)

Ju RT, Wang F, Wan FH, Li B (2011) Effect of host plants on development and reproduction of *Rhynchophorus ferrugineus* (Olivier) (Coleoptera: Curculionidae). J Pest Sci 84:33–39

Lefroy HM (1906) The more important insects injurious to indian agriculture. Gov India Press, Calcutta

Li L, Qin WQ, Ma ZL, Yan W, Huang SC, Peng ZQ (2010) Effect of temperature on the population growth of *Rhynchophorus ferrugineus* (Coleoptera: Curculionidae) on Sugarcane. Environ Entomol 39(3):999–1003

Liu L, Yan W, Wei J, Huang SC, Zhang J, Qin WQ, Cao JH, Peng ZQ (2011) Chemical control of *Rhynchophorus ferrugineus* larvae. Chin J Trop Crops 32(8):1545–1548. (in Chinese)

Montagna M, Chouaia B, Mazza G, Prosdocimi EM, Crotti E, Mereghetti V et al (2015) Effects of the diet on the microbiota of the 空格 red palm weevil (Coleoptera: Dryophthoridae). PLoS ONE 10(1):e0117439

Olson JF, Eaton M, Kells SA, Morin V, Wang C (2013) Cold tolerance of bed bugs and practical recommendations for control. J Econ Entomol 106:2433–2441

Ou SS, Xie EB, Xie YJ, Wang XX, Qin LH, Shen XP (2009) Study on control effects of different medicament on *Rhynchophorus ferrugineus* Fabricius. J Anhui Agric Sci 37(36):18005–18006. (in Chinese)

Parmakelis A, Slotman MA, Marshall JC, Awono-Ambene PH, Nkondjio CA, Simard F, Caccone S, Powell J (2008) The molecular evolution of four antimalarial immune genes in the *Anopheles gambiae* species complex. BMC Evol Biol 8:68–79

Pu YC, Hou YM (2016) Isolation and identification of bacterial strains with insecticidal activities from *Rhynchophorus ferrugineus* Oliver (Coleoptera: Curculionidae). Journal of Applied Entomology, 140: 10.1111/jen.12293

Qin WQ, Li CX, Huang SC (2009) Risk analysis of *Rynchophorus ferrugineus* Olivier (Coleoptera: Curculionidae) in China. Acta Agric Jiangxi 21(9):79–82. (in Chinese)

Roth O, Sadd BM, Schmid-Hempel P, Kurtz J (2009) Strain-specific priming of resistance in the red flour beetle, *Tribolium castaneum*. Proc R Soc Lond Ser B 276:145–151

Roth O, Joop G, Eggert H, Hilbert J, Daniel J, Schmid-Hempel P, Kurtz J (2010) Paternally derived immune priming for offspring in the red flour beetle, *Tribolium castaneum*. J Anim Ecol 79:403–413

Rugman-Jones PF, Hoddle CD, Hoddle MS, Stouthamer R (2013) The lesser of two weevils: molecular-genetics of pest palm weevil populations confirm *Rhynchophorus vulneratus* (Panzer 1798) as a valid species distinct from *R. ferrugineus* (Olivier 1790), and reveal the global extent of both. PLoS ONE 8:e78379

Shi ZH, Sun JH (2010) Immunocompetence of the red turpentine beetle, *Dendroctonus valens* LeConte (Coleoptera: Curculionidae, 空格 Scolytinae): variation between developmental stages and sexes in populations in China. J Insect Physiol 56:1696–1701

Shi ZH, Lin YT, Hou YM (2014) Mother-derived trans-generational immune priming in the red palm weevil, *Rhynchophorus ferrugineus* Olivier (Coleoptera, Dryophthoridae). Bull Entomol Res 104:742–750

Trauer U, Hilker M (2013) Parental legacy in insects: variation of transgenerational immune priming during offspring development. PLoS ONE 8:e63392

Urbański A, Czarniewska E, Baraniak E, Rosiński G (2014) Developmental changes in cellular and humoral responses of the burying beetle *Nicrophorus vespilloides* (Coleoptera, Silphidae). J Insect Physiol 60:98–103

Vacas S, Primo J, Navarro-Llopis V (2013) Advances in the use of trapping systems for *Rhynchophorus ferrugineus* (Coleoptera: Curculionidae): traps and attractants. J Econ Entomol 106:1739–1746

Wang Y (2007) Forest pests and diseases of Shanghai. Shanghai Sci Tech Publishers, Shanghai, p3. (in Chinese)

Wang L, Zhang XW, Pan LL, Liu WF, Wang DP, Zhang GY, Yin YX, Yin A, Jia SG, Yu XG, Sun GY, Hu SN, Al-Mssallem IS, Yu J (2013) A large-scale gene discovery for the red palm weevil *Rhynchophorus ferrugineus* (Coleoptera: Curculionidae). Instr Sci 20:689–702

Wang GH, Zhang X, Hou YM, Tang BZ (2015) Analysis of the population genetic structure of *Rhynchophorus ferrugineus* in Fujian, China, revealed by microsatellite loci and mitochondrial COI sequences. Entomol Exp et Applicata 155:28–38

Wu YS, Dong DZ, Liu DM, Wu GC, Feng RH (1998) Preliminary survey report of the occurrence of *Rhynchophorus ferrugineus* (Olivier) on Palm plants. Guangdong Landscape Architecture 1:38. (in Chinese)

Yan W, Liu L, Huang SC, Zhang J, Qin WQ, Cao JH, Peng ZQ (2011) Application of stepwise regression analysis in predicting red palm weeril (*Rhynchophorus ferrugineus*) movement. Chin J Trop Crops 32(8):1549–1552. (in Chinese)

Yin A, Pan L, Zhang X, Wang L, Yin Y, Jia S, Liu W, Xin C, Liu K, Yu X, Sun G, Al-hudaib K, Hu S, Al-Mssallem IS, Yu J (2015) Transcriptomic study of the red palm weevil *Rhynchophorus ferrugineus* embryogenesis. Instr Sci 22:65–82

Yue F, Zhou Z, Wang LL, Ma ZP, Wang JJ, Wang MQ, Zhang H, Song LS (2013) Maternal transfer of immunity in scallop *Chlamys farreri* and its trans-generational immune protection to offspring against bacterial challenge. Dev Comp Immunol 41:569–577

Zanchi C, Troussard J, Martinaud G, Moreau J, Moret Y (2011) Differential expression and costs between maternally and paternally derived immune priming for offspring in an insect. J Anim Ecol 80:1174–1183

Zhang YF, Tang GL, Wang L, Pei ZW, Tang LF (2008) Bionomics and control of *Rhynchophorus ferrugineus*. Forest Pest Disease 27:12–13. (in Chinese)

Zhang J, Qin WQ, Yan W, Peng ZQ (2011) Isolation and identification of a pathogenic strain of *Rhynchophorus ferrugineus* Oliver. Chin J Trop Crops 32(11):2331–2335. (in Chinese)

Zhang J, Qin WQ, Yan W, Peng ZQ (2012) Detection of pathogenicity of *Meatarhiziums* against *Rhynchophorus ferrugineus* in laboratory. Chin J Trop Crops 33(5):899–905. (in Chinese)

Zhao M, Ju RT (2010) Effects of temperature on the development and fecundity of experimental population of *Rhychophorus ferrugineus*. Acta Phys Sin 37(6):517–521. (in Chinese)

Zhou ZS, Guo JY, Chen HS, Wan FH (2010) Effects of temperature on survival, development, longevity and fecundity of *Ophraella communa* (Coleoptera: Chrysomelidae), a biological control agent against invasive ragweed, *Ambrosia artemisiifolia* L. (Asterales: Asteraceae). Environ Entomol 39:1021–1027

Zhu H, Qin WQ, Huang SC, Yan W, Sun XD (2010) Isolation and identification of an entomopathogenic fungus strain of *Rhynchophorus ferrugineus* Oliver. Acta Phys Sin 37(4):336–340. (in Chinese)

Chapter 14
Nipa Palm Hispid Beetle *Octodonta nipae* (Maulik)

Baozhen Tang and Youming Hou

Abstract In 2001, the nipa palm hispid beetle *Octodonta nipae* (Maulik) (Coleoptera: Chrysomelidae) was first detected in Hainan Province, China, on the California fan palm *Washingtonia filifera* (Linden ex André) H. Wendl. The infested palms displayed serious decline with young leaf fronds with shrivel and curl, and both adults and larvae were found together on the center of young leaves of palms. This beetle now has also invaded other southern regions in China. To understand its invasion mechanisms, we review biological characteristics and ecological adaptation to environments in this chapter. Surveys on biological characteristics showed that females can lay an average of 120 eggs in clusters at the ends of adult feeding scars, and the longevity of males and females can be ~203.5 days and 178.7 days, respectively. Moreover, females and males can mate multiple times with an optimal range throughout their life cycles, within which their reproduction is most successful. When living in harsh environments such as low temperature at 12.5 °C, *O. nipae* can adjust its physiological state to adapt to environmental stressors, however, no evident tolerance to high temperature over 27 °C was observed. This beetle has a wide range of preferred host palm plants except the Chinese fan palm *Livistona chinensis* R. Brown. In addition, upon bacterial challenge, the beetle can make an adaptive immune response with the existence of a sexual dimorphism of immunocompetence. All these traits may together contribute to its wide spread. So far, the potentialities of biological agents, such as *Tetrastichus brontispae* and *Metarhizium anisopliae* var. *anisopliae*, have been validated only in the lab, thus chemical spraying is still the effective method to manage the beetle in the field. Further field investigations are needed to assess control efficiency of these biological agents.

Keywords *Octodonta nipae* (Maulik) • Hispine beetle • *Washingtonia filifera* • Biological control • *Tetrastichus brontispae* Ferrière

---

B. Tang • Y. Hou (✉)
State Key Laboratory of Ecological Pest Control for Fujian and Taiwan Crops and Fujian Provincial Key Laboratory of Insect Ecology, Fujian Agriculture and Forestry University, Fuzhou 350002, China
e-mail: ymhou@fafu.edu.cn

F. Wan et al. (eds.), *Biological Invasions and Its Management in China*,
Invading Nature - Springer Series in Invasion Ecology 11,
DOI 10.1007/978-94-024-0948-2_14

Fig. 14.1 Morphology of *O. nipae* (From eggs to adults)

14.1 Introduction

Octodonta nipae (Maulik) (Fig. 14.1), is a nipa palm hispid beetle belonging to Cryptonychini of Chrysomeloidea, Coleoptera (Maulik 1921; Staines 2012). It is currently a serious invasive pest of palm plants in southern China (Zhang et al. 2008). In China, this beetle was first detected on the California fan palm *Washingtonia*

Fig. 14.2 Field investigation

filifera (Linden ex André) H. Wendl. in 2001 from a forest nursery in Hainan province (Sun et al. 2003), and later it was found in other southern provinces including Fujian, Guangdong and Guangxi (Fig. 14.2) (Liang et al. 2005; Wu et al. 2007; Hou and Weng 2010; Hou et al. 2011). The beetle was ever found on the queen palm *Syagrus romanzoffiana* (Cham.) Glassman (Arecaceae) on the island of Cyprus in 2009, but it has been successfully eradicated in 2010 (Vassiliou et al. 2010).

O. niape attacks young palm leaf fronds, and sustained attack will lead to leaf shrivel, curl, and even death (Hou et al. 2011). Palms are keystone plant species in tropical and semitropical ecosystems, serving as important sources of food and, more importantly, of ornamental industry. As southern China has a wide coverage of host palms and frequent trade of palm plants, this region would be very favorable for this beetle and thus might suffer from serious attacks (Hou et al. 2011). For this reason, in the past decade local governments and researchers have paid much attention on this pest. Now several research programs are on the way in our lab, including risk assessment, negative effects caused by this pest, invasion mechanisms, control methods, and so on.

14.2 Invasive Mechanism of *O. nipae*

14.2.1 Biological Characteristics

O. niape is gregarious, and it is often found in large numbers on one palm. They feed and dwell in the tightly furled fronds and trunk fibers (Fig. 14.3). Females lay an average of 120 eggs in clusters at the ends of adult feeding scars and cover them with regurgitated leaf fiber (Kalshoven 1981). The life cycle from egg to adult takes

Fig. 14.3 Symptoms of Canary island date palm damaged by *O. nipae*

from 5 to 9 weeks, and adults live around 150 days, depending on temperature (Kalshoven 1981). Hou and Weng (2010) reported that the longest longevity of males and females can be ~203.5 days and ~178.7 days, respectively, and the life expectancy for a newborn egg can be up to 171.6 days at 20 °C. The hidden habit and long life span may have been the main reasons for its wide spread.

Both *O. nipae* females and males can mate multiple times throughout their life cycles. According to our observations, a single pair of female and male can mate 25 times during in 48 h. As multiple mating can give both advantages (e.g. material benefits such as supplementation of water, food supply, nuptial gifts and nutrients provided in seminal fluid; genetic benefits) and disadvantages (e.g. harassment, physical and physiological injuries males imposed to females) for insects, we predicted there might be an intermediate optimal female mating rate for the maximization of reproductive success concerning the benefit gains and cost losses attached to increased mating. To test this hypothesis (the existence of an intermediate optimal female mating rate for the maximization of reproductive success), we compared in the laboratory females' fecundity and longevity of this beetle at various mating rates, i.e., 1, 5, 10, 15 and 20 times. Females mating 15 times laid the largest number of eggs (139 eggs), and had a highest egg hatching rate (47.4%) and population growth rate. By contrast, when the mating increased to 20 times, females died earlier and both egg production and egg-hatching rate decreased sharply; the males' longevity was also decreased (Li et al. 2014). This suggests that *O. nipae* females have an optimal range of multiple mating (~ 15 times), within which their reproduction is most successful. That is, the speculated intermediate optimal female mating rate exits in this beetle. The situation under natural conditions is to be investigated. It would be interesting to figure out how the beetle would shift their reproductive strategy and what the outcomes would differ between laboratory population and wild population.

14.2.2 Ecological Adaptation to Physical Environments

In the laboratory, all individuals of *O. nipae* could complete development and females could lay eggs under the temperature from 20 to 30 °C (Yu et al. 2007; Hou and Weng 2010). The optimization temperature for development and females' fecundity was at 25 °C or so (Hou and Weng 2010). Such a temperature range exists in southern China, especially in Hainan, Guangzhou and Fujian, and three generations can be observed in these regions (Yu et al. 2007), thus this should be one of the reasons why *O. nipae* can establish and widely spread in these regions. Furthermore, *O. nipae* can adapt to extreme temperature very well, which has also played a key role in its wide spread. When *O. nipae* (larvae, pupae and adults) were subjected to low-temperature acclimation (12.5, 15, 17.5, and 20 °C) for 10 days under laboratory conditions, contents of glycogen, protein, amino acid, crude fat, glycerol and free water changed significantly, suggesting that *O. nipae* can adjust its physiological state to adapt to low temperatures (Hua et al. 2014). However, *O. nipae* does not have evident tolerance to high temperatures over 27 °C (Hua 2013). As the temperature increased from 29 to 32 °C, the eclosion rate would decrease from 87.5% to 5% (Yu et al. 2007), and the beetle would not develop at 35 °C (Hou and Weng 2010).

Host plants play important roles in the population increases and outbreaks of *O. nipae*. *O. nipae* can infest many palms species, of which some are much favored, such as queen palm *Syagrus romanzoffiana* (Cham.) Glassman, Canary island date palm *Phoenix canariensis* Chabaud (Yu et al. 2009), Chinese windmill palm *Trachycarpus fortune* (Hooker) H. Wendland (Hou et al. 2014a, b; Li et al. 2016), areca palm *Areca catechu* Linn. and coconut palm *Cocos nucifera* (Lever1937; Zhang et al. 2015). This beetle can complete life cycle and lay eggs on pygmy date palm *Phoenix roebelenii* O' Brien, but this plant appears not as suitable as the ones mentioned above (Hou et al. 2014a). *O. nipae* could not complete development on Majestic palm *Ravenea rivularis* Jumelle & Perrier, and females did not lay eggs on the Chinese fan palm *Livistona chinensis* R. Brown (Hou et al. 2014a). These laboratory results were accordant with observations under natural conditions. For example, *O. nipae* often causes serious damages to *P. canariensis*, but it does not damage the adjacent *L. chinensis*. The variance of suitability among hosts are due to the difference of main nutrient components in palms, such as free amino acids, soluble sugar and proteins (Xi et al. 2013). These results will be useful for the design of culture management strategies for *O. nipae*.

14.2.3 Stress-Resistance and Adaptation

The adaptive immune response of *O. nipae* upon bacterial challenges has been surveyed. Feng and Hou (2015) reported that after the injection of lipopolysaccharide (LPS) into *O. nipae*, which is the major component of the outer membrane of Gram-negative bacteria and can elicit strong immune responses in animals, the activity of phenoloxidase (PO) and antibacterial peptide were tremendously increased in a short period of time. Such a response was more rapid in females than in males. Yet, the LPS-induced immunocompetence may lead to costs of reproduction, such as reducing mating rate, mating latency, increasing mounting time, and decreasing copulation time (Feng et al. unpublished data). In brief, there is a sexual dimorphism of immunocompetence in *O. nipae*, and, besides offering the benefits, immune defense may also impose fitness costs, as observed in other insects. Deep investigations are needed to figure out how such an immune response in *O. nipae* may contribute to its population growth and adaptation in new environments.

14.3 Management and Control Strategy of *O. nipae*

Palms should be inspected seriously prior to sales on markets. Because the morphology and biology of *O. nipae* are similar to those of *B. longissima*, they are easily confused when invading a new area. Inspiringly, rapid quarantine decisions can be made with the aid of genetic methods (Zhang et al. 2015). Upon confirmation of its identity, the infested palms should be destroyed by incineration. If unfortunately incursion occurs, eradication measures should be performed immediately. As this beetle has a hidden habit (they live in the tightly furled fronds) and long life cycle, intensive surveys should also be performed in the newly invaded regions.

14.3.1 Biological Control

Tetrastichus brontispae Ferriere (Hymenoptera: Eulophidae) is a gregarious and koinobiont endoparasitoid native to Java, Indonesia. It was introduced from Taiwan into Hainan, mainland China in 2004 to control the coconut hispine beetle *B. longissima* (Gestro). This parasitoid can parasitize both larvae and pupae of *O. nipae* (Howard et al. 2001; Huang et al. 2007; Tang et al. 2014a), mostly preferring one-day-old pupae (Fig. 14.4) (Tang et al. 2014a).

To establish a mass producing and releasing system of *T. brontispae*, suitable conditions for storing this parasitoid and hosts were investigated. *T. brontispae* females could be stored at 13 ± 1 °C (the lower threshold temperature for *T. brontispae*) for 2 days, and naive one-day-old host pupae could be stored for 8 days, without evident loss of quality (this doesn't affect the number of emerged offsprings

Fig. 14.4 *O. nipae* pupa parasitized by *T. brontispae*

per host, and they can develop and survive normally). But storing mummified pupae of *O. nipae* under low temperatures for 7 days may reduce parasitism rate (Tang et al. 2014a).

The physiological interaction between *O. nipae* and *T. brontispae* has been investigated. Expression of immune-related genes in *O. nipae* pupae could be altered after they were parasitized by *T. brontispae* (Tang et al. 2014b). Venom was the major virulent factor that *T. brontispae* injected into host pupae at parasitization, which was thought to be crucial for successful parasitization (Meng et al. 2016; Tang et al. unpublished data). The wasp venoms consist of a cocktail of proteinacious and non-proteinacious components (Moreau and Asgari 2015). This information might be suggestive for the development of immune defense-based strategies for the control of *O. nipae*.

In addition to the parasitoid *T. brontispae*, other natural enemies have also been identified. Xu et al. (2011) found that *Metarhizium anisopliae* var. *anisopliae* was highly virulent against *O. nipae* larvae and adults, with the larvae being more sensitive (Fig. 14.5). Thus, this fungus is a potential biological control agent for this beetle.

14.3.2 Chemical Control

So far chemical control is still the major strategy for the management of *O. nipae* in the nursery stock or palm plants. Normally broad-spectrum insecticides are usually used, such as the combination of imidacloprid and cypermethrin (Vassiliou et al. 2010), carbarl, chlorpyrifos and trichlorphon (Wu et al. 2007). As the beetle lives cryptically, as well as the palm plants normally have a high stem, it is usually difficult for the beetles to be reached by insecticides. To increase the control efficiency, many effective alternative insecticide-applying methods can be developed instead of spraying, for example, yejiaqing eluviation power in a hung bag can be used. The power is a mixture of thiosultap sodium and acetamipri. The bag was hung on palm shoots, which thereby can work well with the aid of rainfall.

Fig. 14.5 Symptoms of *O. nipae* larva (**a**) and adult (**b**) infected with *Metarhizium*

14.4 Conclusion and Prospects

O. nipae is now mainly distributed in Hainan, Guangdong and Fujian, and likely invade other southern regions in China due to their biological characteristics and the appropriate climate in these regions. To uncover their invasion mechanisms, many studies investigated the biological characteristics and the ecological adaptation. So far, chemical spraying is still the effective method to control the beetle in the field. Although the biological control agents such as *T. brontispae* and *M. anisopliae* var. *anisopliae* have been validated under laboratory conditions, their control efficiency needs to be tested in the field. In addition, strict quarantine procedures and surveillance is still not available. Therefore, a series of effective measures should be taken to prevent or control *O. nipae*, including the establishment of management strategies for quarantine, and further investigations on whether the biological control potentialities of *T. brontispae* and *M. anisopliae* var. *anisopliae* also occur in the field.

References

Feng SS, Hou YM (2015) Lipopolysaccharide-induced immune response of *Octodonta nipae* (Coleoptera: Chrysomelidae) adults in relation to their genders. Acta Entomol Sin 58:28–37. (in Chinese)

Hou YM, Weng ZQ (2010) Temperature-dependent development and life table parameters of *Octodonta nipae* (Coleoptera: Chrysomelidae). Environ Entomol 39:1676–1684

Hou YM, Wu ZJ, Wang CF (2011) The status and harm of invasive insects in Fujian, China. In: Xie LH, You MS, Hou YM (eds) Biological invasions: problems and countermeasures. Science Press, Beijing, pp 111–114. (in Chinese)

Hou YM, Miao YX, Zhang ZY (2014a) Study on life parameters of the invasive Species *Octodonta nipae* (Coleoptera: Chrysomelidae) on different palm species, under laboratory conditions. J Econ Entomol 107:1486–1495

Hou YM, Miao YX, Zhang ZY (2014b) Leaf consumption capacity and damage projection of *Octodonta nipae* (Coleoptera: Chrysomelidae) on three palm species. Ann Entomol Soc Am 107:1010–1017

Howard FW, Moore D, Giblin-Davis R, Abad R (2001) Insects on palms. CABI Press, Egham

Hua RX (2013) Effects of acclimation temperature on the cold tolerance and heat tolerance in *Octodonta nipae* (Maulik). Master's Degree Thesis of Fujian Agriculture and Forestry University (in Chinese)

Hua RX, Hou YM, Shi ZH (2014) Changes in the contents of physiologically active substances in *Octodonta nipae* (Coleoptera: Chrysomelidae) after low temperature acclimation. Acta Entomol Sin 57:265–273. (in Chinese)

Huang SC, Tan WQ, Zhou HQ, Ma ZL, Li CX (2007) Study on reproductive biology of *Tetrastichus brontispae*. Entomol J East China 16(168–171):238. (in Chinese)

Kalshoven LGE (1981) Pest of crops in Indonesia. Ichtiar Baru-Van Houve, Jakarta

Lever RJAW (1937) Economic insects and biological control in the British Solomon Islands. Bull Entomol Res 28:325–331

Li JL, Zhang X, Hou YM, Tang BZ (2014) Effects of multiple mating on the fecundity of an invasive pest (*Octodonta nipae*): the existence of an intermediate optimal female mating rate. Physiol Entomol 39:348–354

Li JL, Tang BZ, Hou YM, Xie YX (2016) Molecular cloning and expression of the vitellogenin gene and its correlation with ovarian development in an invasive pest *Octodonta nipae* on different two host plants. Bull Entomol Res 106. doi:10.1017/S0007485316000353

Liang QC, Huang FY, Lai TZ, Xu BD, Li ZW, Chen SY (2005) Capture of an invasive pest, *Octodonta nipae,* in *Ptychosperma macarthurii* from Thailand. Plant Quarantine 19:160–161. (in Chinese)

Maulik S (1921) A new hispid beetle injurious to nipa palm. Ann Mag Nat Hist 9:451–452

Meng E, Tang BZ, Hou YM, Chen XX, Chen JT, Yu XQ (2016) Altered immune response of *Octodonta nipae* (Maulik) to its pupal endoparasitoid, *Tetrastichus brontispae* Ferrière. Comp Biochem Physiol B Biochem Mol Biol 198:100–109

Moreau SJM, Asgari S (2015) Venom proteins from parasitoid wasps and their biological functions. Toxins 7:2385–2412

Staines CL (2012) Catalog of the hispines of the world (Coleoptera: Chrysomelidae: Cassidinae). Tribe Cryptonychini

Sun JH, Yu PY, Zhang YZ, Wang XJ (2003) A new invasive coconut pest in Hainan Province. Entomol Knowl 40:286–287. (in Chinese)

Tang BZ, Xu L, Hou YM (2014a) Effects of rearing conditions on the parasitism of *Tetrastichus brontispae* on its pupal host *Octodonta nipae*. BioControl 59:647–657

Tang BZ, Chen J, Hou YM, Meng E (2014b) Transcriptome immune analysis of the invasive beetle *Octodonta nipae* (Maulik) (Coleoptera: Chrysomelidae) parasitized by *Tetrastichus brontispae* Ferrière (Hymenoptera: Eulophidae). PLoS One 9:e91482

Vassiliou VA, Kazantzis E, Melifronidou-Pantelidou A (2010) First report of the nipa palm hispid *Octodonta nipae* on queen palms in Cyprus. Phytoparasitica 39:51–54

Wu DJ, Du YH, Chen XJ, Yuan H, Li SB (2007) Quarantine and risk assessment of the invasion of *Octodonat nipae* to Shunde. Plant Quarantine 21:25–26. (in Chinese)

Xi B, Zhang ZY, Hou YM, Shi ZH (2013) Effects of host plants on the developmental duration, feeding and reproduction of the nipa palm hispid, *Octodonta nipae* (Coleoptera: Chrysomelidae). Acta Entomol Sin 56:799–806. (in Chinese)

Xu L, Lan JL, Hou YM, Chen YS, Chen ZX, Weng ZQ (2011) Molecular identification and pathogenicity assay on *Metarhizium* against *Octodonta nipae* (Coleoptera: Chrysomelidae). Chin J Appl Entomol 48:922–927. (in Chinese)

Yu FY, Li MZ, Li CX, Huang SC, Li HS, Zhou HQ, Han CW, Tan WQ (2007) Effects of temperature on the development of *Octodonta nipae* (Maulik). Entomol J East China 16:264–267. (in Chinese)

Yu FY, Tan WQ, Ma ZL, Li CX, Huang SC, Han CW (2009) Effects of host plants on development and fecundity of *Octodonta nipae*. Plant Prot 35:72–74. (in Chinese)

Zhang GL, Fu WD, Liu K (2008) Alien invasive species in agriculture. Science Press, Beijing. (in Chinese)

Zhang X, Tang BZ, Hou YM (2015) A rapid diagnostic technique to discriminate between two pests of palms, *Brontispa longissima* and *Octodonta nipae* (Coleoptera: Chrysomelidae), for quarantine applications. J Econ Entomol 108:95–99

Chapter 15
Oriental Fruit Fly *Bactrocera dorsalis* (Hendel)

Dong Wei, Wei Dou, Mingxing Jiang, and Jinjun Wang

Abstract The oriental fruit fly *Bactrocera dorsalis* (Hendel), first reported in Taiwan Island, is one of the most destructive pests of fruits and vegetables around the world, causing huge economic losses each year. In mainland China, it was first recorded in 1930s, and since the 1980s its population size increased rapidly with the distribution expanding to a broad range. Moreover, it is continuing to spread northwards. In order to prevent its spread and reduce damage, great efforts have been made to clarify its ecological and physiological adaptation and to develop control methods. Here, we first review the invasion history of *B. dorsalis*, ecological and physiological mechanisms underlying its invasion, and research advances obtained in major biological features. We focus on problems and factors that are related to its invasion in China. Subsequently, we introduce the major managing approaches that have been or will be potentially employed in China, including quarantine, monitoring, physical control, biological control, sterile insect technique and RNA interference. Finally, we propose directions for future research.

Keywords *Bactrocera dorsalis* • Invasion • Resistance • Integrated management

15.1 Introduction

Fruit flies (Diptera: Tephritidae) are among the most economically important pest species of the world, attacking a wide range of fruits and fleshy vegetables throughout tropical and sub-tropical areas. The tephritid flies of the genus *Bactrocera*,

---

D. Wei • W. Dou • J. Wang (✉)
Key Laboratory of Entomology and Pest Control Engineering, College of Plant Protection, Southwest University, Chongqing 400716, China
e-mail: wangjinjun@swu.edu.cn

M. Jiang
Ministry of Agriculture Key Laboratory of Agricultural Entomology, Institute of Insect Sciences, Zhejiang University, Hangzhou 310058, China

F. Wan et al. (eds.), *Biological Invasions and Its Management in China*,
Invading Nature - Springer Series in Invasion Ecology 11,
DOI 10.1007/978-94-024-0948-2_15

Fig. 15.1 The oriental fruit fly adults on the tangerine (Shen et al. 2010)

which consists of at least 440 species, are distributed primarily in tropical Asia, Australia and the South Pacific. They feed inside fruits and vegetables, leading to both quantitative and qualitative losses, and their damage has enormous threats to fruit and vegetable production of the world (Clarke et al. 2005).

The oriental fruit fly, *Bactrocera dorsalis* (Hendel), is ranked as one of five most damaging and aggressive pest fruit flies in the world. Being highly polyphagous, it has been recorded from over 250 host plant species (Aketarawong et al. 2014). This fly can infest a wide variety of fruit crops, such as citrus, peach and mango, etc., causing significant economic losses through direct fruit damage, fruit drop and export limitations associated to quarantine restrictions (Fig. 15.1). It was first recorded in Taiwan Island, China, in 1912. Due to the species' broad host range, wide climate tolerance and high dispersal capacity, it has spread to the whole Asia-Pacific region in the last century, which ranges from India to Hawaii and encompasses all of South-East Asia (Wan et al. 2012). In China, the oriental fruit fly has three to eleven generations annually with most areas having four to eight generations (Wang et al. 2009). It has the potential of expanding to northern and southern cool areas of this country in the future.

15.2 Spread of *B. dorsalis* in China

After its first detection in Taiwan Island, the oriental fruit fly invaded Hainan Island, China in 1934. It was rarely found in other disjointed areas of southern China before the 1970s. Since the 1980s, however, its population size increased quickly and the distribution area expanded to cover most areas of southern China. In the last decade, the oriental fruit fly expanded across the Yangtze River, reaching 32°N.

Several methods have been used to predict the potential geographical distribution of oriental fruit fly in China. The analysis based on the CLIMEX model showed that the most suitable areas were located in South China, including Guangdong, Hainan, Guangxi Provinces and East China including Fujian, Zhejiang, and Shanghai, and the moderately suitable areas were in Southwest China, including Yunnan, Sichuan, Guizhou Provinces. In comparison, Hunan, Hubei, Jiangxi Provinces were not much suitable in most areas, and the north of the Yangtze River was unsuitable (Hou and Zhang 2005). Similar potential distribution range was suggested in the analyses using the GARP ecological niche modeling (Zhou et al. 2007) and emergence rate model combined with ArcGIS (Wang et al. 2009). By far, the oriental fruit fly has successfully invaded most of the above regions and provinces, even the areas in the north of the Yangtze River, such as Henan and Anhui Provinces. It has the potential of spreading to more northern areas.

Analyses based on 13 microsatellite loci showed that *B. dorsalis* in China originated from Thailand, Laos and Yunnan was the region to be invaded earliest, and Guangxi was probably also among the regions to be invaded early (Wu et al. 2011). High genetic differentiation was detected between various populations in China, suggesting that this insect should have two invasion routes in this country. One was in southeast China, from where it spread to Fujian, Taiwan, Guangdong and Hainan; the other was in southwest China, from where it spread to Sichuan, Yunnan, Guangxi, Guizhou and Hunan (Li et al. 2007; Wang et al. 2014c).

15.3 Major Factors Responsible for *B. dorsalis* Invasions in China

A number of factors are responsible for population growth of *B. dorsalis* in China and thus should have increased its invasive capacity. They factors include wide host range, powerful reproduction, adaptability to environmental physical stress, resistance to insecticides, and the habit of feeding inside hosts at larval stages to keep from natural enemies.

15.3.1 High Reproductivity

Female remating is a widespread phenomenon in insects, through which females can get more supplementary nutrients secreted by male accessory gland, thereby substantially increasing their fitness. In *B. dorsalis* females, there also exists a remating inhibition. Re-mated females with a remating refractory period would produce more offspring (Wei et al. 2015a). This may be one of major reasons for the rapid population development of *B. dorsalis* in the wild.

15.3.2 Adaptation to Environmental Stress

Living in a wide geographical range, *B. dorsalis* can adapt well to extremely temperatures. Eggs have a high level of tolerance to high temperatures (Li et al. 2013), and pupae can be tolerant to low temperatures (Wang et al. 2014b). The fruit fly would become more tolerant to low temperature at the pre-overwintering stage (Wang et al. 2014b). Larval host plant could influence the cold hardiness of next generation (Ren et al. 2006). The mechanisms underlying in response to extreme temperatures have been investigated. One study revealed that many oxidoreductases, binding proteins, and transferases were present abundantly in adults treated with extreme high and/or low temperatures, which gave physiological protection to adults (Wei et al. 2015b). Studies also proved that antioxidant enzymes, such as superoxide dismutase (SOD), probably play an important role in reducing oxidative damage in thermally stressed *B. dorsalis* (Gao et al. 2013; Jia et al. 2011).

B. dorsalis can also adapt to a wide range of humidity. For example, eclosion would not be delayed after third-instar larvae experience desiccation in dry soils (Xie and Zhang 2009). Larvae will reduce its weight within 2 h after treatment to reduce damage from desiccation. Most of pupae would survive and develop healthily at a broad moisture range 10–60% (Hou et al. 2006). Pupae are more resistant to water compared to larvae (Ren et al. 2007; Tian et al. 2005).

15.3.3 Resistance to Chemical Insecticides

The use of chemical insecticides is a primary method for the control of *B. dorsalis*. However, due to the long, highly frequent applications of certain chemicals, this insect has evolved high levels of insecticide resistance, which in turn increases destructive outbreaks (Jin et al. 2011b). For instance, in 2007 and 2008, *B. dorsalis* was detected to have developed a high level of resistance to trichlorfon in Guangdong and a high level of resistance to β-cypermethrin in Jiangsu (Jin et al. 2011a). Resistance to malathion, β-cypermethrin and abamectin has also occurred in recent years (Wang et al. 2013; Chen et al. 2015b). Moreover, high resistance to cyantraniliprole (a new anthranilic diamide insecticide) has been detected in Hubei (Zhang et al. 2014). Insecticide resistance increasingly poses a serious threat to current *B. dorsalis* control effort.

Intense efforts over the past several years have shed light on the insecticide resistance development and underlying biochemical and molecular mechanisms in this pest (Jin et al. 2012, 2014; Wang et al. 2016). The main mechanisms involve the increased detoxification by cytochrome P450 monooxygenases (P450s), glutathione S-transferases (GSTs) and carboxylesterase (CarEs), as well as increased insensitivity of acetylcholinesterase (AChE) (Hu et al. 2011, 2014; Huang et al. 2012, 2013; Jiang et al. 2014; Shen et al. 2012; Wang et al. 2013). Studies also suggest that the physiological resistance, e. g. high digestive enzymes activities, are also involved in resistance development in *B. dorsalis* (Jiang et al. 2014). Moreover, the cuticle thickness, laminated structure of the chitin layers, and interspace of epidermal cells might be correlated with cuticular penetration of insecticides (Lin et al. 2012). More recently, some resistance-related genes were functionally characterized using heterologous expression and RNA interference techniques (Huang et al. 2015; Lu et al. 2016; Wang et al. 2015, 2016). For example, three over-expressed esterases were found to be functional in malathion resistance in *B. dorsalis* (Wang et al. 2015, 2016).

15.4 Research Advances in Biology of *B. dorsalis* in China

In the past several decades, numerous papers have been published on oriental fruit fly's development, reproduction, chemical resistance, communication, and associations with symbionts. Below we present the major findings of these aspects in China.

15.4.1 Development

In order to investigate the regulation mechanism of *B. dorsalis* development, three Halloween genes involved in ecdysone biosynthesis have been cloned from *B. dorsalis*, namely *Cyp302a1*, *Cyp315a1* and *Cyp314a1* (Cong et al. 2015a). An ecdysone receptor gene, *EcR-B1*, has also been cloned from this insect (Cong et al. 2012); this receptor initializes the expression of ecdysone response gene cascades that ultimately affect cell proliferation, differentiation and apoptosis. Some of these genes are potential molecular targets for controlling this pest.

In insects, there are two types of chitin synthase (CHS), CHS1 and CHS2, the crucial enzymes of the final step of chitin bio-synthetic pathway. Both *CHS1* and *CHS2* were cloned in *B. dorsalis* (Chen et al. 2013; Yang et al. 2013a). *CHS1a*, one of two alternative splicing variants of CHS1, is mainly expressed during larval-pupal and pupal-adult transitions; depression its expression by RNAi can induce phenotypic defects and kill most of the treated larvae (Fig. 15.2) (Yang et al. 2013a). *CHS2* transcript was predominately found in larval midgut, and its expression is positively related with total chitin content during development. Genes of other critical enzymes involved in chitin biosynthesis were also cloned and characterized from the fly, such as glucose-6-phosphate isomerase and UDP-N-acetylglucosamine pyrophosphorylase. Depression of their expression by RNAi leads to death and abnormal phenotypes of larvae (Yang et al. 2015). Chitinase required for degradation of glycosidic bonds of chitin has also been identified and characterized in *B. dorsalis* (Yang et al. 2013b).

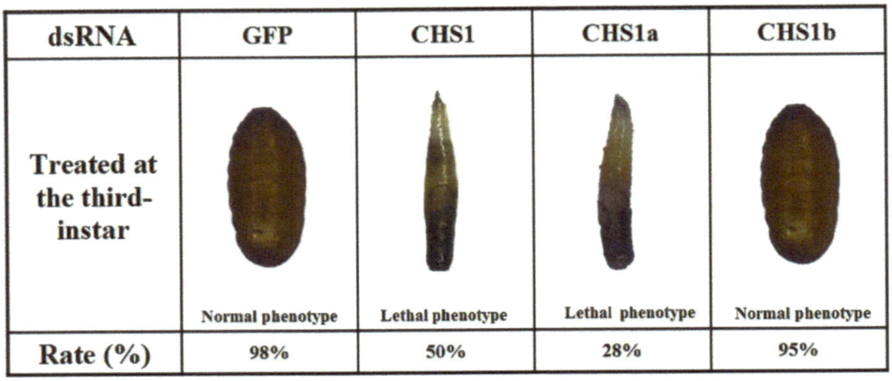

Fig. 15.2 Representative phenotypes of *Bactrocera dorsalis* after injection of day-2 third-instar larvae with *CHS1*, *CHS1a* and *CHS1b* dsRNA (Yang et al. 2013a) (Note: Injection of *CHS1* and *CHS1a* dsRNAs resulted in the same lethal phenotype: the larva was trapped in old cuticle and died without tanning completely, whereas insects injected with *CHS1b* dsRNA seemed normal)

A number of other genes crucial for *B. dorsalis* growth have been studied. For example, importance of the multicopper enzyme phenoloxidase (PO) and its zymogen for *B. dorsalis* growth was evaluated. When PO gene is inhibited, the development of larvae and pupae is significantly suppressed (Bai et al. 2014a, b). Copulated male and female adults have a significant higher PO activity in hemolymph than virgin ones (Shi et al. 2015). Insulin has been identified as a key player in insect metamorphosis, and six insulin signaling pathway components were identified in *B. dorsalis*. Suppression *IRS* expression will inhibit ovarian development (Xu et al. 2015). Three forkhead box (Fox) proteins gene were identified, *FoxL*, *FoxO* and *FoxP* (Zheng et al. 2012a). Fox O transcription factor (*FoxO*) is an important downstream transcription factor capable of regulating insect body size and development (Wu et al. 2016).

To facilitate molecular research into development mechanisms and develop effective control methods, transcriptome of *B. dorsalis* were extensively analyzed (Shen et al. 2011, 2013; Zheng et al. 2012a; Yang et al. 2014b). The digital gene expression libraries were also constructed to identify functional genes associated with *B. dorsalis* development. Now the genome of *B. dorsalis* has been sequenced and accessible in NCBI by far, which can be used combinatively with transcriptome in future studies.

15.4.2 Reproduction

In order to understand the reproductive process of *B. dorsalis* males, the transcriptomes of testis and male accessory glands (MAGs) were sequenced (Wei et al. 2015d, 2016). Many functional genes involved in spermatogenesis were identified, such as *testis-specific serine/threonine-protein kinases*, *Cyclin B*, *Neprilysin 4*, *ferritin 3*, *heat shock protein 70*. The MAGs, which have two types, mesodermal accessory glands and ectodermal accessory glands, are the source of a variety of secreted proteins and peptides that play critical roles in fertilization (Wei et al. 2015a). Many categories of genes involved in immunity and hormone are identified in MAGs (Wei et al. 2015c, 2016). Recently discovered post-transcriptional regulation directed by miRNAs can efficiently fine-tune the expression of target genes in certain types of cells or tissues. A number of specific testicular miRNAs have been identified in *B. dorsalis*, which provides an overview of their expression during spermatogenesis (Tariq et al. 2016b). Moreover, some of the novel miRNAs and target genes are predicted. One miRNA, miR-8-3p, was predicted to bind the 3'UTR of putative *B. dorsalis mitoferrin* and is essential for male fertility (Tariq et al. 2016a). The *maltase D* gene and D-glucose are also critical gene and substrate in male *B. dorsalis* mating process (Cheng et al. 2014).

Reproduction of *B. dorsalis* females has also been studied in terms of vitellogenin biosynthesis and deposition. Vitellogenin receptor is critical for yolk protein absorption and ovary maturation. This gene was cloned and characterized from *B. dorsalis*. Depression of its expression by RNAi leads to delay of ovary development, indicating a vital role in female reproduction (Cong et al. 2015b). Multiple differentially expressed genes, including antimicrobial peptides genes associated with sexual maturation and mating responsive genes involved in female reproduction, are also identified (Shi et al. 2015; Zheng et al. 2016b). These results are expected to facilitate molecular research on *B. dorsalis* reproduction and provide abundant target genes for effective control of this pest.

15.4.3 Responses to Host Volatiles

It has been proved that the fruits of mango, banana, guava, orange, and papaya can attract *B. dorsalis* (Jang et al. 1997; Ren et al. 2008; Shi et al. 2010). In this process, volatiles from ripening fruits, such as methyl eugnol, can enter into the sensillum of antennae (the primary sense organ) *via* numerous pores, combined by odorant binding proteins in the sensillum, and tranfered to the odorant receptor located on the sensory neuron. Using the RNAi method and electrophysiology technique, a diversity of proteins involved in this process have been identified from antennae, including odorant binding proteins (OBPs), chemosensory proteins (CSPs), odorant receptors (ORs), ionotropic receptors (IRs), sensory neuron membrane proteins (SNMPs) and odorant receptor co-receptor (ORCO) (Liu et al. 2016; Wu et al. 2015). As expected, silencing some of these genes (e.g., *OBPs*, *CSPs* and *ORCO*) would significantly decrease electrophysiological response of the antennae (Yi et al. 2014, 2016; Zheng et al. 2012b). Relative to these genes expressing mainly in the antennae, the other genes expressing both in classical olfactory and non-olfactory organs, or expressing only in non-olfactory organs such as heads, legs, and abdomens, is still elusive and should be deeply explored in the future.

15.4.4 Symbiotic Associations with Bacteria

Insects harbor diverse microorganisms in their intestinal tract and other tissues, some of which is vital for their lives. Using culture-dependent and -independent techniques, it was revealed that adult flies harbor a stable bacterial community in gut which is dominated by *Enterobacteriaceae* such as *Klebsiella*, *Citrobacter*, *Enterobacter* and *Pectobacterium*. The bacterial diversity could be influenced by different food supplies (Wang et al. 2011). These bacteria have an indirect contribution to host fitness by preventing the establishment or proliferation of pathogenic bacteria. Some bacteria, such as *Bacillus cereus* and *Enterococcus faecalis*, are able to secrete the substances capable of attracting fruit flies, suggesting that these bacteria might be potential bacterial biocontrol agents (Wang et al. 2014a). Depression of *Duox* gene would lead to an increased bacterial load and a decreased relative abundance of *Enterobacteriaceae* and *Leuconostocaceae* in the gut, suggesting that *Duox* plays a pivotal role in regulating intestinal bacterial community homeostasis of *B. dorsalis* (Yao et al. 2016). While in the reproductive system of female *B. dorsalis*, *Enterobacter sakazakii* and *Klebsiella oxytoca* are the dominant bacterial species (Shi et al. 2012).

15.5 Management

15.5.1 Quarantine

The oriental fruit fly is likely to spread into more northern and southern regions, thus it has been considered as an important quarantine pest in China. Quantitative risk assessment is an important part of integrated management of this pest. As this fly can be dispersed by fruit transportation, the fruits in the infected areas should be strictly quarantined before transported to uninfected regions. Three methods can be applied to detect *B. dorsalis*: morphological identification, acoustic detection, and molecular detection (Yang et al. 2014a). Thermal treatment should be performed for the fruits infected by *B. dorsalis*. After treatment at 60 °C for 2 h or at 45 °C for 5 h, all the fruit flies inside fruits would die (Zhan et al. 2010). This method can be applied at ports for quarantine treatment of infested fruits.

15.5.2 Monitoring and Baits-Based Physical Control

As a part of advanced integrated pest management (IPM) programs of fruits, an automatic infield monitoring system has been established. Methy lugenol, a sex attractant capable of trapping *B. dorsalis* males, is well used in this system (Lu et al. 2006). Many other technologies have also been applied in the system, such as the internet technology, e. g. web database, web servery, and a wireless sensor network, with the aim of developing a system for invasive pest risk analysis and forecast (Li and Zhang 2008; Zhang and Hou 2005; Wen et al. 2013). Such a system can assist farmers and pest control technicians to analyze the relations between population dynamics of the fruit fly and meteorological events and allow them to make a better risk assessment and develop a more optimal decision-making strategy for *B. dorsalis*.

A number of attractants have been developed for the control of this pest, such as cuelure, methoxybutyrophenone, isoeugenol, which are capable of attracting female adults (Wang 2013). This environment-friendly method can reduce the use of chemical pesticides (Jin et al. 2011a). The volatiles from some host plants have significantly high attractive efficacy to the oriental fruit fly. So far, the volatiles from mango (Shi et al. 2010), wax-apple (Jin et al. 2015), durian (Mo et al. 2014) and banana (Zhang et al. 2006) were demonstrated to have potential of being developed into effective attractants for *B. dorsalis* management. Sweeteners, a widely used food additives for human, have also been used to control insects. For *B. dorsalis*, the natural sweetener erythritol and the artificial sweeteners aspartame and saccharin were found to be effective in reducing fly survival (Zheng et al. 2016a). Some defensive chemicals known as repellent have anti-feedant and oviposition deterrence effects on *B. dorsalis*. Rhodojaponin-III, a diterpene compound isolated from the flowers of *Rhododendron molle*, has intense anti-feedant activities against *B. dorsalis* (Yi et al. 2013). Moreover, oil deposits may block the release of plant volatiles that attract female *B .dorsalis* to lay eggs (Ouyang et al. 2008). At present, the use of synergistic and insecticide with various attractants and repellents can significantly increase control efficacy and become the major methods for *B. dorsalis* control in China.

15.5.3 Biological Control

The current *B. dorsalis* management strategies relying on insecticide application leads to the development of resistance in this pest. Since the 1990s, biological control has been suggested as an alternative method for the *B. dorsalis* control (Zhang et al. 2010). An IPM system is currently being developed for *Bactrocera* spp. based on the utilization of parasitoids.

***Fopius arisanus* (Sonan)** is an egg-pupal parasitoid of tephritid fruit flies. It is an important biological control agent of these invasive and damaging pests around the world since it has the ability to parasitize diverse tephritid species during their early developmental stages (Manoukis et al. 2011), and it is the one of only three opiine parasitoids known to infect host eggs (Wang et al. 2004; Vargas et al. 2012). The major impediment to the wide use of *F. arisanus* for fruit fly control is that it is difficult to establish a stable laboratory colony. Yet, recently reliable methods to maintain laboratory populations have been developed (Manoukis et al. 2011).

Spalangia endius (Walker) (Hymenoptera: Pteromalidae) is a solitary pupal endoparasitoid widely distributed in many countries including China. It was first reported in 1939 as a parasitoid of *Musca domestica* L. pupae (Diptera, Muscidae), now it has been found to be capable of parasitizing a great number of Diptera in the families Anthomyiidae, Calliphoridae, Muscidae, Sarcophagidae, and Tephritidae (Zhao et al. 2016). In China, *S. endius* is one of the native parasitoids of *B. dorsalis*, found the earliest in 2009 in the fruit of carambola orchards in the Guangdong and Guangxi Provinces (Tang et al. 2015). It is a potential biological control agent of *B. dorsalis* with the advantages of low cost and easy rearing.

Beauveria bassiana (Sordariomycetes: Clavicipitaceae), an entomopathogenic fungus, is also an effective biological control agent of *B. dorsalis*. In a laboratory assay, mortality of 83–94% was observed in *B. dorsalis* adults when *B. bassiana* spore powder was sprayed on sand (Zhang et al. 2010).

15.5.4 Sterile Insect Technique

Sterile insect technique (SIT) is an environment-friendly, widely used control method against tephritid pests. This technique involves producing a large number of flies in the insectary, sterilizing the males (e.g., by irradiation), and then releasing the sterile males over infested areas where they mate with wild females (Ji et al. 2007b). If the sterile males overwhelmingly outnumber the fertile wild males, the wild fly population can be driven to extinction. For most SIT targets, developing an effective method for large-scale sex separation is a challenging step for the release of only males. In China, a genetic sexing strain for *B. dorsalis* based on pupal color has been developed (Ji et al. 2007a). This strain allows males to be separated from females at the pupal stage. Before the irradiated flies are released to the field, their quality has to be evaluated in terms of adult emergence rate, flight ability, mating performance and longevity of mass-reared males according to the International Quality Control Manual (FAO/IAEA/USDA 2003). Transgenic technology can also be used to produce male-only progeny, so it is a promising strategy for the development of male-only progeny strains (Liu et al. 2015).

15.5.5 RNA Interference

RNA interference (RNAi) is a powerful tool for sequence-specific gene silencing which is triggered by double-stranded RNA (dsRNA). Genetic flaw by RNAi can lead to death, molting deformity and reducing reproduction in some insects (Tariq et al. 2016a; Yang et al. 2013a). Hence, it can be used for the control of insect pests, as an environment-friendly tool. In adult *B. dorsalis*, *rpl19*, *v-ATPase-D*, *noa* and *rab11* could be silenced by feeding with corresponding dsRNA (Li et al. 2011). More potential target genes have been identified in this fly, such as the ones involved in midgut digestion and detoxification (Shen et al. 2013).

However, before RNAi is successfully used for the control of *B. dorsalis*, some difficulties need to be overcome. The biggest challenge is that there lacks an effective approach of delivering dsRNA to *B. dorsalis*. Moreover, there are risks to non-target organisms. For example, *rpl19* is a potential target gene for RNAi controlling of *B. dorsalis*, but the sequence-specific dsRNA may also depress the corresponding gene expression in some natural enemies and other beneficial insects (Chen et al. 2015a), possibly due to the high similarity in *rpl19* sequences between these insects and *B. dorsalis*. Thus, reducing impacts of dsRNA on non-target insects, as well as the fruit, vegetables hosts is one of major goals to be achieved in near future.

15.6 Conclusions and Future Directions

Since the invasion to mainland of China, oriental fruit fly has become one of most important pests in orchards. Moreover, currently it is spreading to more suitable areas of North China. In the past decades, many strategies have been developed to monitor its occurrence and spread and reduce its damage. Meanwhile, many researches have been carried out to elucidate its invasion mechanism. Further studies are needed to understand its adaptation to new environments, resistance to chemicals, and high reproductive ability. Exploration of these fields may give us clues to searching for novel control methods for this pest.

In China, the use of chemical insecticides remains the dominant way to control *B. dorsalis*. However, the chemical control efficacy has substantially reduced due to the development of chemical resistance in this pest. To resolve this problem, we have to take measures to delay development of chemical resistance, at the same time efforts are to be made to screening of new types of insecticides, such as botanical and microbial pesticides. Novel control strategies based on behavior mediation will be largely desired, after we know sufficiently about the physiological process and molecular basis of this pest's mate choice and mating behavior. SIT is a potential strategy to control *B. dorsalis*, but so far it has not been well established in China. This technique can be improved by sexually selected trait, a good indicator of male fitness and knowledge of sexual selection processes. RNAi is a promising tool for sustainable control of this fruit fly in the future; however, there will be a long way for application for the selected target genes, and there are also risks to fruit, vegetables, natural enemies and other non-target insects which should be evaluated further. Nevertheless, we should strive for these potential strategies. The last but most important is the quarantine method. All the fruit and vegetables must be quarantined strictly if they are transported from invaded to un-invaded regions, which is crucial for preventing *B. dorsalis* from spreading to new areas.

Most of the aspects described above are being extensively studied and will be directions of future researches in China. The ultimate goal will be to find safe, sustainable approaches capable of reducing *B. dorsalis* to a low density and thus to reduce its threat to fruit and vegetation production.

References

Aketarawong N, Guglielmino CR, Karam N, Falchetto M, Manni M, Scolari F, Gomulski LM, Gasperi G, Malacrida AR (2014) The oriental fruit fly *Bactrocera dorsalis* s.s. in East Asia: disentangling the different forces promoting the invasion and shaping the genetic make-up of populations. Genetica 142:201–213

Bai PP, Chen EH, Shen GM, Wei D, Wei DD, Wang JJ (2014a) Inhibition of phenoloxidase activity delays development in *Bactrocera dorsalis* (Diptera: Tephritidae). Fla Entomol 97:477–485

Bai PP, Xie YF, Shen GM, Wei DD, Wang JJ (2014b) Phenoloxidase and its zymogen are required for the larval-pupal transition in *Bactrocera dorsalis* (Diptera: Tephritidae). J Insect Physiol 71:137–146

Chen L, Yang WJ, Cong L, Xu K-K, Wang JJ (2013) Molecular cloning, characterization and mRNA expression of a chitin synthase 2 gene from the oriental fruit fly, *Bactrocera dorsalis* (Diptera: Tephritidae). Int J Mol Sci 14:17055–17072

Chen A, Zheng W, Zheng W, Zhang H (2015a) The effects of RNA interference targeting *Bactrocera dorsalis* ds-Bdrpl19 on the gene expression of rpl19 in non-target insects. Ecotoxicology 24:595–603

Chen LJ, Meng QQ, Li ZQ, Zhang SQ, Zeng L, Lu YY (2015b) The resistance monitoring of *Bactrocera dorsalis* in the field of Shenzhen. Chin Plant Protect 35:63–66

Cheng DF, Chen LJ, Yi CY, Liang GW, Xu YJ (2014) Association between changes in reproductive activity and D-glucose metabolism in the Tephritid fruit fly, *Bactrocera dorsalis* (Hendel). Sci Report 4. doi:10.1038/srep07489

Cong L, Yang WJ, Shen GM, Dou W, Wang JJ (2012) Molecular characterization of the cDNA encoding ecdysone receptor isoform β1 and its expression in the oriental fruit fly, *Bactrocera dorsalis* (Diptera: Tephritidae). Fla Entomol 95:650–658

Cong L, Jiang XZ, Yang WJ, Xu KK, Dou W, Ran C, Wang JJ (2015a) Identification of ecdysone synthesis pathway genes and analysis on the impact of food deprivation on larvae development of *Bactrocera dorsalis* Hendel. Sci Agric Sin 48:4469–4482

Cong L, Yang WJ, Jiang XZ, Niu JZ, Shen GM, Ran C, Wang JJ (2015b) The essential role of vitellogenin receptor in ovary development and vitellogenin uptake in *Bactrocera dorsalis* (Hendel). Int J Mol Sci 16:18368–18383

Clarke AR, Armstrong KF, Carmichael AE, Milne JR, Raghu S, Roderick GK, Yeates DK (2005) Invasive phytophagous pests arising through a recent tropical evolutionary radiation: the *Bactrocera dorsalis* complex of fruit flies. Annu Rev Entomol 50:293–319

FAO/IAEA/USDA (2003) Manual for product quality control and shipping procedures for sterile mass-reared tephritid fruit flies, version 5.0. IAEA, Vienna, Austria

Gao XM, Jia FX, Shen GM, Jiang HQ, Dou W, Wang JJ (2013) Involvement of superoxide dismutase in oxidative stress in the oriental fruit fly, *Bactrocera dorsalis*: Molecular cloning and expression profiles. Pest Manag Sci 69:1315–1325

Hou BH, Zhang RJ (2005) Potential distributions of the fruit fly *Bactrocera dorsalis* (Diptera:Tephritidae) in China as predicted by CLIMEX. Acta Ecol Sin 25:1569–1574

Hou B, Xie Q, Zhang R (2006) Depth of pupation and survival of the oriental fruit fly, *Bactrocera dorsalis* (Diptera: Tephritidae) pupae at selected soil moistures. Appl Entomol Zool 41:515–520

Hu F, Dou W, Wang JJ, Jia FX, Wang JJ (2011) Purification and biochemical characterization of glutathione *S*-transferases from four field populations of *Bactrocera dorsalis* (Hendel) (Diptera: Tephritidae). Arch Insect Biochem Physiol 78:201–215

Hu F, Dou W, Wang JJ, Jia FX, Wang JJ (2014) Multiple glutathione *S*-transferase genes: identification and expression in oriental fruit fly, *Bactrocera dorsalis*. Pest Manag Sci 70:295–303

Huang Y, Jiang HB, Shen GM, Dou W, Wang JJ (2012) Molecular characterizations of two cytochrome p450 genes encoding CYP6a41 and CYP6ek1 from the oriental fruit fly, *Bactrocera dorsalis* (Diptera: Tephritidae). Arch Insect Biochem Physiol 79:31–46

Huang Y, Shen GM, Jiang HB, Jiang XZ, Dou W, Wang JJ (2013) Multiple P450 genes: identification, tissue-specific expression and their responses to insecticide treatments in the oriental fruit fly, *Bactrocera dorsalis* (Hendel) (Diptera: Tephritidea). Pestic Biochem Physiol 106:1–7

Huang Y, Lu XP, Wang LL, Wei D, Feng ZJ, Zhang Q, Xiao LF, Dou W, Wang JJ (2015) Functional characterization of NADPH-cytochrome P450 reductase from *Bactrocera dorsalis*: possible involvement in susceptibility to malathion. Sci Report 5. doi:10.1038/srep18394

Jang EB, Carvalho LA, Stark JD (1997) Attraction of female oriental fruit fly, *Bactrocera dorsalis*, to volatile semiochemicals from leaves and extracts of a nonhost plant, panax (*Polyscias guilfoylei*) in laboratory and olfactometer assays. J Chem Ecol 23:1389–1401

Ji QE, Hou WR, Chen JH (2007a) Development of a genetic sexing strain and the sterile male technique of the oriental fruit fly, *Bactrocera dorsalis* (Hendel). Acta Entomol Sin 50:1002–1008

Ji QE, Hou WR, Chen JH (2007b) Sterile insect technique of oriental fruit fly, *Bactrocera dorsalis* (Hendel) - optimal pupal age and dose of irradiation treatment for male pupae. J Nucl Agric Sci 21:523–526

Jia FX, Dou W, Hu F, Wang JJ (2011) Effects of thermal stress on lipid peroxidation and antioxidant enzyme activities of oriental fruit fly, *Bactrocera dorsalis* (Diptera: Tephritidae). Fla Entomol 94:956–963

Jiang JJ, Zhou K, Liang GW, Zeng L, Wen SY (2014) A novel point mutation of acetylcholinesterase in a trichlorfon-resistant strain of the oriental fruit fly *Bactrocera dorsalis* (Diptera: Tephritidae). Appl Entomol Zool 49:129–137

Jin T, Zeng L, Lin YY, Lu YY, Liang GW (2011a) Insecticide resistance of the oriental fruit fly, *Bactrocera dorsalis* (Hendel) (Diptera: Tephritidae), in mainland China. Pest Manag Sci 67:370–376

Jin T, Zeng L, Lu YY, Lin YY, Liang GW (2011b) Changes of the frequency of resistant individuals in populations of the oriental fruit fly, *Bactrocera dorsalis* (Diptera: Tephritidae), with resistant individual flow. Acta Entomol Sin 54:306–311

Jin T, Zeng L, Lin YY, Lu YY, Liang GW (2012) Characteristics of protein variants in trichlorphon-resistant *Bactrocera dorsalis* (Diptera; Tephritidae) larvae. Genet Mol Res 11:2608–2619

Jin T, Liang GW, Zeng L, Lu YY (2014) Detoxification enzymes activities in different *Bactrocera dorsalis* (Hendel) populations and their relationship with the resistant levels. J Environ Entomol 36:58–67

Jin J, Ruan ZY, Huang ZF, Lai GY, Huang SS, Fan XL (2015) Attractions of volatiles from wax-apple fruit to the oriental fruit fly. J South China Agric Univ 36:71–77

Li BN, Zhang RJ (2008) A forecast system for invasive pest risk analysis and forecast based on web and MySQL: case study for *Bactrocera dorsalis*(Hendel). Acta Sci Natur Univ Sunyatseni 47:108–113

Li WF, Yang L, Tang K, Zeng L, Liang GW (2007) Microsatellite polymorphism of *Bactrocera dorsalis* (Hendel) populations in China. Acta Entomol Sin 50:1255–1262

Li XX, Zhang MY, Zhang HY (2011) RNA interference of four genes in adult *Bactrocera dorsalis* by feeding their dsRNAs. PLoS One 6:e17788

Li BS, Zhan GP, Wang YJ, Ren LL, Liu B, Li TX (2013) Thermal death kinetics of fruit flies *Bactrocera dorsalis*, *B. papayae* and *B. correcta* (Diptera: Tephritidae). Acta Entomol Sin 56:1404–1412

Lin Y, Jin T, Zeng L, Lu Y (2012) Cuticular penetration of beta-cypermethrin in insecticide-susceptible and resistant strains of *Bactrocera dorsalis*. Pestic Biochem Physiol 103:189–193

Liu G, Wu Q, Li J, Zhang G, Wan F (2015) RNAi-mediated knock-down of *transformer* and *transformer 2* to generate male-only progeny in the oriental fruit fly, *Bactrocera dorsalis* (Hendel). PLoS One 10:e0128892. doi:10.1371

Liu Z, Smagghe G, Lei Z, Wang JJ (2016) Identification of male- and female-specific olfaction genes in antennae of the oriental fruit fly (*Bactroscera dorsalis*). PLoS One 11: e0147783. doi:10.1371

Lu YY, Zeng L, Liang GW, Lin JT, Yu X, Xu YJ (2006) Improvement of the monitoring technigue of oriental fruit fly, *Bactrocera dorsalis*, males by sex attractant. Chin Bull Entomol 43:123–126

Lu XP, Wang LL, Huang Y, Dou W, Chen CT, Wei D, Wang J (2016) The epsilon glutathione *S*-transferases contribute to the malathion resistance in the oriental fruit fly, *Bactrocera dorsalis* (Hendel). Comp Biochem Physiol C 180:40–48

Manoukis N, Geib S, Seo D, McKenney M, Vargas R, Jang E (2011) An optimized protocol for rearing *Fopius arisanus*, a parasitoid of Tephritid fruit flies. J Vis Exp 53:2901. doi:10.3791/2901

Mo RJ, Ouyang Q, Zhong BE, Wu WJ (2014) Attractiveness of durian fruit volatiles to *Bactrocera dorsalis* (Hendel). Chin J Appl Entomol 51:1336–1342

Ouyang GC, Yang YP, Zhong GL, Xiong JJ, Huang MD, Liang GW (2008) Oviposition repellency and EAG responses of the oriental fruit fly, *Bactrocera dorsalis* (Hendel) to mineral spray oils. Acta Entomol Sin 51:390–394

Ren L, Lu YY, Zeng L, Pang ST (2006) Effect of different hostfruits on the cold hardiness of *Bactrocera dorsalis* (Hendel) (Diptera: Tephritidae). Acta Entomol Sin 49:447–453

Ren L, Lu YY, Zeng L (2007) Effect of sand water content on the pupal survival of *Bactrocera dorsalis* (Hendel). J South China Agric Univ 28:63–66

Ren LL, Qi LY, Jiang QG, Zhou SD, Dai HG (2008) Oviposition preference of oriental fruit fly, *Bactrocera dorsalis*. Chin Bull Entomol 45:593–597

Shen GM, Jiang HB, Wang XN, Wang JJ (2010) Evaluation of endogenous references for gene expression profiling in different tissues of the oriental fruit fly *Bactrocera dorsalis* (Diptera: Tephritidae). BMC Mol Biol 11:76

Shen GM, Dou W, Niu JZ, Jiang HB, Yang WJ, Jia FX, Hu F, Cong L, Wang JJ (2011) Transcriptome analysis of the oriental fruit fly (*Bactrocera dorsalis*). PLoS One 6:e29127

Shen GM, Wang XN, Dou W, Wang JJ (2012) Biochemical and molecular characterisation of acetylcholinesterase in four field populations of *Bactrocera dorsalis* (Hendel) (Diptera: Tephritidae). Pest Manag Sci 68:1553–1563

Shen GM, Dou W, Huang Y, Jiang XZ, Smagghe G, Wang JJ (2013) *In silico* cloning and annotation of genes involved in the digestion, detoxification and RNA interference mechanism in the midgut of *Bactrocera dorsalis* [Hendel (Diptera: Tephritidae)]. Insect Mol Biol 22:354–365

Shi W, Liu H, Ye H (2010) Behavioral response of *Bactrocera dorsalis* to five kinds of odor volatile of mango. Chin Bull Entomol 47:318–321

Shi Z, Wang L, Zhang H (2012) Low diversity bacterial community and the trapping activity of metabolites from cultivable bacteria species in the female reproductive system of the oriental fruit fly, *Bactrocera dorsalis* Hendel (Diptera: Tephritidae). Int J Mol Sci 13:6266–6278

Shi Z, Lin Y, Hou Y, Zhang H (2015) Humoral immunocompetence shifts in response to developmental stage change and mating access in *Bactrocera dorsalis* Hendel (Diptera: Tephritidae). Bull Entomol Res 105:166–172

Tang LD, Lu YY, Zhao HY (2015) Suitability of *Bactrocera dorsalis* (Diptera: Tephritidae) pupae for *Spalangia endius* (Hymenoptera: Pteromalidae). Environ Entomol 44:689–694

Tariq K, Metzendorf C, Peng W, Sohail S, Zhang H (2016a) miR-8-3p regulates mitoferrin in the testes of *Bactrocera dorsalis* to ensure normal spermatogenesis. Sci Rep 6:22565

Tariq K, Peng W, Saccone G, Zhang H (2016b) Identification, characterization and target gene analysis of testicular microRNAs in the oriental fruit fly *Bactrocera dorsalis*. Insect Mol Biol 25:32–43

Tian LJ, Liang GW, Zeng L, Lu YY, Bin SY (2005) Influence of soil moisture on the surviral of oriental fruit fly *Bactrocera (bactrocera)* dorsalis pupa. Chin Bull Entomol 42:416–418

Vargas RI, Leblanc L, Harris EJ, Manoukis NC (2012) Regional suppression of *Bactrocera* fruit flies (Diptera: Tephritidae) in the pacific through biological control and prospects for future introductions into other areas of the world. Insects 3:727–742

Wan XW, Liu YH, Zhang B (2012) Invasion history of the oriental fruit fly, *Bactrocera dorsalis*, in the Pacific-Asia region: two main invasion routes. PLoS One 7:e36176

Wang YL (2013) Study advance in incidence and trap control of *Bactrocera dorsalis* (Hendel). J Environ Entomol 35:253–259

Wang XG, Bokonon-Ganta AH, Ramadan MM, Messing RH (2004) Egg-larval opiine parasitoids (Hym., Braconidae) of tephritid fruit fly pests do not attack the flowerhead-feeder *Trupanea dubautiae* (Dipt., Tephritidae). J Appl Entomol 128:716–722

Wang N, Li Z, Wu J, Rajotte EG, Wan F, Wang Z (2009) The potential geographical distribution of *Bactrocera dorsalis* (Diptera: Tephrididae) in China based on emergence rate model and arcgis. Int Fed Inf Process 293:399–411

Wang H, Jin L, Zhang H (2011) Comparison of the diversity of the bacterial communities in the intestinal tract of adult *Bactrocera dorsalis* from three different populations. J Appl Microbiol 110:1390–1401

Wang JJ, Wei D, Dou W, Hu F, Liu WF, Wang JJ (2013) Toxicities and synergistic effects of several insecticides against the oriental fruit fly (Diptera: Tephritidae). J Econ Entomol 106:970–978

Wang HX, Jin L, Peng T, Zhang HY, Chen QL, Hua YJ (2014a) Identification of cultivable bacteria in the intestinal tract of *Bactrocera dorsalis* from three different populations and determination of their attractive potential. Pest Manag Sci 70:80–87

Wang JH, Zeng L, Han ZJ (2014b) An assessment of cold hardiness and biochemical adaptations for cold tolerance among different geographic populations of the *Bactrocera dorsalis* (Diptera: Tephritidae) in China. J Insect Sci 14

Wang M, Xu L, Zhang RJ, Yu DJ (2014c) Population genetic differentiation analysis of the oriental fruit fly, *Bactrocera dorsalis* (Diptera: Tephritidae) based on mtDNA COI gene. Acta Entomol Sin 57:1424–1438

Wang LL, Huang Y, Lu XP, Jiang XZ, Smagghe G, Feng ZJ, Yuan GR, Wei D, Wang JJ (2015) Overexpression of two α-esterase genes mediates metabolic resistance to malathion in the oriental fruit fly, *Bactrocera dorsalis* (Hendel). Insect Mol Biol 24:467–479

Wang LL, Feng ZJ, Li T, Lu XP, Zhao JJ, Niu JZ, Smagghe G, Wang JJ (2016) Inheritance, realized heritability, and biochemical mechanisms of malathion resistance in *Bactrocera dorsalis* (Diptera: Tephritidae). J Econ Entomol 109:299–306

Wei D, Feng YC, Wei DD, Yuan GR, Dou W, Wang JJ (2015a) Female remating inhibition and fitness of *Bactrocera dorsalis* (Diptera: Tephritidae) associated with male accessory glands. Fla Entomol 98:52–58

Wei D, Jia FX, Tian CB, Tian Y, Smagghe G, Dou W, Wang JJ (2015b) Comparative proteomic analysis of *Bactrocera dorsalis* (Hendel) in response to thermal stress. J Insect Physiol 74:16–24

Wei D, Li HM, Tian CB, Smagghe G, Jia FX, Jiang HB, Dou W, Wang JJ (2015c) Proteome analysis of male accessory gland secretions in oriental fruit flies reveals juvenile hormone-binding protein, suggesting impact on female reproduction. Sci Report 5. doi:10.1038/srep16845

Wei D, Li HM, Yang WJ, Wei DD, Dou W, Huang Y, Wang JJ (2015d) Transcriptome profiling of the testis reveals genes involved in spermatogenesis and marker discovery in the oriental fruit fly, *Bactrocera dorsalis*. Insect Mol Biol 24:41–57

Wei D, Tian CB, Liu SH, Wang T, Smagghe G, Jia FX, Dou W, Wang JJ (2016) Transcriptome analysis to identify genes for peptides and proteins involved in immunity and reproduction from male accessory glands and ejaculatory duct of *Bactrocera dorsalis*. Peptides 80:48–60

Wen T, Hong T, Li L, Li Z, Ye Z, Zhang Y (2013) Experiment and development of monitoring system for *Bactrocera dorsalis* (Hendel) based on wireless sensors network. Nongye Gongcheng Xuebao 29:147–154

Wu ZZ, Li HM, Bin SY, Shen JM, He HL, Luo M, Ma J, Lin JT (2011) Analysis of genetic diversity o fdifferent populations of *Bactroceradorsalis* (Diptera:Tephritidae) using microsatellite marker. Acta Entomol Sin 54:149–156

Wu Z, Zhang H, Wang Z, Bin S, He H, Lin J (2015) Discovery of chemosensory genes in the oriental fruit fly, Bactrocera dorsalis. PLoS One 10. doi:10.1371/journal.pone.0129794

Wu YB, Yang WJ, Xie YF, Xu KK, Tian Y, Yuan GR, Wang JJ (2016) Molecular characterization and functional analysis of *BdFoxO* gene in the oriental fruit fly, *Bactrocera dorsalis* (Diptera: Tephritidae). Gene 578:219–224

Xie Q, Zhang R (2009) Responses of oriental fruit fly (Diptera: Tephritidae) third instars to desiccation and immersion. J Agric Urban Entomol 24:1–11

Xu KK, Yang WJ, Tian Y, Wu YB, Wang JJ (2015) Insulin signaling pathway in the oriental fruit fly: the role of insulin receptor substrate in ovarian development. Gen Comp Endocrinol 126:125–133

Yang WJ, Xu KK, Cong L, Wang JJ (2013a) Identification, mRNA expression, and functional analysis of chitin synthase 1 gene and its two alternative splicing variants in oriental fruit fly, *Bactrocera dorsalis*. Int J Biol Sci 9:331–342

Yang WJ, Xu KK, Zhang RY, Dou W, Wang JJ (2013b) Transcriptional regulation of a chitinase gene by 20-hydroxyecdysone and starvation in the oriental fruit fly, *Bactrocera dorsalis*. Int J Mol Sci 14:20048–20063

Yang L, Huang W, Huang L, Wang W, Jiang J (2014a) Research and development on invaded insect *Bactrocera dorsalis* for detection method. J Agric 4:19–21

Yang WJ, Yuan GR, Cong L, Xie YF, Wang JJ (2014b) *De novo* cloning and annotation of genes associated with immunity, detoxification and energy metabolism from the fat body of the oriental fruit fly, *Bactrocera dorsalis*. PLoS One 9:e94470

Yang WJ, Wu YB, Chen L, Xu KK, Xie YF, Wang JJ (2015) Two chitin biosynthesis pathway genes in *Bactrocera dorsalis* (Diptera: Tephritidae): molecular characteristics, expression patterns, and roles in larval-pupal transition. J Econ Entomol 108:2433–2442

Yao ZC, Wang AL, Li YS, Cai ZH, Lemaitre B, Zhang HY (2016) The dual oxidase gene *BdDuox* regulates the intestinal bacterial community homeostasis of *Bactrocera dorsalis*. ISME J 10:1037–1050

Yi X, Zhao H, Dong X, Wang P, Hu M, Zhong G (2013) BdorCSP2 isimportant for antifeed and oviposition-deterring activities induced by Rhodojaponin-III against *Bactrocera dorsalis*. Plos One 8:e77295

Yi X, Zhao HM, Wang PD, Hu MY, Zhong GH (2014) Bdor\Orco is important for oviposition-deterring behavior induced by both the volatile and non-volatile repellents in *Bactrocera dorsalis* (Diptera: Tephritidae). J Insect Physiol 65:51–56

Yi X, Wang PD, Wang Z, Cai J, Hu MY, Zhong GH (2016) Involvement of a specific chemosensory protein from *Bactrocera dorsalis* in perceiving host plant volatiles. J Chem Ecol 42:461–461

Zhan GH, Chu CR, Chen YF, Fan XH, Gao Y, Hu BH, Lin YJ (2010) Study on quarantine and trapping technique for *Bactrocera dorsalis*. J Southwest For Univ 30:1–4

Zhang RJ, Hou BH (2005) Assessment on the introduction risk of *Bactrocera dorsalis* (Hendel) through imported fruits with fuzzy mathematics. Acta Entomol Sin 48:221–226

Zhang SY, Xiao C, Ye M, Kang M, Wu SR, Hu CH (2006) Attraction of the volatiles from ripen banana pulp for *Dacus dorsalis* (Hendel). J Huazhong Agric Univ 25:512–514

Zhang YP, Huang SH, Li DS, Zhang BX, Chen MY (2010) Control effect of *Beauveria bassiana* B6 Strain on *Bactrocera dorsalis* (Hendel). Chin J Biol Control 26:14–18

Zhang RM, He SY, Chen JH (2014) Monitoring of *Bactrocera dorsalis* (Diptera: Tephritidae) resistance to cyantraniliprole in the south of China. J Econ Entomol 107:1233–1238

Zhao HY, Lu YY, Liang GW (2016) Preference of the parasitoid *Spalangia endius* Walker to its hosts, *Bactrocera dorsalis* and *B. cucurbitae*. J Biosafety 25:35–38

Zheng WW, Peng T, He W, Zhang HY (2012a) High-throughput sequencing to reveal genes involved in reproduction and development in *Bactrocera dorsalis* (Diptera: Tephritidae). PLoS One 7:e36463

Zheng WW, Zhu CP, Peng T, Zhang HY (2012b) *Odorant receptor co-receptor Orco* is upregulated by methyl eugenol in male *Bactrocera dorsalis* (Diptera: Tephritidae). J Insect Physiol 58:1122–1127

Zheng CY, Zeng L, Xu YJ (2016a) Effect of sweeteners on the survival and behaviour of *Bactrocera dorsalis* (Hendel) (Diptera: Tephritidae). Pest Manag Sci 72:990–996

Zheng W, Luo D, Wu F, Wang J, Zhang H (2016b) RNA sequencing to characterize transcriptional changes of sexual maturation and mating in the female oriental fruit fly *Bactrocera dorsalis*. BMC Genomics 17. doi:10.1186/s12864-12016-12532-12866

Zhou GL, Chen C, Ye J, Hu BS, Liu FQ (2007) Predicting potential ecological distribution of *Bactrocera dorsalis* in China using GARP ecological niche modeling. Acta Ecol Sin 27:3362–3369

Chapter 16
Codling Moth *Cydia pomonella* (L.)

Maohua Chen, Xinle Duan, Yuting Li, Qiulei Men, and Fanghao Wan

Abstract The codling moth, *Cydia pomonella* (L.), is one of the highly invasive pests that devastates pome and walnut fruit trees worldwide. It has been listed in the most harmful quarantine species in China. Since the first report of this species in northwestern China in 1957, *C. pomonella* spreads rapidly from Xinjiang Province to Gansu Province, Ningxia Province, and other fruit production regions in the northwest. This species was found in some regions in northeastern China since 2006. *C. pomonella* not only causes serious damages to fruit production, but also influences the export of fruits from China. In this chapter, we review the biology, damage, distribution and invasion history of the codling moth. The invasion sources, population genetics, monitor and control the codling moth in China was also summarized in this chapter.

Keywords Codling moth • Invasive species • Distribution • Population genetic • Control strategy

16.1 Introduction

The codling moth *Cydia pomonella* (L.) is a serious insect pest of deciduous fruits. It occurs in most temperate regions of the world, causing severe damages on apple *Malus domestica* Borkh., *M. pumila* Miller, pear *Pyrus pyrifolia* (Burm. F. Nak.), *Pyrus communis* L., apricot *Prunus armeniaca* (L.), peach *Prunus persica var. persica* Batsch, plum *Prunus salicina,* Nectarine, *Prunus persica* var.*nucipersica* (Suckow), and walnut *Juglans regia* L. (Barnes 1991; Willett et al. 2009; Vreysen et al. 2010).

M. Chen (✉) • X. Duan • Y. Li • Q. Men
College of Plant Protection, Northwest A&F University, Key Laboratory of Crop Pest Integrated Pest Management on the Loess Plateau of Ministry of Agriculture, Yangling, Shaanxi 712100, China
e-mail: maohua.chen@nwsuaf.edu.cn

F. Wan
State Key Laboratory for Biology of Plant Diseases and Insect Pests, Institute of Plant Protection, Chinese Academy of Agricultural Sciences, Beijing 100193, China

F. Wan et al. (eds.), *Biological Invasions and Its Management in China*, Invading Nature - Springer Series in Invasion Ecology 11,
DOI 10.1007/978-94-024-0948-2_16

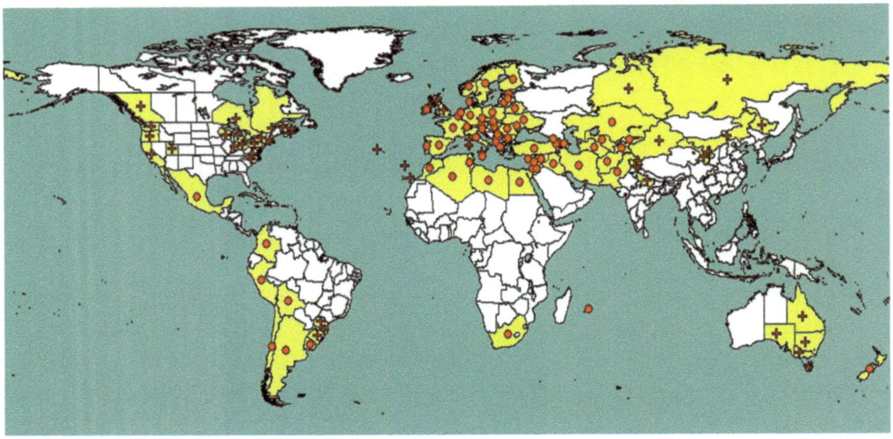

Fig. 16.1 World distribution of *Cydia pomonella* OEPP/EPPO Electronic Knowledge base (http://www.eppo.int/) ● Present (National record), ▲ Present (subnational record), + Transient

The species was considered to be native to southeastern Europe. During the last two centuries, it rapidly spreads to nearly all apple growing areas around the world, largely owing to the quick growth of pome fruit cultivation and human activity-mediated dispersal (Boivin et al. 2004; Franck et al. 2007; Meraner et al. 2008; Thaler et al. 2008). Currently, the codling moth is mainly distributed in the areas between 30°N and 30°S, such as North America, South America, South Africa, Europe, Australia, New Zealand, and other apple and pear growing area, only except Eastern China, Japan and Korean peninsula (Lin and Lin 1996; Franck et al. 2007) (Fig. 16.1).

C. pomonella is listed as a quarantine insect in many countries. In China, it not only causes serious damages to fruits, but also influences the fruit export from China. Consequently, it is considered to be one of most serious invasive species in the country.

16.2 Invasion History and Geographical Distribution of *Cydia pomonella* in China

In China, the moth was first reported in 1957 in Korla region of Xinjiang (Zhang 1957). Since then, it has been the major pest in northwest China. This species spent nearly 50 years to spread the whole fruit production area of Xinjiang (Zhang 1957), and took about 30 years to pass the mountains, deserts and unpopulated areas and reach the neighboring Hexi Corridor of Gansu Province, and invade Dunhuang in 1990, Jiuquan in 1994, Zhangye in 2003, Jinchang in 2005, Wuwei in 2007 and Lanzhou in 2008. Meanwhile, the codling moth penetrated in Inner Mongolia Province and Ningxia Province during 2006–2008 (Qin et al. 2006; Bahatiguli 2009; Wang and Wang 2009). In 2006, new invasion sites were found in Heilongjiang Province, the first sport reported in northeastern China (Qin et al. 2006). Although

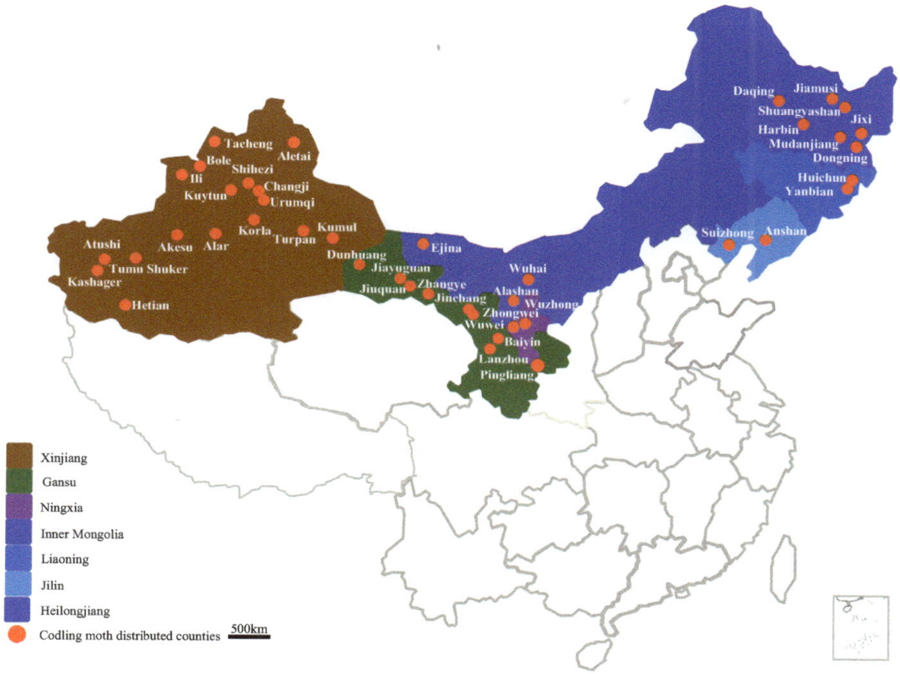

Fig. 16.2 The major distribution regions of the *Cydia pomonella* in China

this species is distributingin the northwest and northeast of China, but no occurrence of it has been recorded in the large areas (about 3000–5000 km away) between the invaded northwestern and northeastern regions despite continuous monitoring.

According to the data derived from the Agricultural Ministry of China, the codling moth is distributed in 136 counties of Xinjiang, Gansu, Ningxia, Inner Mongolia, Liaoning, Jilin and Heilongjiang provinces in China (Fig. 16.2).

16.3 Biology and Damage in China

In China, the codling moth has two to three generations per year (Zhang 1957; Qin et al. 2006, Bahatiguli 2009). The codling moth over-winters as a mature larva in a dense, silken cocoon found under loose bark or in debris underneath the tree. In early spring, the larvae pupate when the temperature reaches 10 °C. Overwintered adults occur in May and the first generation adults occur in June to August depending on climatic conditions (temperature and humidity) and fruit development period. Usually, the first generation causes damages from May to Jul. Females deposit 30–70 eggs on the fruit or nearby leaves and branch. Young larvae bore into the fruit and eat the flesh and pips after eggs hatch, but the rain depresses the development of the population. When the larvae complete the development, they leave the fruit and find the suitable place such as loose scales of bark, soil or debris around the tree base or in trash on the ground near the trunk for pupation. The codling moth enters facultative diapause during which they can tolerate low temperature. The environment of main apple producing areas in China is suitable for the colonization of codling moth (Yang 2008).

On apples, the codling moth often causes damage more seriously than any other insect pests. Fruits are the part mostly affected by the moth, on which it stings and bores deep inside. In addition, it can also feed on leaves and bore into twigs. With the damage of codling moth, the tissue was eaten away and the stings would be greatly enlarged. Finally, larvae would leave small holes for escape; and such holes on the surface of fruits with frass are the distinguishing characteristics of codling moth infestation. Generally, the second generation larvae lead to serious damage. If the control strategies are not timely applied, the moth would cause 30% ~50% product reduction, even 80% lose in certain years in some regions (Kovačević 1952; Ciglar 1998).

China is a major apple producer as well as a major apple consumer in the world (FAO 2007). In 2008, the apple production of China was 29, 846, 600 tons, accounting for 42.88% of the world's total apple production. However, the effected area by the codling moth is nearly 32,700 hectares, resulting in the loss over $157 million. As the codling moth has various host plants and high reproductive ability, as well as the larvae of codling moth bear into the fruit, it is difficult to control it in fruit orchards.

16.4 Invasion Sources and Population Genetics of *Cydia pomonella* in China

Population genetics of new invasive species is related to its invasion history (Grapputo et al. 2005; Bohonak 1999;Watts et al. 2010). The level of genetic diversity and genetic structure for the newly invasive populations are affected by the recent expansion of the species, possible founder effect and dispersal meanings (Ramstad et al. 2004; Dlugosch and Parker 2008; Watts et al. 2010). Information on population genetics of invasive species is important, because it would allow us to understand evolutionary processes after invasions, illuminate mechanisms of population expansion and the gene flow between different environments, which is important for effective control of invasive species (Wright 1951; Zhang and Kang 1999; Miller et al. 2003; Wan et al. 2005; Endersby et al. 2006; Chu et al. 2006).

As a notoriously invasive pest, the codling moth brings serious economic and ecological threats to the production of pomes (primarily apple and pear) in northwestern and northeastern parts of China. Since this species was reported in China, it has caused high economic losses every year (Wan et al. 2005). The invasion sources and population genetics of the pest in these distribution areas of China were still unclear. In Europe and Africa, the genetic diversity and population structure of *C. pomonella* population were reported (Timm et al. 2006; Franck et al. 2007; Fuentes-Contreras et al. 2008; Chen and Dorn 2010; Franck and Timm 2010) and these populations from different countries had different population genetics. In China, the population genetics of codling moth were investigated with microsatellites (Men et al. 2013) and mitochondrial genes (COI, COII and Cytb) (Li et al. 2015).

16.4.1 Invasion Sources

So far, the codling moth is mainly distributed in two provinces in the northwest China (Xinjiang and Gansu) and one province in the northeast China (Heilongjiang). Molecular markers including microsatellites and mitochondrial genes revealed that populations of the moths in northwestern and northeastern China had different population structure and possibly had different invasion sources.

For invasive species, genetic diversity is expected to decrease with range expansion and colonization of new areas via sequential founder populations.(Dlugosch and Parker 2008). Based on microsatellites, the Ili population from Xinjiang Province showed highest genetic diversity in northwest China and Dongning population from Heilongjiang Province had high genetic diversity in northeast China. This indicates that the Ili region and Dongning regions were the primary invasive regions in northwestern and northeast China, respectively (Men et al. 2013). In fact, the geographical distance between Xinjiang and Heilongjiang is 3000–5000 km. Therefore, the natural dispersal of codling moth between northeastern and northwestern distribution region is impossible. Furthermore, the mitochondrial genes revealed that *C. pomonella* populations from northeastern China have more haplotypes and higher nucleotide diversities, proportions of individuals with different haplotypes, proportions of individuals with private haplotypes and proportions of private haplotypes than populations from northwestern China (Li et al. 2015). All these results implied that the northeastern and northwestern populations have different invasion sources.

Based on genetic analyses, northwest populations were quite different from those from northeast China, but the northeast populations had a similar genetic structure with the western European populations (Men et al. 2013; Li et al. 2015). As Ili neighbors to the important central Asia land port, Horgost port, the *C. pomonella* populations in the northwestern China might come from central Asia (Men et al. 2013). Similarly, as there is Dongning port, an import land port near the far eastern region of Russia where the codling moth has been documented (Willett et al. 2009), the invasive sources of this moth in northeastern Heilongjiang may come from the Russian far eastern region (Li et al. 2015).

16.4.2 Population Genetics

The genetic diversity and genetic structure of an invasive species were influenced by not only original populations but also the topography, climate, host plant and other environment factors, as well as the management methods of newly invaded regions (Chen and Dorn 2010). Estimating rates of gene flow among populations can provide an estimate of the rate of invasions or reinvasions of insect populations (Krafsur 2005; Chen and Dorn 2010).

16.4.2.1 Genetic Diversity

Based on the analyses of the number of alleles, the observed heterozygosity, allelic richness and other genetic diversity indices, microsatellite markers revealed Dongning and Mudanjiang populations in northeast China had a higher genetic diversity than the populations in northwest (Xinjiang and Gansu). In Xinjiang, northern populations (Ili, Jinghe, Kuytun and Urumqi) had higher genetic diversity than southern populations (Kashgar and Korla). All the populations in the Hexi Corridor had a similar genetic diversity with the population from the adjacent region of Xinjiang (Kumul) (Men et al. 2013).

Similar to results from microsatellite analyses, results based on mitochondrial genes showed that northeast populations had higher genetic diversity than northwest populations. The northwest populations had similar haplotype number and haplotype diversity, whilst the northeast population had more haplotypes and haplotype diversity. Both microsatellites and mitochondrial genes indicated that the northeast populations had similar genetic diversity with the populations from Europe and Africa countries (Li et al. 2015).

16.4.2.2 Genetic Structure

With the Bayesian clustering approach based on microsatellte data, the codling moth populations in China were divided into four different clusters and the populations in the same cluster were geographical separated (Fig. 16.3). Cluster 1 included the populations from southern Xinjiang and one population from Gansu. Cluster 2 contained the populations from northern Xinjiang. Cluster 3 mainly consisted of two populations from Gansu and one population from the boundary region between Gansu and Xinjiang. Cluster 4 was mostly made-up of populations from northeastern China and the two populations from Western Europe (Men et al. 2013). Similarly, the AMOVA results based on mitochondrial gene data showed no significant genetic variance among the populations from Xinjiang and Gansu, but significant genetic variance was found between the populations from northeastern and northwestern China, as well as between northeastern China, northwestern China and European countries (Li et al. 2015).

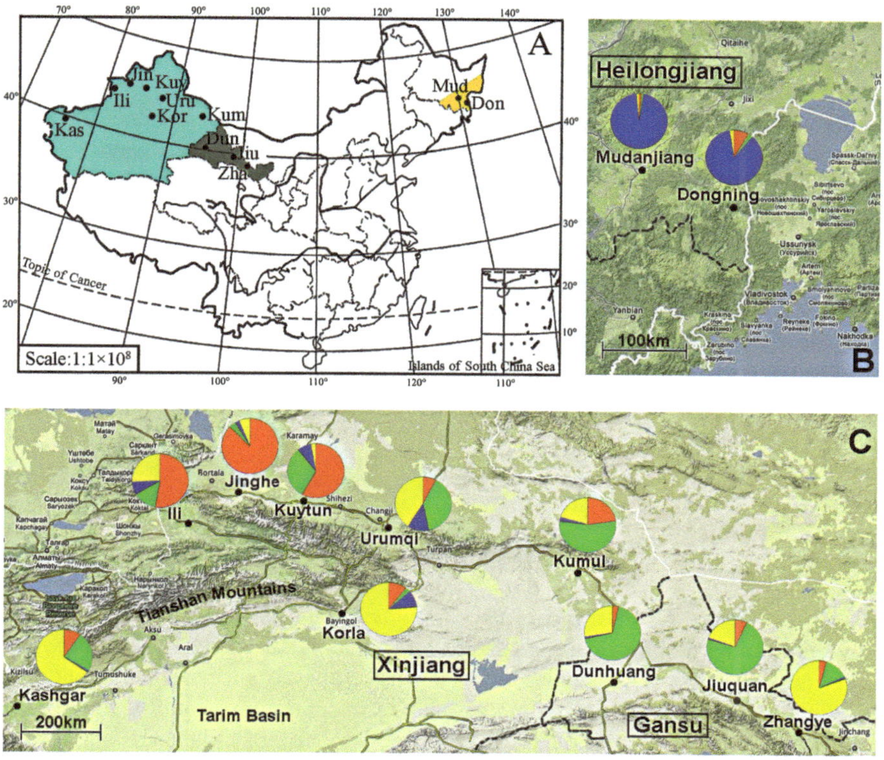

Fig. 16.3 (**a**). Distributed regions and sample locations of 12 *Cydia pomonella* populations; (**b**). Sample locations of the populations from Heilongjiang and the mean membership fractions to each of the four cluster in the populations; (**c**). Sample locations of the populations from Xinjiang and the mean membership fractions to each of the four cluster in the population. Population codes are the same as Table 1 (Refer to Men et al. 2013)

The codling moth is a sedentary insect with low flight capacity (Keil et al. 2001; Timm et al. 2006; Thaler et al. 2008). It completes the entire life cycle on one host plant, the flight distance of the most adult moth was only 50 m and the widest range of movement was only 0.5 ha (Mani and Wildbolz 1977). Therefore, significant genetic differentiation of the codling moth can occur at a small geographic scale, as observed in South Africa (Timm et al. 2006), Italy (Thaler et al. 2008) and Switzerland (Chen and Dorn 2010). However, there was no isolation-by-distance effect among populations from northwest populations and no significant genetic variance was detected among these populations. Similarly, Timm et al. (2006) reported little genetic differentiation among several populations with 150 km geographical distance. It is likely that human activities had led to the similarity of population genetic structure by mixing the individuals from distant geographic regions (Higbee et al. 2001).

16.5 Monitor and Control the Codling Moth in China

As the codling moth has various host plants, high reproductive ability and the larvae bore into fruits, it is difficult to control. Using broad-spectrum insecticides is the major control method. Due to the multiple use of chemicals, the codling moth has developed resistance to various groups of synthetic insecticides all over the world (Franck et al. 2007; Mota-Sanchez et al. 2008; Reyes et al. 2009; Rodríguez et al. 2010; Asser-Kaiser et al. 2011). For this reason, much attention would have to be given to the integrated pest management (IPM) strategy, by combining plant quarantine, filed monitor, cultural control, insecticides and biocontrol. Of these methods, the biocontrol, which is environmental friendly, has been much developed, such as using pheromones (attract-and-kill method and mating disruption), sterile insect technique and nature enemy (Ciglar 1998; Maceljsk 2002; Lacey and Unruh 2005; Shen et al. 2012).

16.5.1 Plant Quarantine and Population Forecasting

The codling moth is one of most important quarantine species in China. This species had low flight ability but the eggs or larvae can be dispersed with transport of fruits, nursery materials or human artifacts over a long distance. Plant quarantine is an effective method to stop the human-aided transportation of the codling moth. Chinese government forbids the import of fresh pome fruits (apple, pear, apricot and quince) and walnut fruits from the distribution areas of the species. The fruits, packaging and filling materials from the distribution areas should be treated with strict quarantine programs. Generally, methyl bromide fumigation, low oxygen-controlled atmosphere, high temperature with low oxygen pulse and other quarantine measures must be used to eradicate the moth (Wan et al. 2005). Meanwhile, the monitor in the distribution areas and threatened areas should be intensive in the apple, pear, apricot and other pome fruit orchards during April and September.

Population forecasting is important for an integrated pest management program for the moth (Balazs et al. 1996; Buban et al. 1996; Laurent 1997). Through pheromone traps and field investigation, the population dynamics of the codling moth was investigated making control strategy of the pest in China (Zhang et al. 2012). In orchard, the degree days, sunset temperatures would be used to calculate the egg hatch and proper spray timing when the codling moth was primarily managed with insecticides and pheromone traps. Compared with the climate, development periods of moth and the phenology of host plants, the duration prediction method and phenological forecasting method could predict the peaking time of adult stage and damage period of the moth, which was helpful to confirm the suitable control time and methods (He et al. 2014).

16.5.2 Cultural Control

The basic principal in orchard management is to keep fruit orchards tidy and clean. Management of the fruit distribution and interplant of the non-host plants of the codling moth between orchards could prohibit the migration of the species. The codling moth-infested fruits often drop from trees early in the season. Removing these fruits from the orchards will prevent larvae maturing in these dropped fruits. Picking infested fruits off is also an effective control measure, but it is only practical in small orchards. Moreover, cloth rings and paper cardboard can be used to trap larvae that overwinter or oversummer. In most apple growing regions of China, the fruit-bagging technology was widely used to control the fruit-borers including the codling moth. In small organic orchards, the infected fruits were often removed and the dropped fruits were collected and removed from the orchards.

16.5.3 Chemical Control

Broad spectrum insecticides are often used to control the codling moth. As this pest has several generations per season and has a high reproductive ability, chemical insecticides have to be applied several times during the growing season (Lacey et al. 2008). Extensive and frequent insecticide application has caused serious resistance of the codling moth to more than sixty insecticides in the world (Croft and Riedl 1991; Thwaite et al. 1993; Blomefield 1994; Sauphanor et al. 1998; Reyes et al. 2004; Ioriatti et al. 2007; Stará and Kocourek 2007; Mota-Sanchez et al. 2008; Schmitt et al. 2013). Based on the data of Insecticide Resistance Action Committee, the codling moth developed resistance to all groups of insecticides including organophosphates, carbamates, pyrethroids, insect growth regulator, abamectin and CpGV and cross-resistance between different types of insecticides (IRAC 2014). To reduce the selective pressure and avoid the development of resistance, insecticides with different action mode can be mixed or rotated in control of the moth. For example, organophosphates or carbamates can be mixed with insect growth regulators (Reyes and Sauphanor 2008). Biological insecticides such as avennectins, Bt and CpGV can also be mixed or rotated with chemical insecticides (Lacey et al. 2004;Yang et al. 2011). Organophosphates, carbamates, pyrethroids are used to control the codling moth in China. So far, monitoring the resistance level of the pest in different regions of China is rare.

16.5.4 Biological Control

Several biological control methods were already applied in the orchards, such as mating disruption, predators, parasitoids and entomopathogens. Mating disruption by synthetic pheromones is the preferred and powerful tool because of its environmental safety. Pheromone traps were used to monitoring the moth, but the mating disruption was not widely used in China (Yan et al. 1999).

Numerous entomopathogens have been reported from the codling moth, such as virus, bacteria, fungus, protozoa, microsporida and nematode. However, only the codling moth granulovirus (CpGV) and entomopathogenic nematodes (EPNs) have been developed as microbial control agents. CpGV was widely used in different countries of the world (Lacey and Unruh 2005), buy not yet in China.

Augmentative releases of the egg parasite *Trichogramma platneri* have been applied to reduce codling moth populations, and three codling moth parasitoids (*Bassus rufipes*, *Mastrus ridibundus* and *Liotryphon caudatus*) were tested to control the codling moth. These parasitoids and many predators could be used for biological control of the codling moth, although there were no commercial products of them at the moment (Nangong et al. 2014).

16.5.5 Genetic Control

Genetic control of pest insects represents a promising alternative to chemical control in terms of safety, specificity, and its limited negative environmental impact (Hoy 2003). The sterile insect technology (SIT) is a highly species-specific and environmental friendly technique available to eradicate insect pests if applied consistently on an area-wide basis, but the control effect would be influenced by the population genetics of different populations (Klassen 2005; Vreysen et al. 2010). This technology is compatible with other pest control methods, and can therefore be effectively integrated with other methods including biological methods, such as parasitoids, predators and insect pathogens. The successful suppression of the codling moth in the Okanagan Valley of British Columbia, Canada in the last 20 years is one of the most successful area-wide IPM programs that integrated the SIT with other control tactics (Klassen 2005; Bloem et al. 2007; Vreysen et al. 2010). SIT had not been used in the control of the codling moth in China (Liu et al. 2012).

16.6 Conclusions

The codling moth had a nearly global distribution due to its high potential of adaptation, being one of the most successful pest insect species known today (Thaler et al. 2008). This invasive species is distributed in seven provinces (Xinjiang, Gansu, Ningxia, Inner Mongolia, Liaoning, Jilin and Heilongjiang) in northeastern and northwestern China, and it is one of the most import quarantine pest in China (Wan et al. 2005). Since the first report in China, the moth has been the key pest of fruit production in the northwest of China (Zhang 1957).

By combining microsatellite and mitochondrial genetic markers, we concluded that the coding moth in China originally came from Central Asia and Russia. The invasion sources of *C. pomonella* populations in the northwestern China might come from central Asia. The origin of *C. pomonella* populations in the northeastern China was not from northwestern China, but likely came from Russia. The research results of population genetics of the codling moth are of important for the control of the codling moth in China. The central and local governments should strengthen the inspection and quarantine of the pest in the potential invasive regions and prevent both international and domestic introductions. Introducing the nature enemies from the origin area of the moth is an effective and eco-friendly way for the management. In addition, the ecotypes and the evolutionary factors in different environments should be considered in the IPM of the pest.

References

Asser-Kaiser S, Radtke P, El-Salamouny S, Winstanley D, Jehle JA (2011) Baculovirus resistance in codling moth (*Cydia pomonella* L.) caused by early block of virus replication. Virology 410(2):360–367

Ayres RM, Pettigrove VJ, Hoffmann AA (2010) Low diversity and high levels of population genetic structuring in introduced eastern mosquitofish (*Gambusia holbrooki*) in the greater Melbourne area, Australia. Biol Invasions 12:3727–3744

Bahatigul I (2009) The damage and control of Cydia pomonella in Xinjiang Province. Prot For Sci Technol 4:118–120

Balazs K, Bujaki G, Farkas (1996) Incorporation of apples clearwing (Synanthedon myopaeformis, Bork) control into the IPM system of apple. Acta Hortic 422:134–139

Barnes MM (1991) Codling moth occurrence, host race formation and damage. In: van der Geest LPS, Evenhuis HH (eds) Tortricid pests: their biology, natural enemies and control, 9 July, 1991. Elsevier, Amsterdam, pp 313–327

Bloem S, Carpenter J, Hofmeyr H (2007) Area-wide control tactics for the false codling moth Thaumatotibia leucotreta in South Africa: a potential invasive species. In: MJB V, Robinson AS, Hendrichs J (eds) Area-wide control of insect pests. From research to field implementation. Springer, Dordrecht, pp 351–359

Blomefield T (1994) Codling moth resistance: is it here, and how do we manage it? Deciduous Fruit Grower 44:130–132

Bohonak AJ (1999) Dispersal, gene flow and population structure. Q Rev Biol 74:21–45

Boivin T, Bouvier JC, Beslay D, Suphanor B (2004) Variability in diapause propensity within populations of a temperate insect species: interactions between insecticide resistance genes and photoperiodism. Biol J Linn Soc 83:341–351

Buban T, Inantsy F, Kajati I, Molnar M, Sallai P, Szoeke L, Lantos J (1996) Experiences with integrated pest management in apple orchard during the initial phase of a long term study. Acta Hortic 422:102–106

Chen MH, Dorn S (2010) Microsatellites reveal genetic differentiation among populations in an insect species with high genetic variability in dispersal, the codling moth, *Cydia pomonella* (L.) (Lepidoptera: Tortricidae). Bull Entomol Res 100:75–85

Chu D, Liu GX, Fan ZX, Tao YL, Zhang YJ (2006) Genetic differentiation of different geographical populations of *Bemisia tabaci* (Gennadius) complex. Sci Agric Sin 39(8):1571–1580

Ciglar I (1998) Integrirana zaštita voćaka i vinove loze. Zrinski, Čakovec

Croft BA, Riedl HW (1991) Chemical control and resistance to pesticides of the codling moth. In: Van Der Geest LPS, Evenhuis HH (eds) World crop pests, Tortricid pests: their biology, natural enemies and control, vol 5. Elsevier, Amsterdam, pp 371–387

Dlugosch KM, Parker IM (2008) Founding events in species invasions: genetic variation, adaptive evolution, and the role of multiple introductions. Mol Ecol 17:43–449

Endersby NM, Mckechnie SW, Ridland PM, Weeks AR (2006) Microsatellites reveal a lack of structure in Australian populations of the diamondback moth, Plutella xylostella (L.). Mol Ecol 15(1):107–108

FAO (2007) FAO statistical database. [Online] Available: http://faostat.fao.org

Franck P, Reyes M, Olivares J, Sauphanor B (2007) Genetic differentiation in the codling moth: comparison between microsatllite and insecticide resistant markers. Mol Ecol 16(17):3554–3564

Franck P, Timm AE (2010) Population genetic structure of Cydia pomonella: a review and case study comparing spatiotemporal variation. J Appl Entomol 134(3):191–200

Fuentes-Contreras E, Espinoza JL, Lavandero B, Ramírez CC (2008) Population genetic structure of codling moth (Lepidoptera: Tortricidae) from apple orchards in central Chile. J Econ Entomol 101(1):190–198

Grapputo A, Boman S, Lindstroem L, Lyytinen A, Mappes J (2005) The voyage of an invasive species across continents: genetic diversity of North American and European Colorado potato beetle populations. Mol Ecol 14(14):4207–4219

He SW, Da SC, Gao YM, Wu D, Wang ZH (2014) Research on the forecasting and control time of codling moth. Mod Agric Technol 5:169

Higbee BS, Calkins CO, Temple CA (2001) Overwintering of codling moth (Lepidoptera: Tortricidae) larvae in apple harvest bins and subsequent moth emergence. J Econ Entomol 94:1511–1517

Hoy MA (2003) Insect molecular genatics, an introduction to principes and applications, 2nd edn. Academic Press, Elsevier Science

Insecticide Resistance Action Committee (2014) [Online] Available: http://www.irac-online.org/pests/cydia-pomonella/

Ioriatti C, Tasin M, Charmillot PJ, Reyes M, Sauphanor B (2007) Early detection of resistance to tebufenozide in field populations of *Cydia pomonella* L.: methods and mechanisms. J Appl Entomol 131(7):453–459

Keil S, Gu HN, Dorn S (2001) Response of *Cydia pomonella* to selection on mobility: laboratory evaluation and field verification. Ecol Entomol 26:495–501

Klassen W (2005) Area-wide integrated pest management and the sterile insect technique. In: Dyck VA, Hendrichs J, Robinson AS (eds) Sterile insect technique. Principles and practice in area-wide integrated pest management. Springer, Dordrecht, pp 43–76

Kovačević Ž (1952) Applied entomology. In: Agricultural pests, 2nd edn. University of Zagreb, Zagreb

Krafsur ES (2005) Role of population genetics in the sterile insect technique. In: Dyck VA, Hendrichs J, Robinson AS (eds) Sterile insect technique. Principles and practice in area-wide integrated pest management. Springer, Dordrecht, pp 389–406

Lacey LA, Unruh TR (2005) Biological control of codling moth (Cydia pomonella, Lepidoptera: Tortricidae) and its role in integrated pest management, with emphasis on entomopathogens. Vedalia 12(1):33–60

Lacey LA, Arthurs SP, Thomson D, Fritts RJ, Granatstein D (2004) Codling moth granulovirus and insect specific nematodes for control of codling moth in the Pacific Northwest. Tilth Prod Q 13:10–12

Lacey LA, Tomson D, Vincent C, Arthurs SP (2008) Codling moth granulovirus: a comprehensive review. Biocontrol Sci Tech 18(7):639–663

Laurent P (1997) Codling moth notes I. To know the codling moth better. Fruit-Belge 65:33–36

Li YT, Duan XL, Qiao XF, Li XY, Wang K, Men QL, Chen MH (2015) Mitochondrial DNA revealed the extent of genetic diversity and invasion origin of populations from two separate invaded areas of a newly invasive pest, *Cydia pomonella* (L.) (Lepidoptera: Tortricidae) in China. Bull Entomol Res. Available on CJO2015. doi:10.1017/S0007485315000334

Lin W, Lin CJ (1996) The function of the ecology index in the geographical distribution of codling moth. Plant Quarantine 1:1–7

Liu W, Xu J, Zhang RZ (2012) Advances in the sterile insect technique for controlling codling moth. Chin J Appl Entomol 49(1):268–274

Maceljski M (2002) Poljoprivredna entomologija, 2nd edn. Zrinski, Čakovec

Mani E, Wildbolz T (1977) The dispersal of male codling moths (*Laspeyresia pomonella* L.) in the Upper Rhine Valley. J Appl Entomol 47:39–48

Men QL, Chen MH, Zhang YL, Feng JN (2013) Genetic structure and diversity of a newly invasive species, the codling moth, Cydia pomonella (L.)(Lepidoptera: Tortricidae) in China. Biol Invasions 15(2):447–458

Meraner A, Brandstätter A, Thaler R, Aray B, Unterlechner M, Niederstätter H, Parson W, Zelger R, Dalla Via J, Dallinger R (2008) Molecular phylogeny and population structure of the codling moth (*Cydia pomonella*) in Central Europe: I. Ancient clade splitting revealed by mitochondrial haplotype markers. Mol Phylogenet Evol 48:825–837

Miller NJ, Birley AJ, Overall ADJ, Tatchell GM (2003) Population genetic structure of the lettuce root aphid, *Pemphigus bursarius* (L.), in relation to geographic distance, gene flow and host plant usage. Heredity 91(3):217–223

Mota-Sanchez D, Wise JC, Poppen RV, Gut LJ, Hollingworth RM (2008) Resistance of codling moth, Cydia pomonella (L.)(Lepidoptera: Tortricidae), larvae in Michigan to insecticides with different modes of action and the impact on field residual activity. Pest Manag Sci 64(9):881–890

Nangong ZY, Song P, Wang YJ, Wang QY (2014) Advances in the research and application of biological control of the codling moth, Cydiapomonella (Lepidoptera: Tortricidae). Chin J Biol Control 30(2):260–265

Qin XH, Ma DC, Zhang Y, Li GH, Wang P (2006) The damage and development of *Cydia pomonella* in the northwest of China. Plant Quarantine 2:95–96

Ramstad KM, Woody CA, Sage GK, Allendorf FW (2004) Founding events influence genetic population structure of sockeye salmon (*Oncorhynchus nerka*) in Lake Clark, Alaska. Mol Ecol 13(2):277–290

Reyes M, Sauphanor B (2008) Resistance monitoring in codling moth: a need for standardization. Pest Manag Sci 64(9):945–953

Reyes M, Bouvier JC, Boivin T, Sauphanor B, Fuentes CE (2004) Susceptibilidad a insecticidas y actividad enzimática en *Cydia pomonella* L. (Lepidoptera: Tortricidae) proveniente de tres huertos de manzano de la regi´ on del Maule, Chile. Agricultura T'ecnica 64:229–237

Reyes M, Franck P, Olivares J, Margaritopoulos J, Knight A, Sauphanor B (2009) Worldwide variability of insecticide resistance mechanisms in the codling moth, *Cydia pomonella* L. (Lepidoptera: Tortricidae). Bull Entomol Res 99(4):359–369

Rodríguez MA, Bosch D, Sauphanor B, Avilla J (2010) Susceptibility to organophosphate insecticides and activity of detoxifying enzymes in Spanish populations of *Cydia pomonella* (Lepidoptera: Tortricidae). J Econ Entomol 103(2):482–491

Sauphanor B, Bouvier JC, Brosse V (1998) Spectrum of insecticide resistance in *Cydia pomonella* (Lepidoptera: Tortricidae) in south-eastern France. J Econ Entomol 91:1225–1231

Schmitt A, Bisutti IL, Ladurner E, Benuzzi M, Sauphanor B, Kienzle J, Jehle JA (2013) The occurrence and distribution of resistance of codling moth to *Cydia pomonella* granulovirus in Europe. J Appl Entomol 137(9):641–649

Shen JR, Liu WX, Wan FH, Zhang FQ (2012) Characterization of the *Cydia pomonella* granulovirus CpGV-CJ01 from northwest China. Chin J Appl Entomol 49(1):96–103

Stara J, Kocourek F (2007) Insecticidal resistance and cross-resistance in populations of *Cydia pomonella* (Lepidoptera: Tortricidae) in Central Europe. J Econ Entomol 100(5):1587–1595

Thaler R, Brandstätter MA, Chabicovski M, Parson M, Zelger R, Dalla Via J, Dallinger R (2008) Molecular phylogeny and population structure of the codling moth (*Cydia pomonella*) in Central Europe: II. AFLP analysis reflects human-aided local adaptation of a global pest species. Mol Phylogenet Evol 48(3):838–849

Thwaite WG, Williams DG, Hately AM (1993) Extent and significance of azinphos-methyl resistance in codling moth in Australia. Pest Control Sustain Agric 93:166–168
Timm AE, Geertsema H, Warnich L (2006) Gene flow among *Cydia pomonella* (Lepidoptera: Tortricidae) geographic and host populations in South Africa. J Econ Entomol 99(2):341–348
Vreysen MJB, Carpenter JE, Marec F (2010) Improvement of the sterile insect technique for codling moth *Cydia pomonella* (Linnaeus) (Lepidoptera Tortricidae) to facilitate expansion of field application. J Appl Entomol 134(3):165–181
Wan FH, Zheng XB, Guo JY (2005) Biology and management of invasive alien species in agriculture and forestry. Science Press, Beijing
Wang CL, Wang FX (2009) Advances in interception and control of the codling moth. Plant Prot 35(2):102–103
Watts PC, Keat S, Thompson DJ (2010) Patterns of spatial genetic structure and diversity at the onset of a rapid range expansion: colonization of the UK by the small red-eyed damselfly Erythromma viridulum. Biol Invasions 12:3887–3903
Willett MJ, Neven L, Miller CE (2009) The occurrence of codling moth in low latitude countries: validation of pest distribution reports. HortTechnology 19(3):633–637
Wright S (1951) The genetical structure of populations. Ann Eugen 15:323–354
Yan F, Bengtsson M, Witzgall P (1999) Behavioral response of female Codling Moths, *Cydia Pomonella*, to apple volatiles. J Chem Ecol 25(6):1343–1351
Yang R (2008) Study on the habitat suitability of Cydia pomonella (L.) in China. Ph. D thesis. Yangling, Northwest A&F University
Yang JQ, Zhao R, Yan YG, Zhang YL, Feng JN (2011) Efficacy of seven kinds of pesticides for controlling codling moth *Cydia pomonella*. Acta Agric Boreali-occidentalis Sinica 20(9):194–196
Zhang XZ (1957) The first record of Cydia pomonella in China. Acta Entomol Sin 4:467–472
Zhang RZ, Kang L (1999) The disaster mechanism and control strategy of the invasive pests. Entomol Knowl 36(3):181–183
Zhang RZ, Wang FX, Zhang YL, Chen HJ, Luo JC, Wang QY, Liu WX, Ainiwaer M, Pu CJ, Yan YG, Guo JM, Liu XY, Chen JG, Zhang ZF, Yang S, Xu JJ, Cui GZ, Xu J (2012) Progress on monitoring and control of the codling moth, Cydia pomonella (L.). Chin J Appl Entomol 49(1):37–42

Chapter 17
Red Imported Fire Ant *Solenopsis invicta* Buren

Lei Wang and Yongyue Lu

Abstract Since it was found in Taiwan and mainland, the spread of red imported fire ant *Solenopsis invicta* Buren into China has been continuous. In order to achieve better control of this pest, science and technology research on fire ants has advanced from many aspects. This paper presents the history of fire ant invasion and spread, estimated origins and social forms, present distribution, and potential expansion in China. Also the invasion biology and ecology is discussed in view of ecological impacts of the invasion. Management of *S. invicta* in China was implemented as soon as it was detected. Besides biological and physical control, monitoring, quarantine, and chemical control were widely studied and employed. Although a lot of work was done, the fire ant spread in China remains and will proceed rapidly in the future, mainly due to poor quarantine.

Keywords Bioinvasions • Red imported fire ant • Myrmecology • Control • Transported goods

17.1 Introduction

The red imported fire ant (RIFA), *Solenopsis invicta* Buren, is a successful invasive ant around the world (Vinson 1997) listed among the 100 World's worst invasive alien species (IUCN 2014). Native to inner South America, *S. invicta* had already invaded about 20 countries and ultramarine territories by the end of January, 2015, including the United States, Mexico, West Indies, China, Australia, and New Zealand (Wetterer 2013; Fig. 17.1).

S. invicta was introduced to the United States between 1933 and 1945, around Mobile, AL, and occupied over 367 million acres in Dec., 2014 in that country

Foundation:National Research and Development Program (2016YFC1201204)

L. Wang • Y. Lu (✉)
Red Imported Fire Ant Research Center, South China Agricultural University,
Guangzhou 510642, China
e-mail: luyongyue@scau.edu.cn

F. Wan et al. (eds.), *Biological Invasions and Its Management in China*,
Invading Nature - Springer Series in Invasion Ecology 11,
DOI 10.1007/978-94-024-0948-2_17

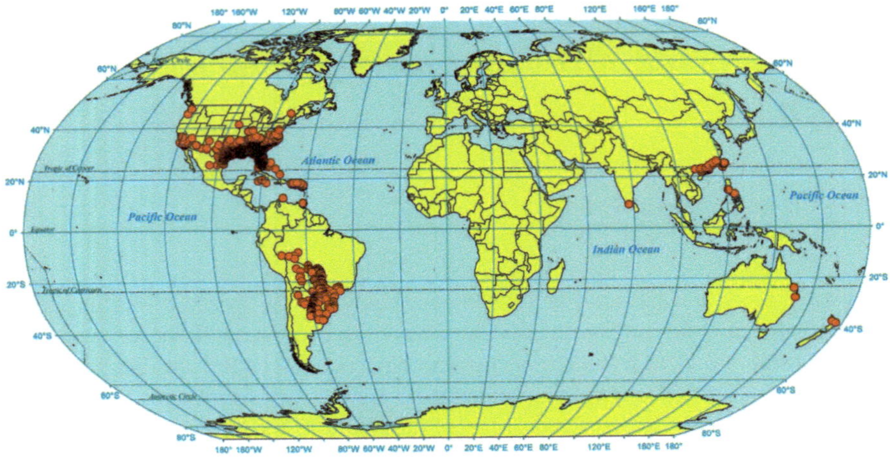

Fig. 17.1 Worldwide distribution of *Solenopsis invicta* (Wetterer 2013) (Note: *Red dots* represent fire ant infestation; records from India and Philippines need to be confirmed)

(USDA, APHIS). The fire ants invasion reduced biological diversity (Wojcik et al. 2001) mainly by destroying ground inhabiting arthropods, and caused heavy agricultural losses (Lard et al. 2002). *S. invicta* is well known for its painful sting, and some people are allergic to its venom (Prahlow and Barnard 1998). It is estimated that more than 200,000 persons per year may require medical treatment for fire ant stings (Williams et al. 2003). The estimated cost for the control, medical treatment, and damage to property in the US is greater than $6 billion annually (Lard et al. 2006).

In 2001 *S. invicta* was detected in Brisbane, Australia, and the infestation area covered over 300 km^2 (Scanlan and Vanderwoude 2006). The National Fire Ant Eradication Program began in September 2001 with a budget of AUD$150 million to eradicate the fire ant by 2006 (Moloney and Vanderwoude 2003). However it was still ongoing in 2014 (Mcnaught et al. 2014) at which the budget had hit 250 million AUD$. It is predicted that *S. invicta* may infest 0.763–4.066 million km^2 by 2035 and reach 200 separate locations around Australia between 2017 and 2027 if no control measures take place (Scanlan and Vanderwoude 2006).

In China, *S. invicta* was first reported in Taoyuan County, Taiwan in October 2003, and subsequent surveillance estimated the infestation area as 40 km^2 (Yang et al. 2013). In Mainland China the pest was confirmed in September 2004 in Wuchuan, Guangdong affecting about 20 km^2 (Zeng et al. 2005). By the end of 2014, the red imported fire ant had already infested 217 counties within 12 provinces of China (Lu 2014; Lu and Zeng 2015, Fig. 17.2), and in 2015 it had expanded to more than 250 counties spanning over 10,000 km^2. More than 1/3 of the human population in the infested areas was eventually stung by fire ants, and around 10% of victims experienced fever, with some reporting dizziness, generalized urticarial, or anaphylactic shock (Xu et al. 2012).

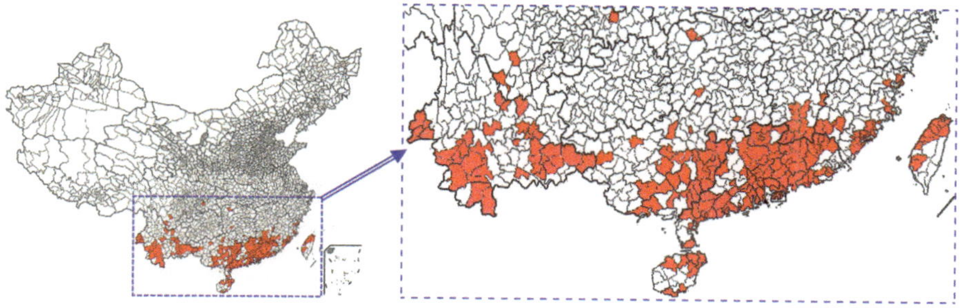

Fig. 17.2 Distribution of *Solenopsis invicta* in China in 2014 (Note: The *red* areas represent fire ant infested territories at a national level)

17.2 Invasion and Spread of *S. invicta* in China

17.2.1 Origin of S. invicta *in China*

Once found in China, learning the introduction route of *S. invicta* became paramount for prevention and management. It was suggested that the *S. invicta* in Taiwan had arrived in two separate introductions, based on microsatellite genotypic data and mtDNA sequences (Yang et al. 2008). Analysis of COI gene of 13 fire ant populations, three haplotypes (I–III) previously reported from Argentina were identified in China, with only Hong Kong population presenting the three haplotypes at the same time (He et al. 2006). Thus, *S. invicta* in China may have come directly from South America near Argentina into Hong Kong as the origin of the spread to other regions of China (He et al. 2006). However, based on the molecular biology data, some claim that the red imported fire ants in Wuchuan, Guangdong seem to originate from the southern United States (Zeng et al. 2005; Yang et al. 2012). Ascunce et al. (2011) also suggested *S. invicta* in China probably came from southern U.S. after assessing the variation of microsatellite markers, inside mtDNA genome and Gp-9 gene among 2144 fire ant colonies from 75 geographic sites both from South America and introduced ranges. Therefore, such results suggest that the origin of *S. invicta* in China appears to be complex and in need of further clarification.

17.2.2 Social Form of S. invicta *in China*

The two social forms of *S. invicta* (monogyne and polygyne colonies) present different biology, such as the number of inseminated queens inside each colony, their modes of colony founding and spread, the potential threat to biodiversity, and territorial organization (Porter and Savignano 1990; Ross and Keller 2003). As such, different monitoring and controlling methods should be adopted according to the different social forms (Drees and Vinson 1990). Both monogyne and polygyne forms of *S. invicta* are found in China (Shao et al. 2008; Chen et al. 2015), with polygynous colonies dominating most of the invaded regions (Zeng et al. 2005; Shao et al. 2008).

17.2.3 Potential Distribution of S. invicta *in China*

The determination of distribution of *S. invicta* in China is important for quarantine. Based on the main biological parameters and climatic variables (Killion and Grant 1995), several models have been used to predict the potential distribution of *S. invicta* in China. The results indicated that all southern regions of the Yangtze River are relative suitable for *S. invicta* (Fig. 17.3). It also predicts there is high possibility of ants reaching the south of Hebei and Shanxi and the north of Anhui and Jiangsu (Xue et al. 2005; Shen et al. 2008; Chen et al. 2006; Du et al. 2007). The range of *S. invicta* in the United States is expected to expand across the next century because of global warming (Korzukhin et al. 2001). In China, a warming trend was confirmed in the past 20 years (Yu et al. 2011), therefore, expansion of *S. invicta* may also continue in the next several decades, and the range limit could be beyond the predicted north line in China.

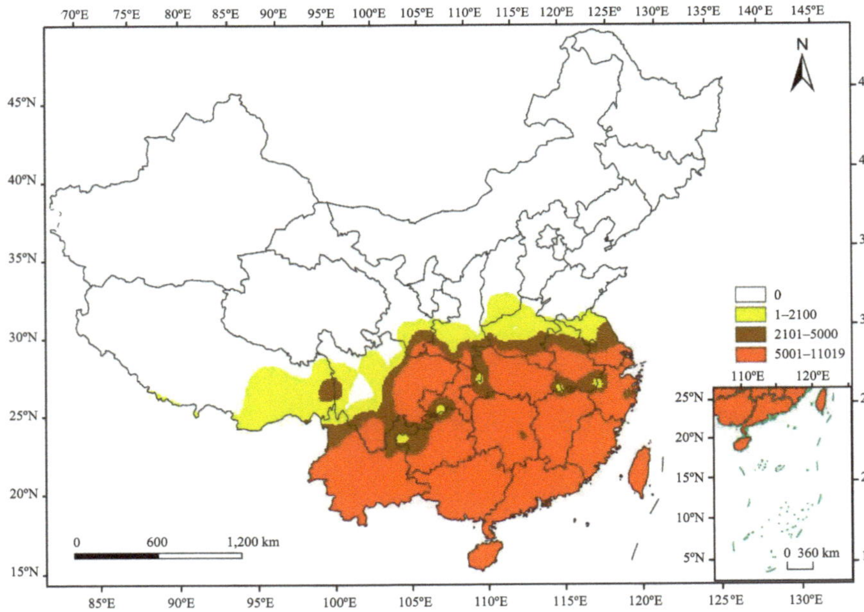

Fig. 17.3 Potential distribution of *Solenopsis invicta* in China based on Climax model (Chen et al. 2006)

17.2.4 Expansion of S. invicta *in China*

S. invicta reaches new ranges via human transportation and also self-dispersal. In a river ecosystem, the natural expansion of *S. invicta* was found to be as slow as 32.7 m–471.2 m per year (Xu et al. 2006). However, the spread can achieve a speed of 26.5–48.1 km per year in China through transportation of turf, nursery stock, waste materials, and others (Lu et al. 2008; Lu 2014; Lu and Zeng 2015). Although it was first identified in September 2003 in China, *S. invicta* has probably been in this country since as early as mid-1990s (Fig. 17.4a) (Lu et al. 2008). In 2005, 40 counties from 7 provinces reported *S. invicta* infestation (Zhang et al. 2007; Lu et al. 2008; Yang et al. 2013; Fig. 17.4b). In 2013, numbers reached more than 200 counties from 12 provinces (Kuo et al. 2010; Lu 2014; Fig. 17.4c), and the ants became widely distributed into Guangdong, south of Fujian and Guangxi (Lu and Zeng 2015). *S. invicta* is expected to spread at a higher speed in the future (Lu 2014; Lu and Zeng 2015, Fig. 17.5) .

17.3 Invasion Biology and Ecology of *S. invicta* in China

17.3.1 Biology of S. invicta

In southern China the population and breeding peaks of *S. invicta* occur in spring and autumn (Xu et al. 2009a). Nuptial flights can take place throughout the year, while 70 ~ 80% take place in spring and summer, being the most frequent just after rainy days in spring (Xu et al. 2009b). Foraging activity of *S. invicta* proved the most

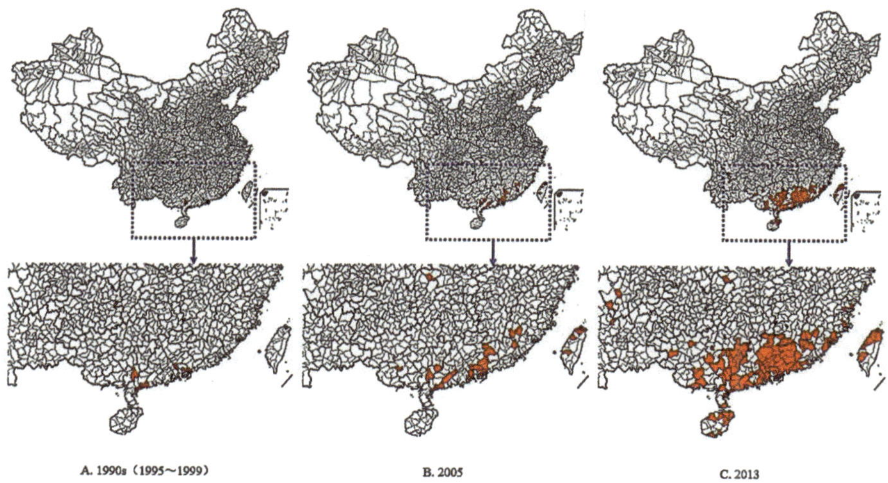

Fig. 17.4 Range of *Solenopsis invicta* in China from 1990s to 2010s (Lu 2014)

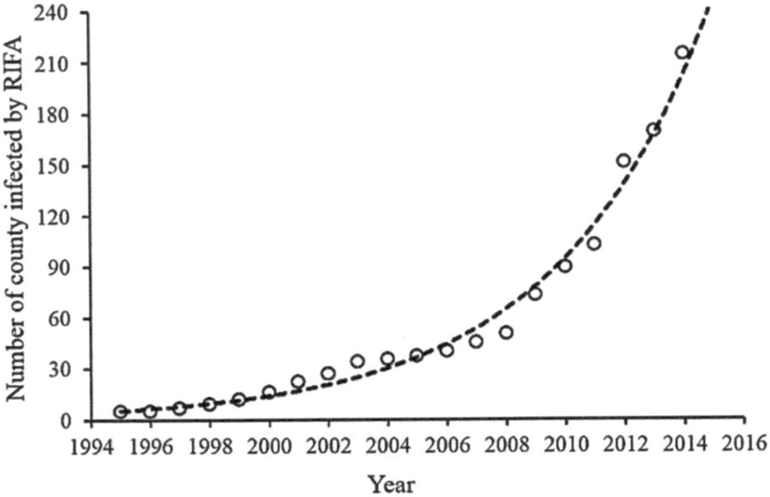

Fig. 17.5 Number of counties infested by *Solenopsis invicta* in China since the mid 1990s (Lu and Zeng 2015)

active during March and June, September and November throughout the year in south China (Li et al. 2008). Daily foraging activity by workers presented two peaks in July and August, and one peak in other months (Chen et al. 2010). The number of defending workers reaches maximum approximately 60–90 s after an initial nest disturbance, being strongly negatively correlated with the distance from the nest core (Xu et al. 2011). Aggressiveness of *S. invicta* is influenced by mound size, colonial structure, temperature, and the other factors, such as pesticides (Gao 2007), but not degree of starvation (Cao et al. 2011) and colony rafting (Huang et al. 2016).

Resistance to unfavorable environmental conditions is one of the main reasons enabling invasive species achieve success. The susceptible lethal temperature for the red imported fire ant was higher than 41.5 °C where exposure time was not longer than 4 h, and heat tolerance could improve after rapid heat hardening (Xu et al., 2009c). Workers can forage when the soil temperatures at 5-cm deep was vary within 16–48 °C (Lu et al. 2012a). Like temperature, drought tolerance was another key factor determining red imported fire ant range. After pre-acclimation for drought tolerance workers survival would rise (Xu et al. 2009d). Starvation had no significant effect on the aggressiveness of *S. invicta* but increased their predation efficiency (Cao et al. 2011). Those biological characteristics may be fundamental to understanding *S. invicta* rapid dispersal.

The main components of *S. invicta* venom are piperidine alkaloids, followed by toxic proteins (Fox 2016). The abundance and composition of fire ant venom correlates not only with castes (Ma et al. 2009), social form (Lai et al. 2008), but also with different colonies (Guan et al. 2013). The contents of *S. invicta* venomous alkaloids decrease when the ants are maintained in laboratory, and from then sustained at a very low level (Liu et al. 2015). In total, eight piperideines, and seven piperidine alkaloids (with *cis* and *trans* isomers for each) were identified (Guan et al. 2013); *cis* and *trans* 2-methyl-6-n-undecenyl piperidine were the two main alkaloids in the venoms of queens (Ma et al. 2009).

Cuticular hydrocarbons play an essential role in ant nestmate recognition and communication. The cuticular compounds of the monogynous and polygynous forms of *S. invicta* vary significantly both quantitatively and qualitatively (Lin et al. 2010). Those results may further evidence the remarkable differences in the social behaviors between the two social forms of *S. invicta*.

It was currently convenient to study *S. invicta* at the molecular level since a whole genome was published based on de novo assembly (Wurm et al. 2011). Apart from numerous deposited expression databases available from NCBI, two genes, e.g. rpl18 and ef1-beta, seem be the most suitable expression reference genes for *S. invicta* (Cheng et al. 2013a). Recently one function of the chemosensory protein 9 (*Si-CSP9*) was found to be related to the development of larvae because fire ant 3rd-instar larvae failed to moult after being fed with *Si-CSP9*-directed siRNA (Cheng et al. 2015). Huang and Wang (2014) revealed that a social supergene of *S. invicta*, consisting of over 600 genes, drives social polymorphism; such chromosome is revealed akin to sex chromosomes in terms of selfish evolutionary behaviors through genetic, genomic, and cytological analysis.

17.3.2 Impacts of S. invicta *Invasion on Biocommunity*

In litchi orchards invaded by *S.invicta*, invertebrate diversity and abundance reduced, mainly arthropods (Xi et al. 2010). In invaded corn fields, spider community structure was also affected (Huang et al. 2012). Native ant diversity can also be negatively affected by *S. invicta* invasion. Ant species richness can be reduced by 46% and 33%, respectively, at *S. invicta*-infested lawns and pastures (Lu et al. 2012b). In Mainland China, most Chinese native ants failed to compete with *S. invicta* at both the individual level and group level (Gao et al. 2011; Chen et al. 2011).

S. invicta also competed strongly with the other ant species for food resources. *S. invicta* can suppress native ant access to honeydew-producing hemipterans, and monopolise most honeydew available (Zhou et al. 2014a; Huang et al. 2010a; Fig. 17.6). Compared to the ghost ant *Tapinoma melanocephalum*, *S. invicta* was more rapid in food searching and recruitment (Lu et al. 2012c). Meanwhile, *S. invicta* can also restrict the foraging behavior and active region of *T. melanocephalum* (Wu et al. 2014a).

As a main pest of agroecosystems, *S. invicta* workers prefer to move and scarify elaiosome-bearing seeds. Respectively, 64%, 56%, and 50% of seeds from *Sesamum indicum*, *Ageratum conyzoides* and *Pennisetum purpureum* seeds will fail to germinate if attacked by fire ants (Huang et al. 2010b). The germination rates of corn seeds (*Zea mays*) decreased from 93 to 80%, while those of mung bean seed (*Vigna radiate*) decreased from 94 to 85% after contact with invading *S. invicta* (Fig. 17.7a)

Fig. 17.6 *Solenopsis invicta* workers tending mealybugs (**a**) and aphids (**b**) (Photo A provided by Aiming Zhou, and photo B by Jun Huang)

Fig. 17.7 Fire ants *S. invicta* feeding on corn seeds (**a**) and preying on a caterpillar *Spodoptera litura* (**b**) (Photos by Jun Huang)

(Huang et al. 2014). Therefore *S. invicta* can also affect crop yield by compromising the crop reproduction (Wu et al. 2014b).

Local mutualism between *S. invicta* and honeydew-producing insects is very common (Huang et al. 2010a; Zhou et al. 2012a). The red imported fire ant will protect aphids from their predators, such as *Menochilus sexmaculatus* (Huang et al. 2011a). This association with aphids can generate further negative effects on rapeseed and mung bean growth, reproduction, and yield (Wu et al. 2014b).

S. invicta is also strongly attracted to the honeydew secreted by the mealybug *Phenacoccus solenopsis*, which is another invasive pest in China (Zhou et al. 2012b). The red imported fire ant not only will defended *P. solenopsis* by attacking predators directly (Zhou et al. 2013; Cheng et al. 2013b), but will also shelter the mealybugs inside structures constructed by the leaf roller *Sylepta derogate*, improving their survival even in predators-containing plants (Zhou et al. 2012c). Moreover, they will assist mealybugs with short-range dispersal (2–3 m) (Zhou et al. 2014b) and suppress competitors on the host plant (Cheng et al. 2013a, b). In order to attract more *S. invicta* workers the mealybugs can regulate metabolism to produce more abundant and lighter honeydew, by increasing the concentration of sugar melezitose and decreasing xylose (Zhou et al. 2015). The benefits to *S. invicta* from the association are considerable: colonies have been shown to grow significantly when feeding from both honeydew and animal protein, and worker longevity seems to increase when fed with honeydew produced by *P. solenopsis* (Zhou et al. 2012a). This way the association between *S. invicta* and *P. solenopsis* has facilitated a successful invasion for both alien species (Zhou et al. 2012a). Considering *P. solenopsis* can infest more than 200 plant species belonging in 57 plant families (Zhang and Wang 2010), it is expected that such mutualism will produce huge economical losses in China agroecosystems in the near future.

In addition to invasive insects, also some invasive plants can benefit from the presence of *S. invicta*. Huang et al. (2011b) found that the local density of an invasive weed found in south China, *Ageratum conyzoides*, increased significantly after the establishment of *S. invicta*. Furthermore, the spatial distribution of *A. conyzoides* seems to have been affected by *S. invicta* (Huang et al. 2011b). Such correlations between weeds and *S. invicta* can cause serious negative impacts to crops.

On the other hand, there is some potential benefit coming from *S. invicta* as a biological control agent. Up to 70% of pupae of the oriental fruit fly *Bactrocera dorsalis* were preyed by *S. invicta* in simulated ground surface in laboratory (Cao et al. 2012). The red imported fire ant was also observed preying *Spodoptera litura* and *Ostrinia furnacalis*. In the corn fields infested with *S. invicta* there were reports of reductions in 50% of egg masses and 94.1% larvae of *O. furnacalis*, as well as a decrease of 92.9% in damaged corn (Huang 2010; Fig. 17.7b). However it has been confirmed that any attempts to use fire ants as a biocontrol agent will always cause more harm than good.

17.4 Management of *S. invicta* in China

17.4.1 Monitoring and Surveillance

Investigating and monitoring *S. invicta* populations is paramount for a successful management. Visual inspection is the primary means of looking for signs of *S. invicta* invasion and infestation (Zeng et al. 2005). Bait traps, mainly using sausage pieces as bait, are effective and reliable in locating colonies and delimiting territories (Zeng et al. 2005; Bao et al. 2011). Dogs can also be trained to detect fire ant colonies in the field (Lin et al. 2011). Based on differential light deflection of fire ant nest relative to other surfaces, it is possible to employ spectroscopic techniques for rapid aerial detection of fire ant nests (Wu et al. 2014c). In order to establish a national technical standard, official Guidelines for quarantine surveillance of *S. invicta* Buren were promulgated and implemented in 2009.

17.4.2 Quarantine

S. invicta was listed as a quarantine pest soon after infestation was confirmed in Taiwan and Mainland China, and quarantine measures implemented in entry and exit ports for domestic goods transportation and trade (Lu and Zeng 2015). The risk of carrying fire ants among transported items was evaluated (Huang et al. 2007b; Ma et al. 2010). Methods such as contact pesticides, fumigation, flooding colonies with boiling water are employed in *S. invicta* quarantine (Lu and Zeng 2015). Although several standards and policies for *S. invicta* quarantine were issued by the Ministry of Agriculture, General Administration of Quality Supervision, Inspection and Quarantine, and State Forestry Administration, the effects of these policies after implementation are still questionable.

17.4.3 Chemical Control

The use of chemicals is the most common method for controlling *S. invicta*. The control efficacy of more than 20 insecticides including flursulamid, spinosad, fipronil, sulfluramid, pyriproxyfen, abamectin, chlorpyrifos, and others, was evaluated in laboratory and field conditions (Li et al. 2007; Huang et al. 2008; Jiang 2008). Huang et al. (2007a) developed a method for fire ant control combining toxic baits and contact insecticides to be used accross a large scale, which has the potential to elminate 94% of fire ant colonies. Currently, one powder contact pesticide and eight toxic baits are approved by the Ministry of Agriculture for the use in red imported fire ant management in China.

Some plants compounds were tested against fire ants in China, and periplocoside X was shown to be highly active, causing severe damage to fire ant midgut epithelium cells (Li et al. 2012; Li and Zeng 2013). Essential oils from some plants exhibited fumigation activity against *S. invicta* workers (Tang et al. 2013; Cheng et al. 2008; Wang et al. 2014b). Over 95% of fire ant workers were killed after exposure to volatile compounds from *Tephrosia vogelii* (Li et al. 2014). In tests using five active components from *Artemisia annua*, over 80% of fire ant workers died, while cineole and D-camphor showed significant repellency (Zhang et al. 2014). Essential oils of *Capsicum annuum* and *Cedrus deodara* also strongly repelled *S. invicta* (Wang et al. 2014a). Kafle and Shih (2013) reported that using *Syzygium aromaticum* powder at 3 and 12 mg/cm^2 could kill all fire ant workers within 6 h, and repel 99% ants within 3 h. These results suggest another method are effective for *S. invicta* management, especially in settings where the use of conventional insecticides is restricted, such as organic agriculture and protected water bodies.

Toxic baits are effective to control fire ants. The comprehensive index system to determine the control effect of toxic baits on the red imported fire ant was developed (Huang et al. 2008). Food preferences of *S. invicta* were tested under laboratory conditions provided the basis for further development of toxic baits (Kafle et al. 2008). Mortality of baits containing either 0.50% priproxyfen, 0.00015% fpronil, or 0.015% spinosad to fire ant colonies exceeded 98% (Hung et al. 2006). Also bait particle size affected the amount being carried back to the nest by workers, indicating the most suitable particle sizes to be 0.8–2.0 mm (Kafle and Shih 2012). Humid conditions and rain always decrease the efficiency of baits, leading to the development of an efficient water-resistant bait containing cypermethrin 0.128% (Kafle et al. 2010). Guan et al. (2014) used electroantennograms coupled to behavioral responses of fire ants to the alarm pheromone component alkylpyrazine and some analogues, indicating alkylpyrazines could be efficient bait attractants for fire ant control.

The above mentioned research conclusions and control methods served as the basis for two further national standards, the Guidelines for Chemical Prevention and Control of *S. invicta* Buren, and Rules for Evaluating Control Effects on *S. invicta* Buren, promulgated in 2013 and 2014, respectively.

17.4.4 Biological Control

Entomopathogenic fungi were tested for the control of *S. invicta* under laboratory conditions (Liu et al. 2010; Yang et al. 2009). A strain of *Paecilomyces lilacinus* was isolated from the red imported fire ant workers in Guangdong, which proved capable of inducing 70% worker mortality at 1×10^8 conidia.ml^{-1} (Liu et al. 2010). Kafle et al. (2011) found a native strain of *Beauveria bassiana* F256 which was effective against *S. invicta* in the laboratory and in the field. As a social insect, *S. invicta* evolved various defensive mechanisms against diseases. For example, some fire ant secretions negatively impact microorganisms (Blum et al. 1958), such as venom applied by workers to eggs and brood (Obin and Vander Meer 1985), and nest parts (Chen 2007). Also grooming behavior helps protect workers from pathogenic fungal infections (Qiu et al. 2014). *S. invicta* can also produce some yet unknown antimicrobial volatiles which can inhibit spore germination and growth of *B. bassiana* (Wang et al. 2015).

17.5 Summary

Much research on *S. invicta* had been completed within the 13 years after the invasion of China was confirmed. Entomologists and politicians united efforts in understanding the invasion biology and developing new technologies to carry out prevention and control. In spite of the control measures applied to many invaded regions, the fire ants have already rapidly spread and continue to expand in China, and this accelerated expansion is expected to progress for 20 years. The main reasons behind this remain the huge transportation rate allied to poor quarantine enforcement (Lu and Zeng 2015). Therefore establishing strict quarantine should be the best way to slow down the spread of fire ants suppressing the infestation. Constructing an effective network for red imported fire ant management should be the next important step: an efficient organization to implement the various measures is necessary. Finally, given sufficient financial support and capable technical expertise, a successful control of red imported fire ants is guaranteed.

References

Ascunce MS, Yang CC, Oakey J, Calcaterra L, Wu WJ, Shih CJ, Goudet J, Ross KG, Shoemaker DD (2011) Global invasion history of the fire ant *Solenopsis invicta*. Science 331(6020):1066

Bao SZ, Kafle L, Cheng JS (2011) A new baited trap for monitoring *Solenopsis invicta* (Formicidae: Hymenoptera) in Taiwan. Appl Entomol Zool 46(2):165–169

Blum MS, Walker JR, Callahan PS, Novak AF (1958) Chemical, insecticidal and antibiotic properties of fire ant venom. Science 128:306–307

Cao L, Kuang BQ, Yang JL, Song YP, Xu YJ (2011) Effect of starvation on the contact-free aggressive behavior and predation activity of *Solenopsis invicta* (Hymenoptera: Formicidae). Sociobiology 57(3):461–469

Chen HT, Luo LZ, Xiong HL (2010) The outing activity of worker ants and nuptial flight of alates in the red imported fire ant, Solenopsis invicta (Hymenoptera: Formicidae) in Guangzhou area, Southern China. Acta Entomol Sin 53(4):449–456

Cao L, Zhou AM, Chen RH, Zeng L, Xu YJ (2012) Predation of the Oriental fruit fly, *Bactrocera dorsalis* puparia by the red imported fire ant, *Solenopsis invicta*: role of host olfactory cues and soil depth. Biocontrol Sci Tech 22(5):551–557

Chen J (2007) Qualitative analysis of red imported fire ant nests constructed in silica gel. J Chem Ecol 33:631–642

Chen L, Korzukhin MD, Cheng DF, Lu QG, Wang FX (2006) The analysis for the red imported fire ant *Solenopsis invicta* (Hymenoptera: Formicidae) distribution in China based on 3S technology, climate model and colony dynamic model. Acta Phys Sin 33(4):384–390

Chen YC, Kafle L, Shih CJ (2011) Interspecific competition between *Solenopsis invicta* and two native ant species, *Pheidole fervens* and *Monomorium chinense*. J Econ Entomol 104(2):614–621

Chen JS, Shen CH, Lee HJ (2015) Monogynous and polygynous red imported fire ants, *Solenopsis invicta* Buren (Hymenoptera: Formicidae), in Taiwan. Environ Entomol 35(1):167–172

Cheng SS, Liu JY, Lin CY, Hsui YR, Lu MC, Wu WJ, Chang ST (2008) Terminating red imported fire ants using *Cinnamomum osmophloeum* leaf essential oil. Bioresour Technol 99:889–893

Cheng DF, Zhang ZL, He XF, Liang GW (2013a) Validation of reference genes in *Solenopsis invicta* in different developmental stages, castes and tissues. PLoS One 8(2):e57718

Cheng SJ, Zeng L, Xu YJ (2013b) Effect of the mutualism between *Solenopsis invicta* (Hymenoptera: Formicidae) and *Phenacoccus solenopsis* Tinsley (Hemiptera: Pseudococcidae) on *Aenasius bambawalei* Hayat and *Echinothrips americanus* Morgan. J Environ Entomol 35(5):555–559

Cheng DF, Lu YY, Zeng L, Liang GW, He XF (2015) *Si-CSP9* regulates the integument and moulting process of larvae in the red imported fire ant, *Solenopsis invicta*. Sci Report 5:9245

Drees BM, Vinson SB (1990) Comparison of the control of monogynous and polygynous forms of the red imported fire ant (Hymenoptera: Formicidae) with a chlorpyrifos mound drench. J Entomol Sci 25(2):317–324

Du YZ, Gu J, Guo JB, Dai L, Ju RT, Hu XN (2007) Study on the potential distribution area of invasive alien pest red imported fire ant, *Solenopsis invicta* Buren in China. Sci Agric Sin 40(1):99–106

Fox EGP (2016) Venom toxins of fire ants. In: Gopalakrishnakone P, Calvete JJ (Eds) Venom genomics and proteomics, pp 149–167. Available at: http://dx.doi.org/10.1007/978-94-007-6416-3_38

Gao YB (2007) Studies on the activity of red imported fire ant *Solenopsis invicta* Buren workers. Master thesis in Anhui Agricultural University

Gao Y, Lv LH, He YR, Qi GJ, Zhang JQ (2011) Interference competition between the red imported fire ant (*Solenopsis invicta* Buren) and two native ant species (Hymenoptera: Formicidae). Acta Entomol Sin 54(5):602–608

Guan D, Liao XL, Chen L (2013) Comparative analysis of venomous alkaloids in workers of the red imported fire ant, *Solenopsis invicta* (Hymenoptera: Formicidae), from China and the United States. Acta Entomol Sin 56(4):365–371

Guan D, Lu YY, Liao XL, Wang L, Chen L (2014) Electroantennogram and behavioral responses of the imported fire ant, *Solenopsis invicta* Buren, to an alarm pheromone component and its analogues. J Agric Food Chem 62:11924–11932

He XF, Lu YY, Zhang WQ, Zeng L (2006) Three haplotypes found in populations of the red imported fire ant invading China. Acta Entomol Sin 49(6):1046–1049

Hung YT, Lin RW, Yang CC, Wu WJ, Shih CJ, Lin CC (2006) A preliminary report on bait use to control the red imported fire ant (*Solenopsis invicta*) in Taiwan. Formosan Entomol 26(1):57–67

Huang J (2010) Influence of *Solenopsis invicta* (Hymenoptera: Formicidae) invasion on insect pests and natural enemies of dry-land corps. PhD thesis in South China Agricultural University

Huang YC, Wang J (2014) Did the fire ant supergene evolve selfishly or socially? BioEssays 36(2):200–208

Huang TF, Xiong ZH, Zeng XN (2007a) Studies on the contact toxicity of insecticides against the worker ants of *Solenopsis invicta*. J South China Agric Univ 28(4):26–29

Huang J, Zeng L, Liang GW, Lu YY, Xu YJ, Gao YB, Zhang Q, Zhang SQ, Yang HZ, Chen ZN (2007b) Technical demonstration of controlling red imported fire ant, *Solenopsis invicta* Buren. China Plant Protect 27(8):41–43

Huang J, Lu YY, Liang GW, Zeng XN, Zeng L (2008) Control effect of four poisonous baits on red imported fire ant, *Solenopsis invicta* Buren. J Environ Entomol 30(2):135–140

Huang J, Xu YJ, Lu YY, Zeng L, Liang GW (2010a) Effects of red imported fire ants (Hymenoptera: Formicidae) on the relationship between native ants and aphids in mung bean fields in China. Sociobiology 55(2):415–426

Huang J, Xu YJ, Zeng L, Liang GW, Lu YY (2010b) Selective feeding of *Solenopsis invicta* on seeds of eight plant species and their influence on germination. J Environ Entomol 32(1):6–10

Huang J, Xu YJ, Lu YY, Liang GW, Zeng L (2011a) Effects of the invasive ant *Solenopsis invicta* (Hymenoptera: Formicidae) on *Menochilus sexmaculatus* (Coleoptera: Coccinellidae) as a predators of *Aphis craccivora* (Hemiptera: Aphididae) in laboratory conditions. Sociobiology 57(3):565–574

Huang J, Xu YJ, Zeng L, Lu YY, Liang GW (2011b) Changes to the spatial distribution of *Ageratum conyzoides* (Asterales: Asteraceae) due to red imported fire ants *Solenopsis invicta* (Hymenoptera: Formicidae) in China. J Insect Behav 24:307–316

Huang J, Xu YJ, Lu YY, Zeng L (2012) Effect of *Solenopsis invicta* invasion on the diversity of spider communities in a corn field. Chin J Appl Ecol 23(4):1111–1116

Huang J, Xu YJ, Liang GW, Lu YY, Zeng L (2014) Effects on the germination of two dry land crop seeds of *Solenopsis invicta* Buren. J Biosaf 23(2):88–92

Huang YQ, Zhu SW, Tan DL, Zeng L, Wang L, Xu YJ (2016) Aggression of Solenopsis invicta towards other ants is not affected by colony rafting. Asian Myrmecol 8:1–4

IUCN View 100 of the World's Worst Invasive Alien Species [EB/OL]. (2010-10-4) [2014-12-20]. http://www.issg.org/worst 100_species.html

Jiang DR (2008) Research progress of red imported fire ant in China. Guangxi Plant Protect 21(3):20–22

Kafle L, Shih CJ (2012) Determining the most effective concentration of cypermethrin and the appropriate carrier particle size for fire ant (Hymenoptera: Formicidae) bait. Pest Manag Sci 68(3):394–398

Kafle L, Shih CJ (2013) Toxicity and repellency of compounds from clove (*Syzygium aromaticum*) to red imported fire ants *Solenopsis invicta* (Hymenoptera: Formicidae). J Econ Entomol 106(1):131–135

Kafle L, Wu WJ, Vander Meer RK, Shih CJ (2008) Simplified approaches to determine the attractant preference of *Solenopsis invicta* (Hymenoptera: Formicidae). Appl Entomol Zool 43:383–390

Kafle L, Wu WJ, Shih CJ (2010) A new fire ant (Hymenoptera: Formicidae) bait base carrier for moist conditions. Pest Manag Sci 66(10):1082–1088

Kafle L, Wu WJ, Kao SS, Shih CJ (2011) Efficacy of *Beauveria bassiana* against the red imported fire ant, Solenopsis invicta (Hymenoptera: Formicidae), in Taiwan. Pest Manag Sci 67(11):1434–1438

Killion MJ, Grant WE (1995) A colony growth model for the imported fire ant: potential geographic range of an invading species. Ecol Model 77:73–84

Korzukhin MD, Porter SD, Thompson LC, Wiley S (2001) Modeling temperature-dependent range limits for the fire ant *Solenopsis invicta* (Hymenoptera: Formicidae) in the United States. Environ Entomol 30(4):645–655

Kuo MH, Lee KL, Hsu CJ, Chiu MC, Liu WT, Wu YT (2010) Monitoring the red imported fire ant *Solenopsis invicta* (Hymenoptera: Formicidae) in Hsinchu county, Taiwan. Formosan Entomol 30(4):263–271

Lai LC, Huang RN, Wu WJ (2008) Venom alkaloids of monogyne and polygyne forms of the red imported fire ant, *Solenopsis invicta*, in Taiwan. Insect Soc 55:443–449

Lard C, Willis D, Salin V, Robison S (2002) Economic assessments of fire ant on Texas urban and agricultural sectors. Southwest Entomol 25:123–137

Lard CF, Schmidt J, Morris B, Estes L, Ryan C, Bergquist D (2006) An economic impact of imported fire ants in the United States of America. Texas A&M University, College Station

Li Y, Zeng X (2013) Effects of periplocoside X on midgut cells and digestive enzymes activity of the soldiers of red imported fire ant. Ecotoxicol Environ Saf 93(4):1–6

Li HL, Wu SH, Li H, Wang L, Liang MS, Wu WJ, Luo YP, Lian HT, Zeng L (2007) Preliminary study on the control effects of ten pesticides to red imported fire ant. Guangdong Agric Sci 34(10):57–59

Li ND, Zeng L, Liang GW, Lu YY (2008) Dynamics of red imported fire ant *Solenopsis invicta* in Wuchuan, Guangdong. Chin Bull Entomol 45(1):54–57

Li Y, Zeng XN, Wang WZ, Luo CH, Yan Q, Tian M (2012) Chemical constituents from the roots of *Periploca sepium* with insecticidal activity. J Asian Nat Prod Res 14(8):811–816

Li WS, Zhou Y, Li H, Wang K, Cheng DM, Zhang ZX (2014) Insecticidal effect of volatile compounds from fresh plant materials of *Tephrosia vogelii* against *Solenopsis invicta* workers. Sociobiology 61(1):28–34

Lin YK, Chang HY, Wu WJ, Ho HY, Lin CC (2010) Different cuticular chemical profiles between the monogynous and polygynous forms of the red imported fire ant, *Solenopsis invicta*, in Taiwan. Sociobiology 56:39–55

Lin HM, Chi WL, Lin CC, Tseng YC, Chen WT, Kung YL, Lien YY, Chen YY (2011) Fire ant-detecting canines: a complementary method in detecting red imported fire ants. J Econ Entomol 104(1):225–231

Liu XY, Lv LH, He YR (2010) Isolation and identification of entmopathgenic fungi naturally infecting *Solenopsis invicta* and their virulence to *S. invicta*. Chin J Biol Control 26(3):373–376

Liu XF, Chen L, Li JK (2015) Dynamic analysis of venomous alkaloids in workers of the red imported fire ant, *Solenopsis invicta* (Hymenoptera: Formicidae) maintained in the laboratory. Acta Entomol Sin 58(1):22–27

Lu YY (2014) Long-distance spreading speed and trend prediction of red imported fire ant, *Solenopsis invicta* Buren, in mainland China. Guangdong Agricultural Science 41(10):70–72. , picture

Lu YY, Zeng L (2015) 10 years after red imported fire ant found to invade China: history, current situation and trend of its infestation. Plant Quarantine 29(2):1–6

Lu YY, Liang GW, Zeng L (2008) Study on expansion pattern of red imported fire ant, *Solenopsis invicta* Buren, in south China. Sci Agric Sin 41(4):1053–1063

Lu YY, Wang L, Zeng L, Xu YJ (2012a) The effects of temperature on the foraging activity of red imported fire ant workers (Hymenoptera: Formicidae) in south China. Sociobiology 59(2):585–588

Lu YY, Wu BQ, Xu YJ, Zeng L (2012b) Effects of red imported fire ants (*Solenopsis invicta*) on the species structure of ant communities in south China. Sociobiology 59(1):275–285

Lu YY, Wu BQ, Zeng L, Xu YJ (2012c) Comparison of foraging ability between *Solenopsis invicta* and *Tapinoma melanocephalum* (Hymenoptera: Formicidae). Sociobiology 59(3):1015–1024

Ma FN, Zeng XN, Pan DQ (2009) Comparison of venom alkaloids in worker ants and queen of *Solenopsis invicta* Buren. Chin Agric Sci Bull 25(17):57–61

Ma J, Hu XN, Wu JJ, Zhao JP, Chen NZ (2010) Ananalysis for the interception dynamic of *Solenopsis invicta* Buren at ports in Guangdong. J Environ Entomol 32(1):122–124

Mcnaught MK, Wylie FR, Harris EJ, Alston CL, Burwell CJ, Jennings C (2014) Effect of broadcast baiting on abundance patterns of red imported fire ants (Hymenoptera: Formicidae) and key local ant genera at long-term monitoring sites in Brisbane, Australia. J Econ Entomol 107(4):1307–1315

Moloney S, Vanderwoude C (2003) Potential ecological impacts of red imported fire ants in eastern Australia. J Agric Urban Entomol 20(3):131–142

Obin MS, Vander Meer RK (1985) Gaster flagging by fire ants (*Solenopsis* spp.): functional significance of venom dispersal behavior. J Chem Ecol 11(12):1757–1768

Porter SD, Savignano DA (1990) Invasion of polygyne fire ants decimates native ants and disrupts arthropod community. Ecology 71(6):2095–2106

Prahlow JA, Barnard JJ (1998) Fatal anaphylaxis due to fire ant stings. Am J Forensic Med Pathol 19(2):137–142

Qiu HL, Lv LH, Shi QX, He YR (2014) Fungus exposed *Solenopsis invicta* ants benefit from grooming. J Insect Behav 27:678–691

Ross KG, Keller L (2003) Ecology and evolution of social organization: insights from fire ants and other highly eusocial insects. Annu Rev Ecol Syst 26(4):631–656

Scanlan JC, Vanderwoude C (2006) Modelling the potential spread of *Solenopsis invicta* Buren (Hymenoptera: Formicidae) (red imported fire ant) in Australia. Aust J Entomol 45(1):1–9

Shao JG, Luo LZ, Chen HT, Jiang XF, Lei CL (2008) Social forms of *Solenopsis invicta* in China, as identified by multiplex PCR of Gp-9 alleles. Acta Entomol Sin 51(1):551–555

Shen WJ, Wang YN, Wan FH (2008) Application of analogy deviation to predict potential distribution of *Solenopsis invicta* Buren (Hymenoptera: Formicidae) and its invasion probability in China. Sci Agric Sin 41(6):1673–1683

Tang L, Sun YY, Zhang QP, Zhou Y, Zhang N, Zhang ZX (2013) Fumigant activity of eight plant essential oils against workers of red imported fire ant, *Solenopsis invicta*. Sociobiology 60(1):35–40

Vinson SB (1997) Invasion of the red imported fire ant (Hymenoptera: Formicidae): spread, biology and impact. Am Entomol 43(1):23–29

Wang L, Elliott B, Jin XX, Zeng L, Chen J (2015) Antimicrobial properties of nest volatiles in red imported fire ants, Solenopsis invicta (Hymenoptera: Formicidae). Sci Nat 102(11):66

Wang J, Qiu X, Zeng L, Xu Y (2014a) Interference of plant essential oils on the foraging behavior of *Solenopsis invicta* (Hymenoptera: Formicidae). Fla Entomol 97(2):454–460

Wang K, Tang L, Zhang N, Zhou Y, Li WS, Li H, Cheng DM, Zhang ZX (2014b) Repellent and fumigant activities of *Eucalyptus globulus* and *Artemisia carvifolia* essential oils against *Solenopsis invicta*. Bull Insectol 67(2):207–211

Wetterer JK (2013) Exotic spread of *Solenopsis invicta* Buren (Hymenoptera: Formicidae) beyond North America. Sociobiology 60(1):50–55

Williams DF, Oi DH, Porter SD, Pereira RM, Briano JA (2003) Biological control of imported fire ants (Hymenoptera: Formicidae). Am Entomol 27(3):203–209

Wojcik DP, Allen CR, Brenner RJ, Forys EA, Jouvenaz DP, Lutz RS (2001) Red imported fire ants: impact on biodiversity. Am Entomol 47(1):16–23

Wu BQ, Wang L, Liang GW, Lu YY, Zeng L (2014a) Food competition mechanism between *Solenopsis invicta* Buren and *Tapinoma melanocephalum* Fabricus. Sociobiology 61(3):265–273

Wu D, Zeng L, Lu YY, Xu YJ (2014b) Effects of *Solenopsis invicta* (Hymenoptera: Formicidae) and its interaction with aphids on the seed productions of mungbean and rapeseed plants. J Econ Entomol 107(5):1758–1764

Wu WB, Hong TS, Zeng L, Xu YJ, Huang DL, Chileshe JM, Chen JT, Yi CD, Liu XY, Lu DH (2014c) Detection of *Solenopsis invicta* (Hymenoptera: Formicidae) nests using spectral data. Fla Entomol 97(3):967–971

Wurm Y, Wang J, Riba-Grognuz O, Corona M, Nygaard S, Hunt BG, Ingramf KK, Falquet L, Nipitwattanaphon M, Gotzek D, Dijkstra MB, Oettler J, Comtesse F, Shih CJ, Wu WJ, Yang CC, Thomas J, Beaudoing E, Pradervand S, Flegel V, Cook ED, Fabbretti R, Stockinger H, Long L, Farmerie WG, Oakey J, Boomsma JJ, Pamilo P, Yi SV, Heinze J, Goodisman MAD, Farinelli L, Harshman K, Hulo N, Cerutti L, Xenarios I, Shoemaker DD, Keller L (2011) The genome of the fire ant *Solenopsis invicta*. Proc Natl Acad Sci 108(14):5679–5684

Xi YB, Lu YY, Liang GW, Zeng L, Xu YJ (2010) Effect of the red imported fire ant (RIFA), *Solenopsis invicta* Buren, on diversity and stability of invertebrate community in litchi orchard. Acta Ecol Sin 30(8):2084–2099

Xu YJ, Lu YY, Zeng L, Xi YB, Huang J (2006) Study on location expansion of *Solenopsis invicta*. J South China Agric Univ 27(1):34–36

Xu YJ, Lu YY, Zeng L (2009a) Colony structure and dynamic of the fire ant *Solenopsis invicta* Buren in several habitats. Acta Phys Sin 36(6):575–576

Xu YJ, Huang J, Lu YY, Zeng L, Liang GW (2009b) Observation of nuptial flights of the red imported fire ant, *Solenopsis invicta* (Hymenoptera: Formicidae) in Mainland China. Sociobiology 54(3):831–840

Xu YJ, Lu YY, Pan ZP, Zeng L, Liang GW (2009c) Heat tolerance of the red imported fire ant, *Solenopsis invicta* (Hymenoptera: Formicidae) in mainland China. Sociobiology 54(1):115–126

Xu YJ, Zeng L, Lu YY, Liang GW (2009d) Effect of soil humidity on the survival of *Solenopsis invicta* Buren workers. Insect Soc 56(4):367–373

Xu YJ, Zeng L, Lu YY (2011) Temporarily defended dispersal area of alarmed workers of *Solenopsis invicta* (Hymenoptera, Formicidae) provoked by physical disturbance. Sociobiology 58(1):119–132

Xu YJ, Huang J, Zhou A, Zeng L (2012) Prevalence of *Solenopsis invicta* (Hymenoptera: Formicidae) venom allergic reactions in mainland China. Fla Entomol 95(4):961–965

Xue DY, Li HM, Han HX, Zhang RZ (2005) A prediction of potential distribution area of *Solenopsis invicta* in China. Chin Bull Entomol 42:57–60

Yang CC, Shoemaker DD, Wu WJ, Shih CJ (2008) Population genetic structure of the red imported fire ant, *Solenopsis invicta*, in Taiwan. Insect Soc 55:54–65

Yang JH, Sun F, Liao KH, Wu WJ, Pang H, Pang Y (2009) Bioassay of 4 strains of *Beauveria bassiana* against *Solenopsis invicta*. J Environ Entomol 31(3):46–51

Yang CC, Ascunce MS, Luo LZ, Shao JG, Shih CJ, Shoemaker D (2012) Propagule pressure and colony social organization are associated with the successful invasion and rapid range expansion of fire ants in China. Mol Ecol 21(4):817–833

Yang JC, Wu WZ, Huang RN (2013) The control of red imported fire ant in Taiwan: present and future. In Proceedings of 2013 international symposium on forest health management. Edited by The Society for Wildlife and Nature, Taiwan: Taibei, pp 1–9

Yu HY, Liu SH, Zhao N, Li D, Yu YT (2011) Characteristics of air temperature and precipitation in different regions of China from 1951 to 2009. J Meteorol Environ 27(4):1–11

Zeng L, Lu YY, He XF, Zhang WQ, Liang GW (2005) Identification of red imported fire ant *Solenopsis invicta* to invade mainland China and infestation in Wuchuan, Guangdong. Chin Bull Entomol 42(2):144–148. 230-231

Zhang RZ, Wang FX (2010) *Phenacoccus solenopsis* (cotton mealybug). Chinese Agriculture Press, Beijing

Zhang RZ, Li Y, Liu N, Porter SD (2007) An overview of the red imported fire ant (Hymenoptera: Formicidae) in Mainland China. Fla Entomol 90:723–731

Zhang N, Tang L, Hu W, Wang K, Zhou Y, Li H, Huang CL, Chun J, Zhang ZX (2014) Insecticidal, fumigant, and repellent activities of sweet wormwood oil and its individual components against red imported fire ant workers (Hymenoptera: Formicidae). J Insect Sci 14(1):201–204

Zhou AM, Lu YY, Zeng L, Xu YJ, Liang GW (2012a) Does mutualism drive the invasion of two alien species? The case of *Solenopsis invicta* and *Phenacoccus solenopsis*. PLoS One 7(7):e41856

Zhou AM, Lu YY, Zeng L, Xu YJ, Liang GW (2012b) Effect of honeydew of *Phenacoccus solenopsis* on foliar foraging by *Solenopsis invicta*. Sociobiology 59(1):71–79

Zhou AM, Zeng L, Lu YY, Xu YJ, Liang GW (2012c) Fire ants protect mealybugs against their natural enemies by utilizing the leaf shelters constructed by the leaf roller *Sylepta derogata*. PLoS One 7(11):e49982

Zhou AM, Lu YY, Zeng L, Xu YJ, Liang GW (2013) *Solenopsis invicta* (Hymenoptera: Formicidae), defend *Phenacoccus solenopsis* (Hemiptera: Pseudococcidae) against its natural enemies. Environ Entomol 42(2):247–252

Zhou AM, Liang GW, Lu YY, Zeng L, Xu YJ (2014a) Interspecific competition between the red imported fire ant, *Solenopsis invicta* Buren and ghost ant, *Tapinoma melanocephalum* Fabricius for honeydew resources produced by an invasive mealybug, *Phenacoccus solenopsis* Tinsiley. Arthropod Plant Interact 8(5):469–474

Zhou AM, Liang GW, Zeng L, Wu D, Xu YJ, Lu YY (2014b) Fire ant (*Solenopsis invicta*) invasion assists the short-range spread of another invasive insect, the cotton mealybug *Phenacoccus solenopsis*. J Biosaf 23(2):93–96

Zhou AM, Wu D, Liang GW, Lu YY, Xu YJ (2015) Effects of tending by *Solenopsis invicta* (Hymenoptera: Formicidae) on the sugar composition and concentration in the honeydew of an invasive mealybug, *Phenacoccus Solenopsis* (Hemiptera: Pseudococcidae). Ethology 121(5):492–500

Appendices

Appendix I

Year of establishment and size of World National Parks and Protected Areas

National park or national forest	Country	Year	Size in hectares
Yellowstone	U.S. (Montana)	1872	899118
Yosemite	U.S. (California)	1890	302687
Sequoia National Park	U.S. (California)	1890	350444
Kings Canyon National Park	U.S. (California)	1940	
Mount Rainier National Park	U.S. (Washington)	1899	95354
Crater Lake	U.S. (Oregon)	1902	74148
Glacier	U.S. (Montana)	1919	410178
Rocky Mountain National Park	U.S. (Colorado)	1915	107595
Haleakalā National Park	U.S. (Hawaii)	1916	13878
Arches National Monument	U.S. (Utah)	1929	30901
Badlands Nat'l Monument	U.S. (South Dakota)	1929	98865
Carlsbad Caverns	U.S. (New Mexico)	1930	18926
Death Valley NP	U.S. (California and Nevada)	1994	1375933
Denali (formerly Mt. Mckinley National Park)	U.S. (Alaska)	1917	2458481
Everglades NP	U.S. (Florida)	1957	624239
Glacier Bay Nat'l Monument	U.S. (Alaska)	1925	1327451
Grand Canyon NP	U.S. (Arizona)	1919	492666
Grand Teton NP	U.S. (Wyoming)	1929	125450
Great Smoky Mountains	U.S. (North Carolina and Tennessee)	1934	211204
Katmai Nat'l Monument	U.S. (Alaska)	1918	1627376
Lafayette NP (renamed Acadia)	U.S. (Maine)	1919	20072

(continued)

F. Wan et al. (eds.), *Biological Invasions and Its Management in China*,
Invading Nature - Springer Series in Invasion Ecology 11,
DOI 10.1007/978-94-024-0948-2

National park or national forest	Country	Year	Size in hectares
Mammoth Cave	U.S. (Kentucky)	1941	21380
Shenandoah	U.S. (Virginia)	1935	79890
Utah NP (renamed Bryce)	U.S. (Utah)	1928	14502
Zion NP	U.S. (Utah)	1919	60190
Everglades National Park	United States (Florida)	1947	610500
Lake Mead National Recreation Area	United States (Arizona and Nevada)	1936	605300
Arctic National Wildlife Refuge	Alaska, United States of America	1960	7800000
Yukon Delta National Wildlife Refuge	Alaska, United States of America	1980	7755000
Noatak National Preserve	United States (Alaska)	1978	2658700
Wrangell–St. Elias National Park and Preserve	United States (Alaska)	1980	5332100
Belovezhskaya National Park	Belarus	1991	87300
Pirin National Park	Bulgaria	1962	27200
České Švýcarsko National Park (Bohemian Switzerland)	Czech Republic	2000	7900
The Broads	England	1989	30300
Dartmoor National Park	England	1951	95300
Exmoor National Park	England	1954	69200
Lake District National Park (England's largest)	England	1951	229200
New Forest National Park	England	2005	56700
Northumberland National Park	England	1956	104900
North York Moors National Park	England	1952	143600
Peak District (Britain's first Nat'l park)	England	1951	143800
South Downs National Park	England	2010	164800
Yorkshire Dales	England	1954	176200
Brecon Beacons	Wales	1957	134700
Loch Lomond and the Trossachs National Park	Scotland	2002	186500
Pembrokeshire Coast National Park	Wales	1952	62000
Snowdonia	Wales	1951	213200
Cairngorms	Scotland	2003	452800
Banff National Park	Alberta, Canada	1885	663960
Cape Breton Highlands National Park	Nova Scotia, Canada	1936	95081
Gros Morne National Park	Newfoundland and Labrador	1973	180452

(continued)

National park or national forest	Country	Year	Size in hectares
Jasper National Park	Alberta (largest NP in Canadian Rockies)	1907	1122563
Mingan Archipelago National Park	Quebec, Canada	1984	9998
Nahanni National Park	Northwest Terretories, Canada	1972	167100
Pacific Rim National Park	British Columbia, Canada	1970	51182
St. Lawrence Island National Park	Ontario, Canada	1904	2400
Wapusk National Park	Manitoba, Canada	1996	1147446
Waterton Lakes National Park	Alberta, Canada	1895	50489
Fundy National Park	New Brunswick Canada	1948	20594
Cinque Terre, Italy	Italy	1999	3900
Oulanka National Park	Finland	1956	27000
Cevennes National Park	France	1970	91300
Pyrénées National Park	France	1967	45700
Bavarian Forest National Park	Germany	1970	24100
Saxon Switzerland National Park	Germany	1990	9300
Killarney National Park	Ireland	1932	10600
Etna National Park	Italy	1987	58100
Durmitor National Park	Montenegro	1952	33900
De Hoge Veluwe National Park	The Netherlands	1935	5400
Peneda Geres National Park	Portugal	1971	69700
Aigüestortes i Estany de Sant Maurici National Park	Spain	1955	10200
Doñana National Park	Spain	1969	54200
Garajonay National Park	Canary Islands, Spain	1981	4000
Northeast Greenland National Park	Greenland	1974	97200000
Hoggar National Park	Algeria/Sahara	1987	45000000
Phoenix Islands Protected Area	Kiribati (Phoenix Islands)	2008	40825000
Papahānaumokuākea Marine National Monument	United States (Hawaii and the Midway Atoll)	2006	36000000
Hawaiian Islands National Wildlife Refuge			
Midway Atoll National Wildlife Refuge			
Battle of Midway National Memorial			
Hawaii State Seabird Sanctuary at Kure Atoll			
Northwestern Hawaiian Islands State Marine Refuge			

(continued)

National park or national forest	Country	Year	Size in hectares
Kavango-Zambezi Transfrontier Conservation Area	Angola, Botswana, Namibia, Zambia, Zimbabwe	2010	38713200
Luiana Game Reserve			
Mavinga Game Reserve			
Chobe National Park			
Makgadikgadi National Park			
Nxai Pan National Park			
Moremi Game Reserve			
Mamili National Park			
Mudumu National Park			
Bwabwata National Park			
Liuwa Plain National Park			
Kafue National Park			
Mosi-oa-Tunya National Park			
Sioma Ngwezi National Park			
Hwange National Park			
Kazuma Pan National Park			
Zambezi National Park			
Victoria Falls National Park			
Marianas Trench National Monument	Mariana Islands, United States of America	2009	25000000
Great Limpopo Transfrontier Park	Mozambique, South Africa, Zimbabwe	2002	9980000
Limpopo National Park			
Makuleke region			
Banhine National Park			
Zinave National Park			
Maputo Special Reserve			
Kruger National Park			
Gonarezhou National Park			
Manjinji Pan Sanctuary			
Malipati Safari Area			
Senge Communial Land Area			
Air and Ténéré Natural Reserves	Niger	1991	7736000
Aïr and Ténéré National Nature Reserve			
Aïr and Ténéré Addax Sanctuary			
Central Kalahari Game Reserve	Botswana	1961	5280000
Namib Naukluft Park	Namibia	1907	5976800
Great Australian Bight Commonwealth Marine Reserve	Australia (south of South Australia)	2012	4592600
Wood Buffalo National Park	Canada (Alberta and Northwest Territories)	1922	4480700

(continued)

National park or national forest	Country	Year	Size in hectares
Selous Game Reserve	Tanzania	1922	4460000
Niassa National Reserve	Mozambique	1954	4200000
Great Arctic State Nature Reserve	Russia	1993	4169200
Gates of the Arctic National Park and Preserve	United States(Alaska)	1980	3946000
Tumucumaque National Park	Brazil	2002	388740
Parima Tapirapecó National Park	Venezuela	1991	3829000
Quittinirpaaq National Park	Canada (Nunavut)	2001	377500
Kgalagadi Transfrontier Park	Botswana, South Africa	2000	3725600
Kalahari Gemsbok National Park			
Gemsbok National Park			
Bernardo O'Higgins National Park	Chile	1969	3525900
Rose Atoll National Monument	American Samoa	1969	3500000
Yukon Flats National Wildlife Refuge	Alaska, United States of America	1969	3500000
Guiana Amazonian Park	French Guiana	2007	33900
Canaima National Park	Venezuela	1962	3000000
Nahanni National Park Reserve	Canada (Northwest Territories)	1976	3000000
Simpson Desert Regional Reserve	Australia (South Australia)	1985	2923900
Boucle du Baoulé National Park	Mali	1982	2533000
Lorentz National Park	West Papua, Indonesia	1997	2505600
Denali National Park and Preserve	United States (Alaska)	1917	2458500
Jaú National Park	Brazil	2000	2300000
Boma National Park	South Sudan	2012	2280000
Kafue National Park	Zambia	1924	2240000
Etosha National Park	Namibia	1907	2227000
Sirmilik National Park	Canada (Nunavut)	2001	2220000
Kluane National Park	Canada (Yukon)	1972	2201300
Auyuittuq National Park	Canada (Nunavut)	2001	2247100
Tsavo National Park	Kenya	1948	2081200
Tsavo East National Park			
Tsavo West National Park			
Ukkusiksalik National Park	Canada (Nunavut)	2003	2050000
Ruaha National Park	Tanzania	1964	2020000
Yellabinna Regional Reserve	Australia (South Australia)	1990	2000800
Bosawás Biosphere Reserve	Nicaragua	1991	2000000
Kakadu National Park	Australia (Northern Territory)	1981	1980400
Nullarbor Regional Reserve	Australia (South Australia)	1989	1919800
Katmai National Park and Preserve	United States (Alaska)	1980	1912299
Yugyd Va National Park	Russia (Komi)	1994	1891700

(continued)

National park or national forest	Country	Year	Size in hectares
Laguna San Rafael National Park	Chile	1959	1742000
Salonga National Park	Democratic Republic of the Congo	1984	1704600
Skeleton Coast National Park	Namibia	1971	1687000
Maloti-Drakensberg Transfrontier Conservation Area	Lesotho, South Africa	2001	1622600
Golden Gate Highlands National Park			
QwaQwa National Park			
Sterkfontein Dam Nature Reserve			
uKhahlamba Drakensberg Park			
Royal Natal National Park			
Sehlabathebe National Park			
Tuktut Nogait National Park	Canada (Northwest Territories)	1996	1634000
Lake Clark National Park and Preserve	United States (Alaska)	1980	1630900
Central Suriname Nature Reserve	Suriname	1998	1600000
Manú National Park	Peru	1987	1532800
Iona National Park	Angola	1964	1515000
Serengeti National Park	Tanzania	1981	1476300
Hwange National Park	Zimbabwe	1930	1465100
Alberto de Agostini National Park	Chile	2005	1460000
Okapi Wildlife Reserve	Democratic Republic of the Congo	1996	1400000
Kerinci Seblat National Park	Sumatra, Indonesia	1982	1379100
Death Valley National Park	United States (California)	1933	1363000
Innamincka Regional Reserve	Australia (South Australia)	1988	1354000
Kati Thanda-Lake Eyre National Park	Australia (South Australia)	1985	1349100
Glacier Bay National Park and Preserve	United States (Alaska)	1980	1328700
Judbarra/Gregory National Park	Australia (Northern Territory)	1990	1288200
Karlamilyi National Park	Australia (Western Australia)	1977	1283700
Chiribiquete National Park	Colombia	1989	1280000
Fiordland National Park	New Zealand	1952	1250000
Aulavik National Park	Canada (Northwest Territories)	1992	1220000
Banc d'Arguin National Park	Mauritania	1978	1200000
Vatnajökull National Park	Iceland	2008	1420000
Upemba National Park	Democratic Republic of the Congo	1939	1173000
Wapusk National Park	Canada (Manitoba)	1996	1147500
Puinawai Natural Reserve	Colombia (Guainía Department)	1989	1092500
Bering Land Bridge National Preserve	United States (Alaska)	1978	1092200

(continued)

National park or national forest	Country	Year	Size in hectares
Jasper National Park	Canada (Alberta)	1907	1087800
Maiko National Park	Democratic Republic of the Congo	1970	1083000
Yaigojé Apaporis Natural Park	Colombia	2009	1055700
Yukon-Charley Rivers National Preserve	United States (Alaska)	1978	1022000
Ivvavik National Park	Canada (Yukon)	1984	1016800
Munga-Thirri National Park	Australia (Queensland)	1967	1012000
Central Karakoram National Park	Skardu District, Gilgit District, Pakistan	1993	1000000
W Transborder Park	Benin, Burkina Faso, Niger	1954	1000000
Quiçama National Park	Angola	1957	996000
Torngat Mountains National Park	Canada (Labrador and Newfoundland)	2005	960000
Laponian Area - UNESCO World Heritage Site	Sweden (Lapland)	1996	940000
Sarek National Park			
Padjelanta National Park			
Stora Sjöfallet National Park			
Muddus National Park			
Sjaunja Nature Reserve (other languages)			
Stubba Nature Reserve (nl/fr)			
Misty Fjords National Monument	United States (Alaska)	1978	924,600
South Luangwa National Park	Zambia	1972	905,000
Yellowstone National Park	United States (Idaho, Montana, and Wyoming)	1872	898,300
Nullarbor Wilderness Protection Area	Australia (South Australia)	2013	894,200
Ngorongoro Conservation Area	Tanzania	1959	828,800
Tadres Reserve	Niger	1940	788,900
Virunga National Park	Democratic Republic of the Congo	1925	780,000
Witjira National Park	Australia (South Australia)	1985	771,100
Grand Staircase-Escalante National Monument	United States (Utah)	1996	757,100
Pechora-Ilych Nature Reserve	Russia	1930	721,300
Nahuel Huapi National Park	Argentina	1934	705,000
Termit Massif Reserve	Niger	1962	700000
Kosciuszko National Park	Australia (New South Wales)	1967	690000
Kobuk Valley National Park	United States (Alaska)	1980	675800
Banff National Park	Canada (Alberta)	1885	664100
Alpine National Park	Australia (Victoria)	1989	646000
Murray-Sunset National Park	Australia (Victoria)	1991	633000
Karijini National Park	Australia (Western Australia)	1991	627400
Mojave National Preserve	United States (California)	1994	621100
Southwest National Park	Australia (Tasmania)	1955	618300

(continued)

National park or national forest	Country	Year	Size in hectares
Ai-Ais/Richtersveld Transfrontier Park	Namibia, South Africa	2003	604500
Ai-Ais Hot Springs Game Park			
Richtersveld National Park			
Kahuzi-Biega National Park	Democratic Republic of the Congo	1980	600000
Lake Torrens National Park	Australia (South Australia)	1991	567700
Waterton-Glacier International Peace Park	Canada (Alberta), United States (Montana)	1995	560600
Glacier National Park			
Waterton Lakes National Park			
Vindelfjällen Nature Reserve (other languages)	Sweden (Lapland)	1995	560000
Saoyú-ʔehdacho National Historic Site	Canada (Northwest Territories)	1997	558700
Lake Gairdner National Park	Australia (South Australia)	1991	548100
Lakefield National Park	Australia (Queensland)	1979	537000
Glen Canyon National Recreation Area	United States (Arizona and Utah)	1972	507600
Diamantina National Park	Australia (Queensland)	1993	507000
Conkouati-Douli National Park	Republic of Congo	1993	505000
Wollemi National Park	Australia (New South Wales)	1979	501700
Yellabinna Wilderness Protection Area	Australia (South Australia)	2005	500700
Bafing National Park	Mali	2000	500000
Sioma Ngwezi National Park	Zambia	2007	500000
Garamba National Park	Democratic Republic of the Congo	1938	492000
Staaten River National Park	Australia (Queensland)	1977	470000
Oyala Thumotang National Park	Australia (Queensland)	1994	457000
Cairngorms National Park	Scotland	2003	452800
Kahurangi National Park	New Zealand	1996	452000
Drysdale River National Park	Australia (Western Australia)	1974	448300
Katavi National Park	Tanzania	1974	447100
Franklin-Gordon Wild Rivers National Park	Australia (Tasmania)	1908	457000
Los Glaciares National Park	Argentina	1981	445900
Vuntut National Park	Canada (Yukon)	1995	434500
Taman Negara National Park	Malaysia	1938	434300
Nouabalé-Ndoki National Park	Republic of Congo	1993	400000
Sighisoara, Tarnava Mare, Podisul Hartibaciului	Romania	2008	360000

Data were collected from Website of National Parks (http://travel.nationalgeographic.com/travel/national-parks/)

Appendix II

Year of establishment, geographic and climatic information, and the number of invasive species reported for 24 natural conservation districts in China

Natural conservation districts	Year	elevation (m)	Latitude (Low)	Latitude (High)	Area (hm^2)	Distance to city (km)	No. of plant species	Annual temp. (°C)	Annual Precipitation (mm)	No.of invasive species
Bannahe	1991	1000	22.07	22.28	26660.0	538	1954	20.0	1500	28
Bawangling	1980	1390	18.95	19.18	29980.0	226	2103	21.3	1657	27
Dashahe	1984	1400	29.00	29.22	26990.0	331	1591	16.0	1280	28
Datian	1986	60	19.08	19.28	1314.0	222	602	23.0	1019	34
Dayudao	1995	25	24.46	24.47	0.2	2	147	20.7	1143	15
Dinghushan	1956	507	23.16	23.19	1133.0	86	2500	20.9	1927	51
Dongzhaigang	1980	20	19.85	20.02	3 337. 6	30	317	23.8	1700	36
Fangchengjinhua	1986	1400	21.73	21.83	9159.0	164	1387	22.5	2800	25
Ganshiling	1985	347	18.33	18.35	1715.5	262	1334	24 .5	1200	44
Hengshui	2000	21	37.53	37.70	187.9	200	382	13.0	518.9	28
Hongshulin	2000	20	22.05	23.12	533. 33	90	49	22.0	2085	20
Jinfoshan	1979	1400	28.83	29.15	41850.0	107	4883	8.2	1434	43
Jiuduansha	2000	3	31.05	31.28	42320.0	47	48	15.7	1145	18
Lushan	1981	1100	29.50	29.68	30466.0	122	2269	13.4	1939	19
Poyanghu	1983	12	29.08	29.25	22400.0	29	526	17.1	1600	19
Shedao-Laotieshan	1980	25	38.95	38.95	17073.0	432	422	10.0	614.5	18
Shiwandashan	1982	1000	21.67	22.07	58 277. 1	135	2233	21.8	2700	22
Songshan	1985	1300	40.49	38.56	4671.0	90	783	8.5	493	23
Taohongling	1981	400	29.70	29.88	12500.0	210	1200	16.5	1300	49
Tianmushan	1956	900	30.31	30.42	4284.0	87	2665	11.8	2179	46
Tongguling	1983	42.5	19.62	19.69	4400.0	107	904	23.9	2361	34
Xingdoushan	1988	1200	29.95	30.18	68339.0	52	3906	14.3	1300	25
Yalujiangkou	1987	2	39.68	40.83	108057.0	285	289	9.8	786	7
Yiwulvshan	1986	530	41.43	41.77	11459.0	224	718	8.0	600	3

Appendix III

Species reported in nature reserves in China, and their life form and growth form

Species	Family	Life form	Growth form	Natural reservation	Major references
Conyza canadensis	Compositae	A	H	Taohongling	Xu et al. (2012)
Euphorbia maculate	Eurphorbiaceae	A	H	Bawangling	Hu et al. (2011)
Lantana camara	Verbenaceae	P	S	Bawangling	Hu et al. (2011)
Lantana camara	Verbenaceae	P	S	Tongguling	Jiang et al. (2007) and Qin et al. (2008)
Leucaena leucocephala	Leguminosae	P	T	Bawangling	Hu et al. (2011)
Manihot esculenta	Euphorbiaceae	P	T	Bawangling	Hu et al. (2011)
Nerium indicum	Apocynaceae	P	S	Bawangling	Hu et al. (2011)
Ricinus communis	Euphorbiaceae	A	H	Bawangling	Hu et al. (2011)
Tamarindus indica	Leguminosae	P	T	Bawangling	Hu et al. (2011)
Abutilon crispum	Malvaceae	P	H	Bawangling	Hu et al. (2011)
Abutilon crispum	Malvaceae	P	H	Datian	Jiang et al. (2007) and Qin et al. (2008)
Abutilon crispum	Malvaceae	P	H	Dongzhaigang	Jiang et al. (2007) and Qin et al. (2008)
Acanthospermum australe	Compositae	A	H	Dinghushan	Song et al. (2009), He and Huang (2004) and Jiang et al. (2007)
Acanthospermum australe	Compositae	A	H	Dongzhaigang	Jiang et al. (2007) and Qin et al. (2008)
Aegilops squarrosa	Gramineae	A	H	Hengshui	Li (2008)
Ageratum conyzoides	Compositae	A	H	Bawangling	Hu et al. (2011)
Ageratum conyzoides	Compositae	A	H	Dashahe	Lin et al. (2008)
Ageratum conyzoides	Compositae	A	H	Datian	Jiang et al. (2007) and Qin et al. (2008)
Ageratum conyzoides	Compositae	A	H	Dinghushan	Song et al. (2009), He and Huang (2004) and Jiang et al. (2007)
Ageratum conyzoides	Compositae	A	H	Dinghushan	Song et al. (2009), He and Huang (2004) and Jiang et al. (2007)
Ageratum conyzoides	Compositae	A	H	Dinghushan	Song et al. (2009), He and Huang (2004) and Jiang et al. (2007)
Ageratum conyzoides	Compositae	A	H	Dongzhaigang	Jiang et al. (2007) and Qin et al. (2008)
Ageratum conyzoides	Compositae	A	H	Ganshiling	Zhang and Xing (2011)

(continued)

Species	Family	Life form	Growth form	Natural reservation	Major references
Ageratum conyzoides	Compositae	A	H	Hongshulin	Cao et al. (2007)
Ageratum conyzoides	Compositae	A	H	Shiwandashan	Wei et al. (2006), Jiang et al. (2007) and Ye et al. (2008)
Ageratum conyzoides	Compositae	A	H	Tianmushan	Chen et al. (2011), Li et al. (2011) and Bai et al. (2013)
Ageratum conyzoides	Compositae	A	H	Tongguling	Jiang et al. (2007) and Qin et al. (2008)
Ageratum houstonianum	Compositae	A	H	Nabanhe	Liu et al. (2008)
Ageratum houstonianum	Compositae	A	H	Shiwandashan	Wei et al. (2006), Jiang et al. (2007) and Ye et al. (2008)
Agropyron cristatum	Gramineae	P	H	Songshan	Liu et al. (2012)
Alternanthera philoxeroides	Amaranthaceae	P	H	Bawangling	Hu et al. (2011)
Alternanthera philoxeroides	Amaranthaceae	P	H	Songshan	Liu et al. (2012)
Alternanthera philoxeroides	Amaranthaceae	P	H	Dashahe	Lin et al. (2008)
Alternanthera philoxeroides	Amaranthaceae	P	H	Dayudao	Zhu et al. (2006)
Alternanthera philoxeroides	Amaranthaceae	P	H	Dinghushan	Song et al. (2009), He and Huang (2004) and Jiang et al. (2007)
Alternanthera philoxeroides	Amaranthaceae	P	H	Dinghushan	Song et al. (2009), He and Huang (2004) and Jiang et al. (2007)
Alternanthera philoxeroides	Amaranthaceae	P	H	Dinghushan	Song et al. (2009), He and Huang (2004) and Jiang et al. (2007)
Alternanthera philoxeroides	Amaranthaceae	P	H	Dongzhaigang	Jiang et al. (2007) and Qin et al. (2008)
Alternanthera philoxeroides	Amaranthaceae	P	H	Hengshui	Li (2008)
Alternanthera philoxeroides	Amaranthaceae	P	H	Jinfoshan	Lin et al. (2007) and Sun et al. (2009)
Alternanthera philoxeroides	Amaranthaceae	P	H	Lushan	Deng et al. (2009), Zheng et al. (2011) and Tang et al. (2012)

(continued)

Species	Family	Life form	Growth form	Natural reservation	Major references
Alternanthera philoxeroides	Amaranthaceae	P	H	Lushan	Deng et al. (2009), Zheng et al. (2011) and Tang et al. (2012)
Alternanthera philoxeroides	Amaranthaceae	P	H	Nabanhe	Liu et al. (2008)
Alternanthera philoxeroides	Amaranthaceae	P	H	Poyanghu	Ge et al. (2010) and Zeng et al. (2013)
Alternanthera philoxeroides	Amaranthaceae	P	H	Shiwandashan	Wei et al. (2006), Jiang et al. (2007) and Ye et al. (2008)
Alternanthera philoxeroides	Amaranthaceae	P	H	Taohongling	Xu et al. (2012)
Alternanthera philoxeroides	Amaranthaceae	P	H	Tianmushan	Chen et al. (2011), Li et al. (2011) and Bai et al. (2013)
Alternanthera pungens	Amaranthaceae	A	H	Bawangling	Hu et al. (2011)
Alternanthera pungens	Amaranthaceae	A	H	Datian	Jiang et al. (2007) and Qin et al. (2008)
Alternanthera pungens	Amaranthaceae	A	H	Dongzhaigang	Jiang et al. (2007) and Qin et al. (2008)
Alternanthera pungens	Amaranthaceae	A	H	Tongguling	Jiang et al. (2007) and Qin et al. (2008)
Amaranthus blitoides	Amaranthaceae	A	H	Hengshui	Li (2008)
Amaranthus blitoides	Amaranthaceae	A	H	Shedao-Laotieshan	Jiang et al. (2007) and Wu et al. (2010)
Amaranthus caudatus	Amaranthaceae	A	H	Lushan	Deng et al. (2009), Zheng et al. (2011) and Tang et al. (2012)
Amaranthus caudatus	Amaranthaceae	A	H	Taohongling	Xu et al. (2012)
Amaranthus hybridus	Amaranthaceae	A	H	Poyanghu	Ge et al. (2010) and Zeng et al. (2013)
Amaranthus hybridus	Amaranthaceae	A	H	Shedao-Laotieshan	Jiang et al. (2007) and Wu et al. (2010)
Amaranthus lividus	Amaranthaceae	A	H	Dinghushan	Song et al. (2009), He and Huang (2004) and Jiang et al. (2007)
Amaranthus lividus	Amaranthaceae	A	H	Tianmushan	Chen et al. (2011), Li et al. (2011) and Bai et al. (2013)

(continued)

Species	Family	Life form	Growth form	Natural reservation	Major references
Amaranthus paniculatus L.	Amaranthaceae	A	H	Tianmushan	Chen et al. (2011), Li et al. (2011) and Bai et al. (2013)
Amaranthus polygonoides	Amaranthaceae	A	H	Shedao-Laotieshan	Jiang et al. (2007) and Wu et al. (2010)
Amaranthus retroflexus	Amaranthaceae	A	H	Songshan	Liu et al. (2012)
Amaranthus retroflexus	Amaranthaceae	A	H	Hengshui	Li (2008)
Amaranthus retroflexus	Amaranthaceae	A	H	Jinfoshan	Lin et al. (2007) and Sun et al. (2009)
Amaranthus retroflexus	Amaranthaceae	A	H	Jiuduansha	Jiang et al. (2007) and Qin et al. (2007)
Amaranthus retroflexus	Amaranthaceae	A	H	Lushan	Deng et al. (2009), Zheng et al. (2011) and Tang et al. (2012)
Amaranthus retroflexus	Amaranthaceae	A	H	Shiwandashan	Wei et al. (2006), Jiang et al. (2007) and Ye et al. (2008)
Amaranthus retroflexus	Amaranthaceae	A	H	Taohongling	Xu et al. (2012)
Amaranthus retroflexus	Amaranthaceae	A	H	Tianmushan	Chen et al. (2011), Li et al. (2011) and Bai et al. (2013)
Amaranthus retroflexus	Amaranthaceae	A	H	Xingdoushan	Lu et al. (2005)
Amaranthus retroflexus	Amaranthaceae	A	H	Yalujiangkou	Jiang et al. (2007) and Wu et al. (2010)
Amaranthus retroflexus	Amaranthaceae	A	H	Yiwulvshan	Wu et al. (2010)
Amaranthus spinosus	Amaranthaceae	A	H	Bawangling	Hu et al. (2011)
Amaranthus spinosus	Amaranthaceae	A	H	Dashahe	Lin et al. (2008)
Amaranthus spinosus	Amaranthaceae	A	H	Datian	Jiang et al. (2007) and Qin et al. (2008)
Amaranthus spinosus	Amaranthaceae	A	H	Dayudao	Zhu et al. (2006)
Amaranthus spinosus	Amaranthaceae	A	H	Dinghushan	Song et al. (2009), He and Huang (2004) and Jiang et al. (2007)
Amaranthus spinosus	Amaranthaceae	A	H	Dinghushan	Song et al. (2009), He and Huang (2004) and Jiang et al. (2007)
Amaranthus spinosus	Amaranthaceae	A	H	Dongzhaigang	Jiang et al. (2007) and Qin et al. (2008)

(continued)

Species	Family	Life form	Growth form	Natural reservation	Major references
Amaranthus spinosus	Amaranthaceae	A	H	Ganshiling	Zhang and Xing (2011)
Amaranthus spinosus	Amaranthaceae	A	H	Hengshui	Li (2008)
Amaranthus spinosus	Amaranthaceae	A	H	Jinfoshan	Lin et al. (2007) and Sun et al. (2009)
Amaranthus spinosus	Amaranthaceae	A	H	Nabanhe	Liu et al. (2008)
Amaranthus spinosus	Amaranthaceae	A	H	Poyanghu	Ge et al. (2010) and Zeng et al. (2013)
Amaranthus spinosus	Amaranthaceae	A	H	Shiwandashan	Wei et al. (2006), Jiang et al. (2007) and Ye et al. (2008)
Amaranthus spinosus	Amaranthaceae	A	H	Taohongling	Xu et al. ((2012))
Amaranthus spinosus	Amaranthaceae	A	H	Tianmushan	Chen et al. (2011), Li et al. (2011) and Bai et al. (2013)
Amaranthus spinosus	Amaranthaceae	A	H	Tongguling	Jiang et al. (2007) and Qin et al. (2008)
Amaranthus spinosus	Amaranthaceae	A	H	Xingdoushan	Lu et al. (2005)
Amaranthus tricolor	Amaranthaceae	A	H	Songshan	Liu et al. (2012)
Amaranthus tricolor	Amaranthaceae	A	H	Datian	Jiang et al. (2007) and Qin et al. (2008)
Amaranthus tricolor	Amaranthaceae	A	H	Dongzhaigang	Jiang et al. (2007) and Qin et al. (2008)
Amaranthus tricolor	Amaranthaceae	A	H	Hongshulin	Cao et al. (2007)
Amaranthus tricolor	Amaranthaceae	A	H	Tianmushan	Chen et al. (2011), Li et al. (2011) and Bai et al. (2013)
Amaranthus tricolor	Amaranthaceae	A	H	Tongguling	Jiang et al. (2007) and Qin et al. (2008)
Amaranthus tricolor	Amaranthaceae	A	H	Xingdoushan	Lu et al. (2005)
Amaranthus viridis	Amaranthaceae	A	H	Songshan	Liu et al. (2012)
Amaranthus viridis	Amaranthaceae	A	H	Dashahe	Lin et al. (2008)
Amaranthus viridis	Amaranthaceae	A	H	Dayudao	Zhu et al. (2006)
Amaranthus viridis	Amaranthaceae	A	H	Dinghushan	Song et al. (2009), He and Huang (2004) and Jiang et al. (2007)
Amaranthus viridis	Amaranthaceae	A	H	Dinghushan	Song et al. (2009), He and Huang (2004) and Jiang et al. (2007)
Amaranthus viridis	Amaranthaceae	A	H	Hengshui	Li (2008)

(continued)

Species	Family	Life form	Growth form	Natural reservation	Major references
Amaranthus viridis	Amaranthaceae	A	H	Jinfoshan	Lin et al. (2007) and Sun et al. (2009)
Amaranthus viridis	Amaranthaceae	A	H	Nabanhe	Liu et al. (2008)
Amaranthus viridis	Amaranthaceae	A	H	Shiwandashan	Wei et al. (2006), Jiang et al. (2007) and Ye et al. (2008)
Amaranthus viridis	Amaranthaceae	A	H	Taohongling	Xu et al. (2012)
Amaranthus viridis	Amaranthaceae	A	H	Tongguling	Jiang et al. (2007) and Qin et al. (2008)
Amaranthus viridis	Amaranthaceae	A	H	Xingdoushan	Lu et al. (2005)
Ambrosia artemisiifolia	Compositae	A	H	Lushan	Deng et al. (2009), Zheng et al. (2011) and Tang et al. (2012)
Ambrosia artemisiifolia	Compositae	A	H	Poyanghu	Ge et al. (2010) and Zeng et al. (2013)
Ambrosia artemisiifolia	Compositae	A	H	Shedao-Laotieshan	Jiang et al. (2007) and Wu et al. (2010)
Ambrosia artemisiifolia	Compositae	A	H	Taohongling	Xu et al. (2012)
Ambrosia artemisiifolia	Compositae	A	H	Tianmushan	Chen et al. (2011), Li et al. (2011) and Bai et al. (2013)
Ambrosia trifida	Compositae	A	H	Yalujiangkou	Jiang et al. (2007) and Wu et al. (2010)
Anredera cordifolia	Basellaceae	P	C	Dashahe	Lin et al. (2008)
Anredera cordifolia	Basellaceae	P	C	Dayudao	Zhu et al. (2006)
Anredera cordifolia	Basellaceae	P	C	Jinfoshan	Lin et al. (2007) and Sun et al. (2009)
Arctium lappa	Compositae	P	H	Tianmushan	Chen et al. (2011), Li et al. (2011) and Bai et al. (2013)
Aster subulatus	Compositae	A	H	Dashahe	Lin et al. (2008)
Aster subulatus	Compositae	A	H	Jiuduansha	Jiang et al. (2007) and Qin et al. (2007)
Aster subulatus	Compositae	A	H	Xingdoushan	Lu et al. (2005)
Atropa belladonna	Solanaceae	P	H	Dinghushan	Song et al. (2009), He and Huang (2004) and Jiang et al. (2007)
Avena fatua	Gramineae	A	H	Dashahe	Lin et al. (2008)

(continued)

Species	Family	Life form	Growth form	Natural reservation	Major references
Avena fatua	Gramineae	A	H	Dinghushan	Song et al. (2009), He and Huang (2004) and Jiang et al. (2007)
Avena fatua	Gramineae	A	H	Jinfoshan	Lin et al. (2007) and Sun et al. (2009)
Avena fatua	Gramineae	A	H	Jiuduansha	Jiang et al. (2007) and Qin et al. (2007)
Avena fatua	Gramineae	A	H	Poyanghu	Ge et al. (2010) and Zeng et al. (2013)
Avena fatua	Gramineae	A	H	Taohongling	Xu et al. (2012)
Avena fatua	Gramineae	A	H	Tianmushan	Chen et al. (2011), Li et al. (2011) and Bai et al. (2013)
Avena fatua	Gramineae	A	H	Xingdoushan	Lu et al. (2005)
Axonopus compressus	Gramineae	P	H	Bawangling	Hu et al. (2011)
Axonopus compressus	Gramineae	P	H	Songshan	Liu et al. (2012)
Axonopus compressus	Gramineae	P	H	Dinghushan	Song et al. (2009), He and Huang (2004) and Jiang et al. (2007)
Axonopus compressus	Gramineae	P	H	Dinghushan	Song et al. (2009), He and Huang (2004) and Jiang et al. (2007)
Axonopus compressus	Gramineae	P	H	Dongzhaigang	Jiang et al. (2007) and Qin et al. (2008)
Axonopus compressus	Gramineae	P	H	Ganshiling	Zhang and Xing (2011)
Axonopus compressus	Gramineae	P	H	Hongshulin	Cao et al. (2007)
Axonopus compressus	Gramineae	P	H	Nabanhe	Liu et al. (2008)
Axonopus compressus	Gramineae	P	H	Shiwandashan	Wei et al. (2006), Jiang et al. (2007) and Ye et al. (2008)
Axonopus compressus	Gramineae	P	H	Tongguling	Jiang et al. (2007) and Qin et al. (2008)
Basella alba	Basellaceae	A	H	Ganshiling	Zhang and Xing (2011)
Basella alba	Basellaceae	A	H	Taohongling	Xu et al. (2012)
Bidens frondosa	Compositae	A	H	Songshan	Liu et al. (2012)
Bidens frondosa	Compositae	A	H	Songshan	Liu et al. (2012)
Bidens frondosa	Compositae	A	H	Dinghushan	Song et al. (2009), He and Huang (2004) and Jiang et al. (2007)

(continued)

Species	Family	Life form	Growth form	Natural reservation	Major references
Bidens frondosa	Compositae	A	H	Dinghushan	Song et al. (2009), He and Huang (2004) and Jiang et al. (2007)
Bidens frondosa	Compositae	A	H	Dongzhaigang	Jiang et al. (2007) and Qin et al. (2008)
Bidens frondosa	Compositae	A	H	Hongshulin	Cao et al. (2007)
Bidens frondosa	Compositae	A	H	Yalujiangkou	Jiang et al. (2007) and Wu et al. (2010)
Bidens pilosa	Compositae	A	H	Dashahe	Lin et al. (2008)
Bidens pilosa	Compositae	A	H	Datian	Jiang et al. (2007) and Qin et al. (2008)
Bidens pilosa	Compositae	A	H	Dayudao	Zhu et al. (2006)
Bidens pilosa	Compositae	A	H	Dinghushan	Song et al. (2009), He and Huang (2004) and Jiang et al. (2007)
Bidens pilosa	Compositae	A	H	Dinghushan	Song et al. (2009), He and Huang (2004) and Jiang et al. (2007)
Bidens pilosa	Compositae	A	H	Hongshulin	Cao et al. (2007)
Bidens pilosa	Compositae	A	H	Jinfoshan	Lin et al. (2007) and Sun et al. (2009)
Bidens pilosa	Compositae	A	H	Nabanhe	Liu et al. (2008)
Bidens pilosa	Compositae	A	H	Poyanghu	Ge et al. (2010) and Zeng et al. (2013)
Bidens pilosa	Compositae	A	H	Shiwandashan	Wei et al. (2006), Jiang et al. (2007) and Ye et al. (2008)
Bidens pilosa	Compositae	A	H	Taohongling	Xu et al. (2012)
Bidens pilosa	Compositae	A	H	Tianmushan	Chen et al. (2011), Li et al. (2011) and Bai et al. (2013)
Bidens pilosa	Compositae	A	H	Xingdoushan	Lu et al. (2005)
Bryophyllum pinnatum	Crassulaceae	P	H	Bawangling	Hu et al. (2011)
Bryophyllum pinnatum	Crassulaceae	P	H	Ganshiling	Zhang and Xing (2011)
Canna indica	Cannaceae	P	H	Bawangling	Hu et al. (2011)
Cannabis sativa	Moraceae	A	H	Songshan	Liu et al. (2012)
Cannabis sativa	Moraceae	A	H	Hengshui	Li (2008)
Capsella bursa-pastoris	Cruciferae	A	H	Taohongling	Xu et al. (2012)
Capsicum annuum	Solanaceae	A	H	Ganshiling	Zhang and Xing (2011)

(continued)

Species	Family	Life form	Growth form	Natural reservation	Major references
Capsicum annuum	Solanaceae	A	H	Taohongling	Xu et al. (2012)
Cassia leschenaultiana	Leguminosae	P	S	Lushan	Deng et al. (2009), Zheng et al. (2011) and Tang et al. (2012)
Cassia leschenaultiana	Leguminosae	P	S	Yalujiangkou	Jiang et al. (2007) and Wu et al. (2010)
Cassia mimosoides	Leguminosae	P	H	Datian	Jiang et al. (2007) and Qin et al. (2008)
Cassia mimosoides	Leguminosae	P	H	Dongzhaigang	Jiang et al. (2007) and Qin et al. (2008)
Cassia occidentalis	Leguminosae	P	S	Bawangling	Hu et al. (2011)
Cassia occidentalis	Leguminosae	P	S	Datian	Jiang et al. (2007) and Qin et al. (2008)
Cassia occidentalis	Leguminosae	P	S	Dongzhaigang	Jiang et al. (2007) and Qin et al. (2008)
Cassia occidentalis	Leguminosae	P	S	Tianmushan	Chen et al. (2011), Li et al. (2011) and Bai et al. (2013)
Cassia occidentalis	Leguminosae	P	S	Tongguling	Jiang et al. (2007) and Qin et al. (2008)
Catharanthus roseus	Apocynaceae	P	H	Ganshiling	Zhang and Xing (2011)
Catharanthus roseus	Apocynaceae	P	H	Ganshiling	Zhang and Xing (2011)
Catharanthus roseus	Apocynaceae	P	H	Taohongling	Xu et al. (2012)
Celosia argentea	Amaranthaceae	A	H	Bawangling	Hu et al. (2011)
Celosia cristata	Amaranthaceae	A	H	Taohongling	Xu et al. (2012)
Cenchrus pauciflorus	Gramineae	A	H	Tongguling	Jiang et al. (2007) and Qin et al. (2008)
Centrosema pubescens	Leguminosae	P	C	Ganshiling	Zhang and Xing (2011)
Chaerophyllum villosum	Umbelliferae	A	H	Jinfoshan	Lin et al. (2007) and Sun et al. (2009)
Chenopodium album	Chenopodiaceae	A	H	Hengshui	Li (2008)
Chenopodium album	Chenopodiaceae	A	H	Tianmushan	Chen et al. (2011), Li et al. (2011) and Bai et al. (2013)

(continued)

Species	Family	Life form	Growth form	Natural reservation	Major references
Chenopodium ambrosioides	Chenopodiaceae	P	H	Bawangling	Hu et al. (2011)
Chenopodium ambrosioides	Chenopodiaceae	A	H	Dashahe	Lin et al. (2008)
Chenopodium ambrosioides	Chenopodiaceae	P	H	Dinghushan	Song et al. (2009), He and Huang (2004) and Jiang et al. (2007)
Chenopodium ambrosioides	Chenopodiaceae	A	H	Hengshui	Li (2008)
Chenopodium ambrosioides	Chenopodiaceae	P	H	Jinfoshan	Lin et al. (2007) and Sun et al. (2009)
Chenopodium ambrosioides	Chenopodiaceae	P	H	Jiuduansha	Jiang et al. (2007) and Qin et al. (2007)
Chenopodium ambrosioides	Chenopodiaceae	P	H	Lushan	Deng et al. (2009), Zheng et al. (2011) and Tang et al. (2012)
Chenopodium ambrosioides	Chenopodiaceae	P	H	Nabanhe	Liu et al. (2008)
Chenopodium ambrosioides	Chenopodiaceae	P	H	Poyanghu	Ge et al. (2010) and Zeng et al. (2013)
Chenopodium ambrosioides	Chenopodiaceae	P	H	Shiwandashan	Wei et al. (2006), Jiang et al. (2007) and Ye et al. (2008)
Chenopodium ambrosioides	Chenopodiaceae	P	H	Taohongling	Xu et al. (2012)
Chenopodium ambrosioides	Chenopodiaceae	P	H	Tianmushan	Chen et al. (2011), Li et al. (2011) and Bai et al. (2013)
Chenopodium hybridum	Chenopodiaceae	A	H	Dashahe	Lin et al. (2008)
Chenopodium hybridum	Chenopodiaceae	A	H	Jinfoshan	Lin et al. (2007) and Sun et al. (2009)
Chromolaena odorata	Compositae	P	H	Bawangling	Hu et al. (2011)
Chromolaena odorata	Compositae	P	H	Datian	Jiang et al. (2007) and Qin et al. (2008)
Chromolaena odorata	Compositae	P	H	Dongzhaigang	Jiang et al. (2007) and Qin et al. (2008)
Chromolaena odorata	Compositae	P	H	Ganshiling	Zhang and Xing (2011)

(continued)

Species	Family	Life form	Growth form	Natural reservation	Major references
Chromolaena odorata	Compositae	P	H	Jiuduansha	Jiang et al. (2007) and Qin et al. (2007)
Chromolaena odorata	Compositae	P	H	Tongguling	Jiang et al. (2007) and Qin et al. (2008)
Chromolaena odorata	Compositae	P	H	Bawangling	Hu et al. (2011)
Chromolaena odorata	Compositae	P	S	Datian	Jiang et al. (2007) and Qin et al. (2008)
Chromolaena odorata	Compositae	P	H	Dinghushan	Song et al. (2009), He and Huang (2004) and Jiang et al. (2007)
Chromolaena odorata	Compositae	P	H	Ganshiling	Zhang and Xing (2011)
Chromolaena odorata	Compositae	P	S	Nabanhe	Liu et al. (2008)
Chromolaena odorata	Compositae	P	H	Shiwandashan	Wei et al. (2006), Jiang et al. (2007) and Ye et al. (2008)
Chromolaena odorata	Compositae	P	S	Tongguling	Jiang et al. (2007) and Qin et al. (2008)
Cissampelopsis volubilis	Compositae	P	H	Dinghushan	Song et al. (2009), He and Huang (2004) and Jiang et al. (2007)
Cocos nucifera	Palmae	P	T	Ganshiling	Zhang and Xing (2011)
Coix lacryma-jobi	Poaceae	A	H	Tianmushan	Chen et al. (2011), Li et al. (2011) and Bai et al. (2013)
Conyza bonariensis	Compositae	A	H	Ganshiling	Zhang and Xing (2011)
Conyza bonariensis	Compositae	A	H	Hongshulin	Cao et al. (2007)
Conyza bonariensis	Compositae	A	H	Tianmushan	Chen et al. (2011), Li et al. (2011) and Bai et al. (2013)
Conyza bonariensis	Compositae	A	H	Xingdoushan	Lu et al. (2005)
Conyza canadensis	Compositae	A	H	Dashahe	Lin et al. (2008)
Conyza canadensis	Compositae	A	H	Dinghushan	Song et al. (2009), He and Huang (2004) and Jiang et al. (2007)
Conyza canadensis	Compositae	A	H	Dinghushan	Song et al. (2009), He and Huang (2004) and Jiang et al. (2007)
Conyza canadensis	Compositae	A	H	Ganshiling	Zhang and Xing (2011)

(continued)

Species	Family	Life form	Growth form	Natural reservation	Major references
Conyza canadensis	Compositae	A	H	Jinfoshan	Lin et al. (2007) and Sun et al. (2009)
Conyza canadensis	Compositae	A	H	Shedao-Laotieshan	Jiang et al. (2007) and Wu et al. (2010)
Conyza canadensis	Compositae	A	H	Shiwandashan	Wei et al. (2006), Jiang et al. (2007) and Ye et al. (2008)
Conyza canadensis	Compositae	A	H	Xingdoushan	Lu et al. (2005)
Conyza canadensis (L.) *Cronq.*	Compositae	A	H	Hengshui	Li (2008)
Conyza canadensis (L.) *Cronq.*	Compositae	A	H	Datian	Jiang et al. (2007) and Qin et al. (2008)
Conyza canadensis (L.) *Cronq.*	Compositae	A	H	Tianmushan	Chen et al. (2011), Li et al. (2011) and Bai et al. (2013)
Conyza canadensis (L.) *Cronq.*	Compositae	A	H	Tongguling	Jiang et al. (2007) and Qin et al. (2008)
Conyza sumatrensis	Compositae	A	H	Dashahe	Lin et al. (2008)
Conyza sumatrensis	Compositae	A	H	Dashahe	Lin et al. (2008)
Conyza sumatrensis	Compositae	A	H	Jinfoshan	Lin et al. (2007) and Sun et al. (2009)
Corchorus capsularis	Tiliaceae	A	H	Taohongling	Xu et al. (2012)
Coreopsis grandiflora	Compositae	P	H	Lushan	Deng et al. (2009), Zheng et al. (2011) and Tang et al. (2012)
Coreopsis grandiflora	Compositae	P	H	Tianmushan	Chen et al. (2011), Li et al. (2011) and Bai et al. (2013)
Coreopsis lanceolata	Compositae	P	H	Tianmushan	Chen et al. (2011), Li et al. (2011) and Bai et al. (2013)
Coriandrum sativum	Umbelliferae	A	H	Taohongling	Xu et al. (2012)
Coronopus didymus	Cruciferae	A	H	Dayudao	Zhu et al. (2006)
Coronopus didymus	Cruciferae	A	H	Jiuduansha	Jiang et al. (2007) and Qin et al. (2007)
Coronopus didymus	Cruciferae	A	H	Poyanghu	Ge et al. (2010) and Zeng et al. (2013)
Coronopus didymus	Cruciferae	A	H	Taohongling	Xu et al. (2012)

(continued)

Species	Family	Life form	Growth form	Natural reservation	Major references
Coronopus didymus	Cruciferae	A	H	Tianmushan	Chen et al. (2011), Li et al. (2011) and Bai et al. (2013)
Crassocephalum crepidioides	Compositae	A	H	Dayudao	Zhu et al. (2006)
Crassocephalum crepidioides	Compositae	A	H	Dinghushan	Song et al. (2009), He and Huang (2004) and Jiang et al. (2007)
Crassocephalum crepidioides	Compositae	A	H	Dinghushan	Song et al. (2009), He and Huang (2004) and Jiang et al. (2007)
Crassocephalum crepidioides	Compositae	A	H	Hongshulin	Cao et al. (2007)
Crassocephalum crepidioides	Compositae	A	H	Jinfoshan	Lin et al. (2007) and Sun et al. (2009)
Crassocephalum crepidioides	Compositae	A	H	Nabanhe	Liu et al. (2008)
Crassocephalum crepidioides	Compositae	A	H	Taohongling	Xu et al. (2012)
Crassocephalum crepidioides	Compositae	A	H	Tianmushan	Chen et al. (2011), Li et al. (2011) and Bai et al. (2013)
Crassocephalum crepidioides	Compositae	A	H	Xingdoushan	Lu et al. (2005)
Cuphea balsamona	Lythraceae	A	H	Dinghushan	Song et al. (2009), He and Huang (2004) and Jiang et al. (2007)
Cyperus rotundus	Cyperaceae	P	H	Taohongling	Xu et al. (2012)
Datura metel	Solanaceae	A	H	Taohongling	Xu et al. (2012)
Datura metel	Solanaceae	A	H	Tianmushan	Chen et al. (2011), Li et al. (2011) and Bai et al. (2013)
Datura stramonium	Solanaceae	A	H	Dongzhaigang	Jiang et al. (2007) and Qin et al. (2008)
Datura stramonium	Solanaceae	A	H	Hengshui	Li (2008)
Datura stramonium	Solanaceae	A	H	Jinfoshan	Lin et al. (2007) and Sun et al. (2009)
Datura stramonium	Solanaceae	A	H	Nabanhe	Liu et al. (2008)
Datura stramonium	Solanaceae	A	H	Shedao-Laotieshan	Jiang et al. (2007) and Wu et al. (2010)

(continued)

Species	Family	Life form	Growth form	Natural reservation	Major references
Datura stramonium	Solanaceae	A	H	Shiwandashan	Wei et al. (2006), Jiang et al. (2007) and Ye et al. (2008)
Datura stramonium	Solanaceae	A	H	Tongguling	Jiang et al. (2007) and Qin et al. (2008)
Datura stramonium	Solanaceae	A	H	Xingdoushan	Lu et al. (2005)
Daucus carota	Umbelliferae	B	H	Dashahe	Lin et al. (2008)
Daucus carota	Umbelliferae	B	H	Dinghushan	Song et al. (2009), He and Huang (2004) and Jiang et al. (2007)
Daucus carota	Umbelliferae	B	H	Jinfoshan	Lin et al. (2007) and Sun et al. (2009)
Daucus carota	Umbelliferae	B	H	Jiuduansha	Jiang et al. (2007) and Qin et al. (2007)
Daucus carota	Umbelliferae	B	H	Taohongling	Xu et al. (2012)
Daucus carota	Umbelliferae	B	H	Xingdoushan	Lu et al. (2005)
Eichhornia crassipes	Pontederiaceae	P	H	Dashahe	Lin et al. (2008)
Eichhornia crassipes	Pontederiaceae	P	H	Hongshulin	Cao et al. (2007)
Eichhornia crassipes	Pontederiaceae	P	H	Poyanghu	Ge et al. (2010) and Zeng et al. (2013)
Eichhornia crassipes	Pontederiaceae	P	H	Tianmushan	Chen et al. (2011), Li et al. (2011) and Bai et al. (2013)
Eichhornia crassipes	Pontederiaceae	P	H		
Eleusine indica	Gramineae	A	H	Bawangling	Hu et al. (2011)
Eleusine indica	Gramineae	A	H	Hongshulin	Cao et al. (2007)
Eleusine indica	Gramineae	A	H	Jinfoshan	Lin et al. (2007) and Sun et al. (2009)
Eleusine indica	Gramineae	A	H	Taohongling	Xu et al. (2012)
Eleusine indica	Gramineae	A	H	Xingdoushan	Lu et al. (2005)
Eleusine indica	Gramineae	A	H	Yiwulvshan	Wu et al. (2010)
Erechtites hieracifolia	Compositae	A	H	Shiwandashan	Wei et al. (2006), Jiang et al. (2007) and Ye et al. (2008)
Erechtites valerianaefolia (Wolf) DC.	Compositae	A	H	Ganshiling	Zhang and Xing (2011)
Erigeron annuus	Compositae	A	H	Bawangling	Hu et al. (2011)
Erigeron annuus	Compositae	A	H	Songshan	Liu et al. (2012)
Erigeron annuus	Compositae	A	H	Dashahe	Lin et al. (2008)
Erigeron annuus	Compositae	A	H	Datian	Jiang et al. (2007) and Qin et al. (2008)

(continued)

Species	Family	Life form	Growth form	Natural reservation	Major references
Erigeron annuus	Compositae	A	H	Dinghushan	Song et al. (2009), He and Huang (2004) and Jiang et al. (2007)
Erigeron annuus	Compositae	A	H	Jinfoshan	Lin et al. (2007) and Sun et al. (2009)
Erigeron annuus	Compositae	A	H	Lushan	Deng et al. (2009), Zheng et al. (2011) and Tang et al. (2012)
Erigeron annuus	Compositae	A	H	Poyanghu	Ge et al. (2010) and Zeng et al. (2013)
Erigeron annuus	Compositae	A	H	Shedao-Laotieshan	Jiang et al. (2007) and Wu et al. (2010)
Erigeron annuus	Compositae	A	H	Taohongling	Xu et al. (2012)
Erigeron annuus	Compositae	A	H	Tianmushan	Chen et al. (2011), Li et al. (2011) and Bai et al. (2013)
Erigeron annuus	Compositae	A	H	Xingdoushan	Lu et al. (2005)
Erigeron annuus	Compositae	A	H	Yalujiangkou	Jiang et al. (2007) and Wu et al. (2010)
Erigeron annuus	Compositae	A	H	Yiwulvshan	Wu et al. (2010)
Eryngium foetidum	Umbelliferae	P	H	Nabanhe	Liu et al. (2008)
Eryngium foetidum	Umbelliferae	P	H	Shiwandashan	Wei et al. (2006), Jiang et al. (2007) and Ye et al. (2008)
Eupatorium adenophorum	Compositae	P	H	Nabanhe	Liu et al. (2008)
Eupatorium catarium	Compositae	A	H	Bawangling	Hu et al. (2011)
Eupatorium catarium	Compositae	A	H	Datian	Jiang et al. (2007) and Qin et al. (2008)
Eupatorium catarium	Compositae	A	H	Dongzhaigang	Jiang et al. (2007) and Qin et al. (2008)
Eupatorium catarium	Compositae	A	H	Ganshiling	Zhang and Xing (2011)
Eupatorium catarium	Compositae	A	H	Hongshulin	Cao et al. (2007)
Eupatorium catarium	Compositae	A	H	Tongguling	Jiang et al. (2007) and Qin et al. (2008)
Euphorbia helioscopia	Euphorbiaceae	A	H	Taohongling	Xu et al. (2012)
Euphorbia helioscopia	Euphorbiaceae	A	H	Tianmushan	Chen et al. (2011), Li et al. (2011) and Bai et al. (2013)

(continued)

Species	Family	Life form	Growth form	Natural reservation	Major references
Euphorbia helioscopia	Euphorbiaceae	A	H	Xingdoushan	Lu et al. (2005)
Euphorbia hirta	Euphorbiaceae	A	H	Bawangling	Hu et al. (2011)
Euphorbia hirta	Euphorbiaceae	A	H	Dashahe	Lin et al. (2008)
Euphorbia hirta	Euphorbiaceae	A	H	Datian	Jiang et al. (2007) and Qin et al. (2008)
Euphorbia hirta	Euphorbiaceae	A	H	Dinghushan	Song et al. (2009), He and Huang (2004) and Jiang et al. (2007)
Euphorbia hirta	Euphorbiaceae	A	H	Dinghushan	Song et al. (2009), He and Huang (2004) and Jiang et al. (2007)
Euphorbia hirta	Euphorbiaceae	A	H	Dinghushan	Song et al. (2009), He and Huang (2004) and Jiang et al. (2007)
Euphorbia hirta	Euphorbiaceae	A	H	Dongzhaigang	Jiang et al. (2007) and Qin et al. (2008)
Euphorbia hirta	Euphorbiaceae	A	H	Dongzhaigang	Jiang et al. (2007) and Qin et al. (2008)
Euphorbia hirta	Euphorbiaceae	A	H	Ganshiling	Zhang and Xing (2011)
Euphorbia hirta	Euphorbiaceae	A	H	Hongshulin	Cao et al. (2007)
Euphorbia hirta	Euphorbiaceae	A	H	Jinfoshan	Lin et al. (2007) and Sun et al. (2009)
Euphorbia hirta	Euphorbiaceae	A	H	Nabanhe	Liu et al. (2008)
Euphorbia hirta	Euphorbiaceae	A	H	Poyanghu	Ge et al. (2010) and Zeng et al. (2013)
Euphorbia hirta	Euphorbiaceae	A	H	Shiwandashan	Wei et al. (2006), Jiang et al. (2007) and Ye et al. (2008)
Euphorbia hirta	Euphorbiaceae	A	H	Tongguling	Jiang et al. (2007) and Qin et al. (2008)
Euphorbia maculata	Euphorbiaceae	A	H	Dashahe	Lin et al. (2008)
Euphorbia maculata	Euphorbiaceae	A	H	Jinfoshan	Lin et al. (2007) and Sun et al. (2009)
Euphorbia maculata	Euphorbiaceae	A	H	Lushan	Deng et al. (2009), Zheng et al. (2011) and Tang et al. (2012)

(continued)

Species	Family	Life form	Growth form	Natural reservation	Major references
Euphorbia maculata	Euphorbiaceae	A	H	Shedao-Laotieshan	Jiang et al. (2007) and Wu et al. (2010)
Euphorbia maculata	Euphorbiaceae	A	H	Taohongling	Xu et al. (2012)
Euphorbia maculata	Euphorbiaceae	A	H	Tianmushan	Chen et al. (2011), Li et al. (2011) and Bai et al. (2013)
Fagopyrum esculentum	Polygonaceae	A	H	Songshan	Liu et al. (2012)
Flaveria bidentis (L.) *Kuntze*	Compositae	A	H	Hengshui	Li (2008)
Galinsoga parviflora	Compositae	A	H	Dashahe	Lin et al. (2008)
Galinsoga parviflora	Compositae	A	H	Jinfoshan	Lin et al. (2007) and Sun et al. (2009)
Galinsoga parviflora	Compositae	A	H	Lushan	Deng et al. (2009), Zheng et al. (2011) and Tang et al. (2012)
Galinsoga parviflora	Compositae	A	H	Taohongling	Xu et al. (2012)
Galinsoga parviflora	Compositae	A	H	Tongguling	Jiang et al. (2007) and Qin et al. (2008)
Galinsoga parviflora	Compositae	A	H	Yalujiangkou	Jiang et al. (2007) and Wu et al. (2010)
Galinsoga quadriradiata	Compositae	A	H	Xingdoushan	Lu et al. (2005)
Gaura lindheimeri	Onagraceae	P	H	Lushan	Deng et al. (2009), Zheng et al. (2011) and Tang et al. (2012)
Gaura parviflora	Onagraceae	P	H	Shedao-Laotieshan	Jiang et al. (2007) and Wu et al. (2010)
Gentianopsis contorta	Geraniaceae	A	H	Songshan	Liu et al. (2012)
Geranium carolinianum	Geraniaceae	A	H	Dashahe	Lin et al. (2008)
Geranium carolinianum	Geraniaceae	A	H	Jiuduansha	Jiang et al. (2007) and Qin et al. (2007)
Geranium carolinianum	Geraniaceae	A	H	Poyanghu	Ge et al. (2010) and Zeng et al. (2013)
Geranium carolinianum	Geraniaceae	A	H	Tianmushan	Chen et al. (2011), Li et al. (2011) and Bai et al. (2013)
Geranium carolinianum	Geraniaceae	A	H	Xingdoushan	Lu et al. (2005)
Gomphrena celosioides	Amaranthaceae	A	H	Datian	Jiang et al. (2007) and Qin et al. (2008)

(continued)

Species	Family	Life form	Growth form	Natural reservation	Major references
Gomphrena celosioides	Amaranthaceae	A	H	Dongzhaigang	Jiang et al. (2007) and Qin et al. (2008)
Gomphrena celosioides	Amaranthaceae	A	H	Tongguling	Jiang et al. (2007) and Qin et al. (2008)
Gralinsoga parviflora	Compositae	A	H	Tianmushan	Chen et al. (2011), Li et al. (2011) and Bai et al. (2013)
Helianthus annuus	Compositae	A	H	Taohongling	Xu et al. (2012)
Helianthus tuberosus	Compositae	P	H	Hongshulin	Cao et al. (2007)
Helianthus tuberosus	Compositae	P	H	Lushan	Deng et al. (2009), Zheng et al. (2011) and Tang et al. (2012)
Helianthus tuberosus	Compositae	P	H	Taohongling	Xu et al. (2012)
Helianthus tuberosus	Compositae	P	H	Tianmushan	Chen et al. (2011), Li et al. (2011) and Bai et al. (2013)
Hemerocallis fulva	Liliaceae	P	H	Songshan	Liu et al. (2012)
Hibiscus trionum	Malvaceae	A	H	Hengshui	Li (2008)
Hibiscus trionum	Malvaceae	A	H	Poyanghu	Ge et al. (2010) and Zeng et al. (2013)
Hibiscus trionum	Malvaceae	A	H	Shedao-Laotieshan	Jiang et al. (2007) and Wu et al. (2010)
Humulus scandens	Malvaceae	P	H	Tianmushan	Chen et al. (2011), Li et al. (2011) and Bai et al. (2013)
Hyptis brevipes Poit.	Labiatae	A	H	Ganshiling	Zhang and Xing (2011)
Hyptis rhomboidea	Labiatae	A	H	Datian	Jiang et al. (2007) and Qin et al. (2008)
Hyptis rhomboidea	Labiatae	A	H	Ganshiling	Zhang and Xing (2011)
Hyptis suaveolens	Labiatae	A	H	Datian	Jiang et al. (2007) and Qin et al. (2008)
Hyptis suaveolens	Labiatae	A	H	Dongzhaigang	Jiang et al. (2007) and Qin et al. (2008)
Hyptis suaveolens	Labiatae	A	H	Ganshiling	Zhang and Xing (2011)
Impatiens balsamina L.	Balsaminaceae	A	H	Taohongling	Xu et al. (2012)

(continued)

Species	Family	Life form	Growth form	Natural reservation	Major references
Impatiens balsamina L.	Balsaminaceae	A	H	Tianmushan	Chen et al. (2011), Li et al. (2011) and Bai et al. (2013)
Ipomoea cairica	Convolvulaceae	P	H	Bawangling	Hu et al. (2011)
Ipomoea cairica	Convolvulaceae	P	H	Datian	Jiang et al. (2007) and Qin et al. (2008)
Ipomoea cairica	Convolvulaceae	P	H	Dinghushan	Song et al. (2009), He and Huang (2004) and Jiang et al. (2007)
Ipomoea cairica	Convolvulaceae	P	H	Dinghushan	Song et al. (2009), He and Huang (2004) and Jiang et al. (2007)
Ipomoea cairica	Convolvulaceae	P	H	Dongzhaigang	Jiang et al. (2007) and Qin et al. (2008)
Ipomoea cairica	Convolvulaceae	P	H	Ganshiling	Zhang and Xing (2011)
Ipomoea cairica	Convolvulaceae	P	H	Hongshulin	Cao et al. (2007)
Ipomoea cairica	Convolvulaceae	P	H	Shiwandashan	Wei et al. (2006), Jiang et al. (2007) and Ye et al. (2008)
Ipomoea cairica	Convolvulaceae	P	H	Tongguling	Jiang et al. (2007) and Qin et al. (2008)
Ipomoea nil	Convolvulaceae	A	H	Songshan	Liu et al. (2012)
Ipomoea nil	Convolvulaceae	P	H	Shiwandashan	Wei et al. (2006), Jiang et al. (2007) and Ye et al. (2008)
Ipomoea nil	Convolvulaceae	P	H	Xingdoushan	Lu et al. (2005)
Ipomoea purpurea	Convolvulaceae	A	H	Bawangling	Hu et al. (2011)
Ipomoea purpurea	Convolvulaceae	A	H	Songshan	Liu et al. (2012)
Ipomoea purpurea	Convolvulaceae	A	H	Datian	Jiang et al. (2007) and Qin et al. (2008)
Ipomoea purpurea	Convolvulaceae	A	H	Dinghushan	Song et al. (2009), He and Huang (2004) and Jiang et al. (2007)
Ipomoea purpurea	Convolvulaceae	A	H	Dongzhaigang	Jiang et al. (2007) and Qin et al. (2008)
Ipomoea purpurea	Convolvulaceae	A	H	Hengshui	Li (2008)
Ipomoea purpurea	Convolvulaceae	A	H	Jinfoshan	Lin et al. (2007) and Sun et al. (2009)

(continued)

Species	Family	Life form	Growth form	Natural reservation	Major references
Ipomoea purpurea	Convolvulaceae	A	H	Jiuduansha	Jiang et al. (2007) and Qin et al. (2007)
Ipomoea purpurea	Convolvulaceae	A	H	Nabanhe	Liu et al. (2008)
Ipomoea purpurea	Convolvulaceae	A	H	Shedao-Laotieshan	Jiang et al. (2007) and Wu et al. (2010)
Ipomoea purpurea	Convolvulaceae	A	H	Tianmushan	Chen et al. (2011), Li et al. (2011) and Bai et al. (2013)
Ipomoea purpurea	Convolvulaceae	A	H	Xingdoushan	Lu et al. (2005)
Ipomoea purpurea	Convolvulaceae	A	H	Yiwulvshan	Wu et al. (2010)
Jacquemontia tamnifolia	Convolvulaceae	A	H	Taohongling	Xu et al. (2012)
Jussiaealinifolia Vahl.	Onagraceae	A	H	Songshan	Liu et al. (2012)
Kochia scoparia	Chenopodiaceae	A	H	Tianmushan	Chen et al. (2011), Li et al. (2011) and Bai et al. (2013)
Kochia scoparia	Chenopodiaceae	A	H	Taohongling	Xu et al. (2012)
Lablab purpureus (L.) *Sweet.*	Leguminosae	P	C	Jinfoshan	Lin et al. (2007) and Sun et al. (2009)
Lactuca sativa	Compositae	A	H	Taohongling	Xu et al. (2012)
Lantana camara	Verbenaceae	P	S	Dayudao	Zhu et al. (2006)
Lantana camara	Verbenaceae	P	S	Dinghushan	Song et al. (2009), He and Huang (2004) and Jiang et al. (2007)
Lantana camara	Verbenaceae	P	S	Dinghushan	Song et al. (2009), He and Huang (2004) and Jiang et al. (2007)
Lantana camara	Verbenaceae	P	S	Ganshiling	Zhang and Xing (2011)
Lantana camara	Verbenaceae	P	S	Hongshulin	Cao et al. (2007)
Lantana camara	Verbenaceae	P	S	Shiwandashan	Wei et al. (2006), Jiang et al. (2007) and Ye et al. (2008)
Lantana camara L.	Verbenaceae	P	S	Dongzhaigang	Jiang et al. (2007) and Qin et al. (2008)
Lepidium sativum	Cruciferae	A	H	Dashahe	Lin et al. (2008)
Lepidium virginicum	Cruciferae	B	H	Jiuduansha	Jiang et al. (2007) and Qin et al. (2007)
Lepidium virginicum	Cruciferae	B	H	Lushan	Deng et al. (2009), Zheng et al. (2011) and Tang et al. (2012)

(continued)

Species	Family	Life form	Growth form	Natural reservation	Major references
Lepidium virginicum	Cruciferae	B	H	Poyanghu	Ge et al. (2010) and Zeng et al. (2013)
Lepidium virginicum	Cruciferae	B	H	Taohongling	Xu et al. (2012)
Lepidium virginicum	Cruciferae	B	H	Tianmushan	Chen et al. (2011), Li et al. (2011) and Bai et al. (2013)
Lepidium virginicum	Cruciferae	B	H	Xingdoushan	Lu et al. (2005)
Lepidium virginicum	Cruciferae	B	H	Yalujiangkou	Jiang et al. (2007) and Wu et al. (2010)
Lepidium virginicum	Cruciferae	B	H		
Leucaena glauca (L.) *Benth.*	Mimosaceae	P	T	Datian	Jiang et al. (2007) and Qin et al. (2008)
Leucaena glauca (L.) *Benth.*	Mimosaceae	P	T	Tongguling	Jiang et al. (2007) and Qin et al. (2008)
Leucaena leucocep halavar. salvador	Leguminosae	P	T	Dinghushan	Song et al. (2009), He and Huang (2004) and Jiang et al. (2007)
Leucaena leucocephala	Leguminosae	P	T	Ganshiling	Zhang and Xing (2011)
Lolium multiflorum	Leguminosae	A	H	Hengshui	Li (2008)
Lolium perenne	Gramineae	P	H	Songshan	Liu et al. (2012)
Lolium temulentum	Gramineae	A	H	Hengshui	Li (2008)
Ludwigia prostrata	Onagraceae	A	H	Songshan	Liu et al. (2012)
Lycopersicon esculentum	Solanaceae	A	H	Jinfoshan	Lin et al. (2007) and Sun et al. (2009)
Malvastrum coromandelianum	Malvaceae	P	H	Bawangling	Hu et al. (2011)
Malvastrum coromandelianum	Malvaceae	P	H	Nabanhe	Liu et al. (2008)
Malvastrum coromandelianum	Malvaceae	P	H	Shiwandashan	Wei et al. (2006), Jiang et al. (2007) and Ye et al. (2008)
Malvastrum coromandelianum	Malvaceae	P	H	Tongguling	Jiang et al. (2007) and Qin et al. (2008)
Medicago sativa L.	Leguminosae	P	H	Hengshui	Li (2008)
Medicago sativa L.	Leguminosae	P	H	Jiuduansha	Jiang et al. (2007) and Qin et al. (2007)
Medicago polymorpha L.	Leguminosae	A	H	Poyanghu	Ge et al. (2010) and Zeng et al. (2013)
Medicago polymorpha L.	Leguminosae	A	H	Xingdoushan	Lu et al. (2005)

(continued)

Species	Family	Life form	Growth form	Natural reservation	Major references
Melilotus albus Desr.	Leguminosae sp.	A	H	Jiuduansha	Jiang et al. (2007) and Qin et al. (2007)
Mimosa bimucronata	Leguminosae	P	S	Ganshiling	Zhang and Xing (2011)
Mimosa invisa	Leguminosae sp.	P	H	Ganshiling	Zhang and Xing (2011)
Mimosa sepiaria	Leguminosae	P	S	Dinghushan	Song et al. (2009), He and Huang (2004) and Jiang et al. (2007)
Mimosa pudica Linn.	Leguminosae sp.	P	H	Bawangling	Hu et al. (2011)
Mimosa pudica Linn.	Leguminosae sp.	P	H	Dinghushan	Song et al. (2009), He and Huang (2004) and Jiang et al. (2007)
Mimosa pudica Linn.	Leguminosae sp.	P	H	Dinghushan	Song et al. (2009), He and Huang (2004) and Jiang et al. (2007)
Mimosa pudica Linn.	Leguminosae sp.	P	H	Dongzhaigang	Jiang et al. (2007) and Qin et al. (2008)
Mimosa pudica Linn.	Leguminosae sp.	P	H	Ganshiling	Zhang and Xing (2011)
Mimosa pudica Linn.	Leguminosae sp.	P	H	Hongshulin	Cao et al. (2007)
Mimosa pudica Linn.	Leguminosae sp.	P	H	Nabanhe	Liu et al. (2008)
Mimosa pudica Linn.	Leguminosae sp.	P	H	Shiwandashan	Wei et al. (2006) and Jiang et al. (2007) and Ye et al. (2008)
Mimosa pudica Linn.	Leguminosae sp.	P	H	Taohongling	Xu et al. (2012)
Mimosa pudica Linn.	Leguminosae sp.	P	H	Tongguling	Jiang et al. (2007) and Qin et al. (2008)
Mirabilis jalapa L.	Nyctaginaceae	A	H	Dashahe	Lin et al. (2008)
Mirabilis jalapa L.	Nyctaginaceae	A	H	Dinghushan	Song et al. (2009), He and Huang (2004) and Jiang et al. (2007)
Mirabilis jalapa L.	Nyctaginaceae	A	H	Dongzhaigang	Jiang et al. (2007) and Qin et al. (2008)
Mirabilis jalapa L.	Nyctaginaceae	A	H	Jinfoshan	Lin et al. (2007) and Sun et al. (2009)
Mirabilis jalapa L.	Nyctaginaceae	A	H	Nabanhe	Liu et al. (2008)
Mirabilis jalapa L.	Nyctaginaceae	A	H	Taohongling	Xu et al. (2012)
Mirabilis jalapa L.	Nyctaginaceae	A	H	Tianmushan	Chen et al. (2011), Li et al. (2011) and Bai et al. (2013)

(continued)

Species	Family	Life form	Growth form	Natural reservation	Major references
Monarda fistulosa L.	Labiatae	A	H	Lushan	Deng et al. (2009), Zheng et al. (2011) and Tang et al. (2012)
Nicandra physalodes	Solanaceae	A	H	Shedao-Laotieshan	Jiang et al. (2007) and Wu et al. (2010)
Oenothera biennis	Onagraceae	B	H	Shedao-Laotieshan	Jiang et al. (2007) and Wu et al. (2010)
Oenothera parviflora L.	Onagraceae	B	H	Lushan	Deng et al. (2009), Zheng et al. (2011) and Tang et al. (2012)
Opuntia stricta var dillenii	Cactaceae	P	T	Dongzhaigang	Jiang et al. (2007) and Qin et al. (2008)
Opuntia stricta var dillenii	Cactaceae	P	T	Taohongling	Xu et al. (2012)
Opuntia stricta var dillenii	Cactaceae	P	T	Tongguling	Jiang et al. (2007) and Qin et al. (2008)
Oxalis corymbosa	Oxalidaceae	P	H	Jinfoshan	Lin et al. (2007) and Sun et al. (2009)
Oxalis corymbosa DC.	Oxalidaceae	P	H	Bawangling	Hu et al. (2011)
Oxalis corymbosa DC.	Oxalidaceae	P	H	Dashahe	Lin et al. (2008)
Oxalis corymbosa DC.	Oxalidaceae	P	H	Dayudao	Zhu et al. (2006)
Oxalis corymbosa DC.	Oxalidaceae	P	H	Dinghushan	Song et al. (2009), He and Huang (2004) and Jiang et al. (2007)
Oxalis corymbosa DC.	Oxalidaceae	P	H	Dinghushan	Song et al. (2009), He and Huang (2004) and Jiang et al. (2007)
Oxalis corymbosa DC.	Oxalidaceae	P	H	Ganshiling	Zhang and Xing (2011)
Oxalis corymbosa DC.	Oxalidaceae	P	H	Lushan	Deng et al. (2009), Zheng et al. (2011) and Tang et al. (2012)
Oxalis corymbosa DC.	Oxalidaceae	P	H	Shiwandashan	Wei et al. (2006), Jiang et al. (2007) and Ye et al. (2008)
Oxalis corymbosa DC.	Oxalidaceae	P	H	Taohongling	Xu et al. (2012)
Pachyrhizus erosus (Linn.) Urb.	Leguminosae sp.	P	C	Jinfoshan	Lin et al. (2007) and Sun et al. (2009)

(continued)

Species	Family	Life form	Growth form	Natural reservation	Major references
Panicum maximum	Gramineae	P	H	Ganshiling	Zhang and Xing (2011)
Panicum repens L.	Gramineae	P	H	Datian	Jiang et al. (2007) and Qin et al. (2008)
Panicum repens L.	Gramineae	P	H	Dayudao	Zhu et al. (2006)
Panicum repens L.	Gramineae	P	H	Dinghushan	Song et al. (2009), He and Huang (2004) and Jiang et al. (2007)
Panicum repens L.	Gramineae	P	H	Shiwandashan	Wei et al. (2006), Jiang et al. (2007) and Ye et al. (2008)
Panicum repens L.	Gramineae	P	H	Taohongling	Xu et al. (2012)
Panicum repens L.	Gramineae	P	H	Tongguling	Jiang et al. (2007) and Qin et al. (2008)
Parthenium hysterophorus	Compositae	A	H	Datian	Jiang et al. (2007) and Qin et al. (2008)
Parthenium hysterophorus	Compositae	A	H	Dinghushan	Song et al. (2009), He and Huang (2004) and Jiang et al. (2007)
Parthenium hysterophorus	Compositae	A	H	Dongzhaigang	Jiang et al. (2007) and Qin et al. (2008)
Parthenocissus quinquefo lia (L.) *Planch.*	Vitaceae	P	C	Tongguling	Jiang et al. (2007) and Qin et al. (2008)
Paspalum conjugatum	Gramineae	P	H	Bawangling	Hu et al. (2011)
Paspalum conjugatum	Gramineae	P	H	Dinghushan	Song et al. (2009), He and Huang (2004) and Jiang et al. (2007)
Paspalum conjugatum	Gramineae	P	H	Dongzhaigang	Jiang et al. (2007) and Qin et al. (2008)
Paspalum conjugatum	Gramineae	P	H	Nabanhe	Liu et al. (2008)
Paspalum conjugatum	Gramineae	P	H	Tongguling	Jiang et al. (2007) and Qin et al. (2008)
Paspalum dilatatum	Gramineae	P	H	Hongshulin	Cao et al. (2007)
Passiflora edulis Sims	Passifloraceae	P	C	Ganshiling	Zhang and Xing (2011)
Passiflora foetida L.	Passifloraceae	P	C	Bawangling	Hu et al. (2011)
Passiflora foetida L.	Passifloraceae	P	C	Datian	Jiang et al. (2007) and Qin et al. (2008)

(continued)

Species	Family	Life form	Growth form	Natural reservation	Major references
Passiflora foetida L.	Passifloraceae	P	C	Dongzhaigang	Jiang et al. (2007) and Qin et al. (2008)
Passiflora foetida L.	Passifloraceae	P	C	Ganshiling	Zhang and Xing (2011)
Passiflora foetida L.	Passifloraceae	P	C	Hongshulin	Cao et al. (2007)
Passiflora foetida L.	Passifloraceae	P	C	Jinfoshan	Lin et al. (2007) and Sun et al. (2009)
Passiflora foetida L.	Passifloraceae	P	C	Tongguling	Jiang et al. (2007) and Qin et al. (2008)
Peperomia pellucida (L.) *Kunth*	Piperaceae	A	H	Dinghushan	Song et al. (2009), He and Huang (2004) and Jiang et al. (2007)
Peperomia pellucida (L.) *Kunth*	Piperaceae	A	H	Dinghushan	Song et al. (2009), He and Huang (2004) and Jiang et al. (2007)
Peperomia pellucida (L.) *Kunth*	Piperaceae	A	H	Dongzhaigang	Jiang et al. (2007) and Qin et al. (2008)
Peperomia pellucida (L.) *Kunth*	Piperaceae	A	H	Nabanhe	Liu et al. (2008)
Pharbitis nil (L.) *Choisy*	Convolvulaceae	A	H	Songshan	Liu et al. (2012)
Pharbitis nil (L.) *Choisy*	Convolvulaceae	A	H	Dinghushan	Song et al. (2009), He and Huang (2004) and Jiang et al. (2007)
Pharbitis nil (L.) *Choisy*	Convolvulaceae	A	H	Dinghushan	Song et al. (2009), He and Huang (2004) and Jiang et al. (2007)
Pharbitis nil (L.) *Choisy*	Convolvulaceae	A	H	Shedao-Laotieshan	Jiang et al. (2007) and Wu et al. (2010)
Pharbitis nil (L.) *Choisy*	Convolvulaceae	A	H	Tianmushan	Chen et al. (2011), Li et al. (2011) and Bai et al. (2013)
Phaseolus coccineus L.	Leguminosae sp.	P	H	Songshan	Liu et al. (2012)
Phaseolus lunatus Linn.	Leguminosae sp.	P	H	Jinfoshan	Lin et al. (2007) and Sun et al. (2009)
Phaseolus vulgaris L.	Leguminosae sp.	A	H	Jinfoshan	Lin et al. (2007) and Sun et al. (2009)

(continued)

Species	Family	Life form	Growth form	Natural reservation	Major references
Phleum pratense L.	Gramineae	P	H	Lushan	Deng et al. (2009), Zheng et al. (2011) and Tang et al. (2012)
Phy to lacca americana	Phytolaccaceae	P	H	Dashahe	Lin et al. (2008)
Physalis angulata L.	Solanaceae	A	H	Jinfoshan	Lin et al. (2007) and Sun et al. (2009)
Physalis pubescens	Solanaceae	A	H	Jinfoshan	Lin et al. (2007) and Sun et al. (2009)
Phytolacca americana	Phytolaccaceae	P	H	Jinfoshan	Lin et al. (2007) and Sun et al. (2009)
Phytolacca americana	Phytolaccaceae	P	H	Lushan	Deng et al. (2009), Zheng et al. (2011) and Tang et al. (2012)
Phytolacca americana	Phytolaccaceae	P	H	Nabanhe	Liu et al. (2008)
Phytolacca americana	Phytolaccaceae	P	H	Poyanghu	Ge et al. (2010) and Zeng et al. (2013)
Phytolacca americana	Phytolaccaceae	P	H	Shedao-Laotieshan	Jiang et al. (2007) and Wu et al. (2010)
Phytolacca americana	Phytolaccaceae	P	H	Taohongling	Xu et al. (2012)
Phytolacca americana	Phytolaccaceae	P	H	Tianmushan	Chen et al. (2011), Li et al. (2011) and Bai et al. (2013)
Pilea microphy lla	Urticaceae	A	H	Ganshiling	Zhang and Xing (2011)
Pilea microphylla	Urticaceae	A	H	Dinghushan	Song et al. (2009), He and Huang (2004) and Jiang et al. (2007)
Pilea microphylla	Urticaceae	A	H	Dinghushan	Song et al. (2009), He and Huang (2004) and Jiang et al. (2007)
Pilea microphylla	Urticaceae	A	H	Dinghushan	Song et al. (2009), He and Huang (2004) and Jiang et al. (2007)
Pilea microphylla	Urticaceae	A	H	Jinfoshan	Lin et al. (2007) and Sun et al. (2009)
Pilea microphylla	Urticaceae	A	H	Nabanhe	Liu et al. (2008)
Pilea microphylla	Urticaceae	A	H	Tongguling	Jiang et al. (2007) and Qin et al. (2008)
Pistia stratiotes	Araceae	P	H		
Pistis stratiotes	Araceae	P	H	Tianmushan	Chen et al. (2011), Li et al. (2011) and Bai et al. (2013)

(continued)

Species	Family	Life form	Growth form	Natural reservation	Major references
Plantago aristata	Plantaginaceae	A	H	Taohongling	Xu et al. (2012)
Plantago lanceolata	Plantaginaceae	P	H	Shedao-Laotieshan	Jiang et al. (2007) and Wu et al. (2010)
Plantago virginica	Plantaginaceae	A	H	Hongshulin	Cao et al. (2007)
Plantago virginica	Plantaginaceae	A	H	Poyanghu	Ge et al. (2010) and Zeng et al. (2013)
Plantago virginica	Plantaginaceae	A	H	Tianmushan	Chen et al. (2011), Li et al. (2011) and Bai et al. (2013)
Psidium guajava Linn.	Myrtaceae	P	T	Ganshiling	Zhang and Xing (2011)
Raphanus raphanistrum	Cruciferae	A	H	Poyanghu	Ge et al. (2010) and Zeng et al. (2013)
Reseda lutea L.	Resedaceae	A	H	Shedao-Laotieshan	Jiang et al. (2007) and Wu et al. (2010)
Rhus typhina	Anacardiaceae	P	T	Songshan	Liu et al. (2012)
Rhynchelytrum repens	Gramineae	P	H	Bawangling	Hu et al. (2011)
Rhynchelytrum repens	Gramineae	P	H	Ganshiling	Zhang and Xing (2011)
Rhynchelytrum repens	Gramineae	P	H	Nabanhe	Liu et al. (2008)
Richardia brasiliensis	Rubiaceae	A	H	Ganshiling	Zhang and Xing (2011)
Ricinus communis	Euphorbiaceae	A	H	Dashahe	Lin et al. (2008)
Ricinus communis	Euphorbiaceae	A	H	Dongzhaigang	Jiang et al. (2007) and Qin et al. (2008)
Ricinus communis	Euphorbiaceae	A	H	Ganshiling	Zhang and Xing (2011)
Ricinus communis	Euphorbiaceae	A	H	Jinfoshan	Lin et al. (2007) and Sun et al. (2009)
Ricinus communis	Euphorbiaceae	A	H	Nabanhe	Liu et al. (2008)
Ricinus communis	Euphorbiaceae	A	H	Shiwandashan	Wei et al. (2006), Jiang et al. (2007) and Ye et al. (2008)
Ricinus communis	Euphorbiaceae	A	H	Taohongling	Xu et al. (2012)
Ricinus communis	Euphorbiaceae	A	H	Tianmushan	Chen et al. (2011), Li et al. (2011) and Bai et al. (2013)
Ricinus communis	Euphorbiaceae	A	H	Tongguling	Jiang et al. (2007) and Qin et al. (2008)
Robinia pseudoacacia	Leguminosae sp.	P	T	Songshan	Liu et al. (2012)
Robinia pseudoacacia	Leguminosae sp.	P	T	Jiuduansha	Jiang et al. (2007) and Qin et al. (2007)

(continued)

Species	Family	Life form	Growth form	Natural reservation	Major references
Robinia pseudoacacia	Leguminosae sp.	P	T	Tianmushan	Chen et al. (2011), Li et al. (2011) and Bai et al. (2013)
Robinia pseudoacacia	Leguminosae sp.	P	T	Xingdoushan	Lu et al. (2005)
S.verbascifolium L.	Sonneratiaceae	P	T	Tongguling	Jiang et al. (2007) and Qin et al. (2008)
Scoparia dulcis L.	Scrophulariaceae	P	H	Datian	Jiang et al. (2007) and Qin et al. (2008)
Scoparia dulcis L.	Scrophulariaceae	P	H	Dinghushan	Song et al. (2009), He and Huang (2004) and Jiang et al. (2007)
Scoparia dulcis L.	Scrophulariaceae	P	H	Dinghushan	Song et al. (2009), He and Huang (2004) and Jiang et al. (2007)
Scoparia dulcis L.	Scrophulariaceae	P	H	Dongzhaigang	Jiang et al. (2007) and Qin et al. (2008)
Scoparia dulcis L.	Scrophulariaceae	P	H	Ganshiling	Zhang and Xing (2011)
Scoparia dulcis L.	Scrophulariaceae	P	H	Hongshulin	Cao et al. (2007)
Scoparia dulcis L.	Scrophulariaceae	P	H	Nabanhe	Liu et al. (2008)
Setaria palmifolia	Gramineae	P	H	Dinghushan	Song et al. (2009), He and Huang (2004) and Jiang et al. (2007)
Setaria palmifolia	Gramineae	P	H	Ganshiling	Zhang and Xing (2011)
Setaria palmifolia	Gramineae	P	H	Jinfoshan	Lin et al. (2007) and Sun et al. (2009)
Setaria palmifolia	Gramineae	P	H	Taohongling	Xu et al. (2012)
Setaria parviflora	Gramineae	P	H	Taohongling	Xu et al. (2012)
Silybum marianum	Compositae	A	H	Jiuduansha	Jiang et al. (2007) and Qin et al. (2007)
Sisymbrium altissimum	Cruciferae	A	H	Taohongling	Xu et al. (2012)
Solanum aculeatissimum	Solanaceae	A	H	Bawangling	Hu et al. (2011)
Solanum aculeatissimum	Solanaceae	A	H	Dashahe	Lin et al. (2008)
Solanum aculeatissimum	Solanaceae	A	H	Datian	Jiang et al. (2007) and Qin et al. (2008)

(continued)

Species	Family	Life form	Growth form	Natural reservation	Major references
Solanum aculeatissimum	Solanaceae	A	H	Jinfoshan	Lin et al. (2007) and Sun et al. (2009)
Solanum aculeatissimum	Solanaceae	A	H	Nabanhe	Liu et al. (2008)
Solanum aculeatissimum	Solanaceae	A	H	Tongguling	Jiang et al. (2007) and Qin et al. (2008)
Solanum americanum	Solanaceae	A	H	Ganshiling	Zhang and Xing (2011)
Solanum capsicoides	Solanaceae	P	S	Datian	Jiang et al. (2007) and Qin et al. (2008)
Solanum capsicoides	Solanaceae	P	S	Tongguling	Jiang et al. (2007) and Qin et al. (2008)
Solanum erianthum	Solanaceae	P	S	Ganshiling	Zhang and Xing (2011)
Solanum torvum Swartz	Solanaceae	P	S	Datian	Jiang et al. (2007) and Qin et al. (2008)
Solanum torvum Swartz	Solanaceae	P	S	Dinghushan	Song et al. (2009), He and Huang (2004) and Jiang et al. (2007)
Solanum torvum Swartz	Solanaceae	P	S	Dinghushan	Song et al. (2009), He and Huang (2004) and Jiang et al. (2007)
Solanum torvum Swartz	Solanaceae	P	S	Ganshiling	Zhang and Xing (2011)
Solanum torvum Swartz	Solanaceae	P	S	Nabanhe	Liu et al. (2008)
Solanum torvum Swartz	Solanaceae	P	S	Shiwandashan	Wei et al. (2006), Jiang et al. (2007) and Ye et al. (2008)
Solidago canadensis L.	Compositae	P	H	Jiuduansha	Jiang et al. (2007) and Qin et al. (2007)
Solidago canadensis L.	Compositae	P	H	Tianmushan	Chen et al. (2011), Li et al. (2011) and Bai et al. (2013)
Soliva anthemifolia	Compositae	A	H	Dinghushan	Song et al. (2009), He and Huang (2004) and Jiang et al. (2007)
Soliva anthemifolia	Compositae	A	H	Poyanghu	Ge et al. (2010) and Zeng et al. (2013)
Sonchus asper (L.) *Hill.*	Compositae	A	H	Dinghushan	Song et al. (2009), He and Huang (2004) and Jiang et al. (2007)
Sonchus asper (L.) *Hill.*	Compositae	A	H	Tianmushan	Chen et al. (2011), Li et al. (2011) and Bai et al. (2013)

(continued)

Species	Family	Life form	Growth form	Natural reservation	Major references
Sonchus oleraceus L.	Compositae	A	H	Bawangling	Hu et al. (2011)
Sonchus oleraceus L.	Compositae	A	H	Hengshui	Li (2008)
Sonchus oleraceus L.	Compositae	A	H	Jinfoshan	Lin et al. (2007) and Sun et al. (2009)
Sonchus oleraceus L.	Compositae	A	H	Taohongling	Xu et al. (2012)
Sonchus oleraceus L.	Compositae	A	H	Tianmushan	Chen et al. (2011), Li et al. (2011) and Bai et al. (2013)
Sonneratia apetala	Solanaceae	P	S	Hongshulin	Cao et al. (2007)
Sorghum halepense (L.) *Pers.*	Gramineae	P	H	Datian	Jiang et al. (2007) and Qin et al. (2008)
Sorghum halepense (L.) *Pers.*	Gramineae	P	H	Dongzhaigang	Jiang et al. (2007) and Qin et al. (2008)
Sorghum halepense (L.) *Pers.*	Gramineae	P	H		
Spartina alterniflora	Gramineae	P	H	Dayudao	Zhu et al. (2006)
Spartina alterniflora	Gramineae	P	H	Jiuduansha	Jiang et al. (2007) and Qin et al. (2007)
Spermacoce latifolia	Rubiaceae	P	H	Nabanhe	Liu et al. (2008)
Stachy tarpheta jamaicensis	Verbenaceae	P	H	Ganshiling	Zhang and Xing (2011)
Stachytarpheta jamaicensis (L.) *Vahl.*	Verbenaceae	P	H	Bawangling	Hu et al. (2011)
Stachytarpheta jamaicensis (L.) *Vahl.*	Verbenaceae	P	H	Datian	Jiang et al. (2007) and Qin et al. (2008)
Stachytarpheta jamaicensis (L.) *Vahl.*	Verbenaceae	P	H	Tongguling	Jiang et al. (2007) and Qin et al. (2008)
Stylosanthes guianensis (Aubl.) Sw.	Leguminosae sp.	P	H	Ganshiling	Zhang and Xing (2011)
Synedrella nodiflora	Compositae	A	H	Bawangling	Hu et al. (2011)
Synedrella nodiflora	Compositae	A	H	Datian	Jiang et al. (2007) and Qin et al. (2008)
Synedrella nodiflora	Compositae	A	H	Dinghushan	Song et al. (2009), He and Huang (2004) and Jiang et al. (2007)
Synedrella nodiflora	Compositae	A	H	Dinghushan	Song et al. (2009), He and Huang (2004) and Jiang et al. (2007)

(continued)

Species	Family	Life form	Growth form	Natural reservation	Major references
Synedrella nodiflora	Compositae	A	H	Dinghushan	Song et al. (2009), He and Huang (2004) and Jiang et al. (2007)
Synedrella nodiflora	Compositae	A	H	Dongzhaigang	Jiang et al. (2007) and Qin et al. (2008)
Synedrella nodiflora	Compositae	A	H	Nabanhe	Liu et al. (2008)
Synedrella nodiflora	Compositae	A	H	Tongguling	Jiang et al. (2007) and Qin et al. (2008)
Tagetes erecta	Compositae	A	H	Datian	Jiang et al. (2007) and Qin et al. (2008)
Tagetes erecta	Compositae	A	H	Tianmushan	Chen et al. (2011), Li et al. (2011) and Bai et al. (2013)
Tagetes patula	Compositae	A	H	Jinfoshan	Lin et al. (2007) and Sun et al. (2009)
Tagetes patula	Compositae	A	H	Taohongling	Xu et al. (2012)
Talinum paniculatum	Portulacaceae	P	H	Ganshiling	Zhang and Xing (2011)
Talinum paniculatum	Portulacaceae	P	H	Jinfoshan	Lin et al. (2007) and Sun et al. (2009)
Talinum paniculatum	Portulacaceae	P	H	Tianmushan	Chen et al. (2011), Li et al. (2011) and Bai et al. (2013)
Thalictrum aquilegifolium Linn.	Ranunculaceae	P	H	Songshan	Liu et al. (2012)
Tithonia diversifolia	Compositae	A	H	Dinghushan	Song et al. (2009), He and Huang (2004) and Jiang et al. (2007)
Tithonia diversifolia	Compositae	A	H	Nabanhe	Liu et al. (2008)
Tridax procumbens	Compositae	P	H	Datian	Jiang et al. (2007) and Qin et al. (2008)
Tridax procumbens	Compositae	P	H	Dinghushan	Song et al. (2009), He and Huang (2004) and Jiang et al. (2007)
Tridax procumbens	Compositae	P	H	Dongzhaigang	Jiang et al. (2007) and Qin et al. (2008)
Tridax procumbens	Compositae	P	H	Ganshiling	Zhang and Xing (2011)
Tridax procumbens	Compositae	P	H	Tongguling	Jiang et al. (2007) and Qin et al. (2008)

(continued)

Species	Family	Life form	Growth form	Natural reservation	Major references
Trifolium pratense	Leguminosae sp.	P	H	Xingdoushan	Lu et al. (2005)
Trifolium repens	Leguminosae sp.	P	H	Jinfoshan	Lin et al. (2007) and Sun et al. (2009)
Trifolium repens	Leguminosae sp.	P	H	Jiuduansha	Jiang et al. (2007) and Qin et al. (2007)
Trifolium repens	Leguminosae sp.	P	H	Lushan	Deng et al. (2009), Zheng et al. (2011) and Tang et al. (2012)
Trifolium repens	Leguminosae sp.	P	H	Xingdoushan	Lu et al. (2005)
Trifolium repens	Leguminosae sp.	P	H	Yalujiangkou	Jiang et al. (2007) and Wu et al. (2010)
Trifoliumrepens	Leguminosae sp.	P	H	Jinfoshan	Lin et al. (2007) and Sun et al. (2009)
Vaccaria segetalis	Caryophyllaceae	A	H	Hengshui	Li (2008)
Verbena tenera Spreng.	Verbenaceae	P	H	Dongzhaigang	Jiang et al. (2007) and Qin et al. (2008)
Veronica polita	Scrophulariaceae	A	H	Dashahe	Lin et al. (2008)
Veronica polita	Scrophulariaceae	A	H	Jinfoshan	Lin et al. (2007) and Sun et al. (2009)
Veronica polita	Scrophulariaceae	A	H	Taohongling	Xu et al. (2012)
Veronica polita	Scrophulariaceae	A	H	Xingdoushan	Lu et al. (2005)
Veronica arvensis	Scrophulariaceae	A	H	Jinfoshan	Lin et al. (2007) and Sun et al. (2009)
Veronica arvensis	Scrophulariaceae	A	H	Jiuduansha	Jiang et al. (2007) and Qin et al. (2007)
Veronica arvensis	Scrophulariaceae	A	H	Taohongling	Xu et al. (2012)
Veronica arvensis L.	Scrophulariaceae	A	H	Tianmushan	Chen et al. (2011), Li et al. (2011) and Bai et al. (2013)
Veronica persica	Scrophulariaceae	A	H	Dashahe	Lin et al. (2008)
Veronica persica	Scrophulariaceae	A	H	Jinfoshan	Lin et al. (2007) and Sun et al. (2009)
Veronica persica	Scrophulariaceae	A	H	Jiuduansha	Jiang et al. (2007) and Qin et al. (2007)
Veronica persica	Scrophulariaceae	A	H	Taohongling	Xu et al. (2012)
Viola tricolor L.	Violaceae	P	H	Taohongling	Xu et al. (2012)

(continued)

Species	Family	Life form	Growth form	Natural reservation	Major references
Waltheria indica L. (W. americana L.)	Cucurbitaceae	P	H	Bawangling	Hu et al. (2011)
Waltheria indica L. (W. americana L.)	Cucurbitaceae	P	H	Datian	Jiang et al. (2007) and Qin et al. (2008)
Waltheria indica L. (W. americana L.)	Cucurbitaceae	P	H	Dongzhaigang	Jiang et al. (2007) and Qin et al. (2008)
Wedelia prostrata	Compositae	A	H	Dinghushan	Song et al. (2009), He and Huang (2004) and Jiang et al. (2007)
Wedelia trilobata (Hook. et Arn.) Hemsl.	Compositae	P	H	Dinghushan	Song et al. (2009), He and Huang (2004) and Jiang et al. (2007)
Wedelia trilobata (Hook. et Arn.) Hemsl.	Compositae	P	H	Ganshiling	Zhang and Xing (2011)
Wedelia trilobata (Hook. et Arn.) Hemsl.	Compositae	P	H	Hongshulin	Cao et al. (2007)
Wedelia trilobata (Hook. et Arn.) Hemsl.	Compositae	P	H	Tongguling	Jiang et al. (2007) and Qin et al. (2008)
Xanthium italicum	Compositae	A	H	Shedao-Laotieshan	Jiang et al. (2007) and Wu et al. (2010)
Xanthium spinosum	Compositae	A	H	Shedao-Laotieshan	Jiang et al. (2007) and Wu et al. (2010)

A annual, *P* Perennial, *B* Biennial, *H* Herb, *H* Shrub, *T* Tree, *C* Climbing vine

References

Bai SB, Zhou GM, Wang YX, Liang QQ, Chen J, Chen YY, Shen R (2013) Plant species diversity and dynamics in forests invaded by Moso bamboo (*Phyllostachys edulis*) in Tianmu Mountain Nature Reserve. Biodivers Sci 21(3):288–295. (in Chinese with English abstract)

Cao F, Song XL, He YH, Qiang S, Qin WH, Jiang MK (2007) Investigation of alien invasive plants in Huizhou Mangrove Natural Reserve. J Plant Resour Environ 16(4):61–66. (in Chinese with English abstract)

Chen J, Xu P, Yao ZS (2011) Analysis and control of invasive plants of Tianmushan Nature Reserve. China Medical Herald 8(9):137–139. (in Chinese with English abstract)

Deng SS, Zhong AY, Lin H, Song FD, Tang YL (2009) Invasive Risk and Countermeasure of *Hylobitelus xiaoi* Zhang in Lushan Nature Reserve. Jiangxi Plant Prot 32(2):74–76. (in Chinese with English abstract)

Ge G, Li EX, Wu HP, Wu ZQ (2010) Invasive plants in the National Nature Reserve of Lake Poyang. J Lake Sci 22(1):93–97. (in Chinese with English abstract)

He WQ, Huang ZL (2004) Dynamics and impacts of invasion by nonnative plant species to Dinghushan nature reserve. Sci Technol Guangdong Province 20(3):42–45. (in Chinese with English abstract)

Hu XH, Xiao YA, Zen JJ, Zhou B (2011) Investigation and analysis on alien plants in Bawangling National Natural Reserve in Hainan Province. J Jinggangshan University (Nat Sci) 32(1):131–136. (in Chinese with English abstract)

Jiang MK, Qin WG, Wang Z, Xu WG, Cai L, Zhang LR, Li J, Qiang S (2007) An investigation of exotic species in costal nature reserves in China. 13:37–43 (in Chinese)

Li HX (2008) A preliminary study on invasive plants of Hebei Hengshui Lake Natural Reserve. Wetl Sci Manage 4(2):51–53. (in Chinese with English abstract)

Li XH, Wang C, Chen X, Wang T, Hu JL (2011) Avian seed dispersal of *Phytolacca americana* in Tianmu Mountain Nature Reserve of Zhejiang Province. Sichuan J Zool 30(3):421–423, 428 (in Chinese with English abstract)

Lin MX, Han F, Liu ZY, Ren MB, Liu X, Zhang J (2007) Weed Sci 4:26–28. (in Chinese)

Lin MX, Liu ZY, Han F, Ren MB, Zhang J (2008) Weed Sci 1:31–32. (in Chinese)

Liu F, Tao GD, Wang DS (2008) Investigation on the alien invasive plants in Nabanhe Nature Reserve and precautionary measures. For Invent Plan 33(6):112–117. (in Chines with English abstract)

Liu J, Yao K, Zhang R, Xia Y, Wang HH, Zhang ZM, Wu JG, Jiang WJ (2012) The research of alien invasive plant in Beijing Songshan Nature Reserve. Chin Agric Sci Bull 28(31):91–95. (in Chinese with English abstract)

Lu SF, Liu SX, Fang YP (2005) A preliminary research of invasive species around Xingdoushan National Nature Reserve, Hubei. J Huanggang Normal Univ 25(3):48–52. (in Chinese with English abstract)

Qin WH, Yu SP, Jiang MK, Wang Z, Chen XZ, Xu WG (2007) Weed Sci 1:29–33

Qin WH, Wang Z, Xu WG, Jiang MK (2008) Investigation and analysis on alien invasive plants in three national nature reserves in Hainan. J Plant Resour Environ 17(2):44–49

Song L, Cao F, He YH, Qiang S, Qin WH, Jiang MK (2009) A survey of invasive alien plant species in Dinghushan National Nature Reserve. J Zhejiang For Coll 26(4):538–543. (in Chinese with English abstract)

Sun J, Yang GF, Chen YC, Teng YQ, Yan T, Yi SR (2009) Acta Prataculturae Sinica 18(3):34–42. (in Chinese)

Tang MZ, Ren MX, Zheng JM, Ding JQ (2012) Seasonal dynamics of plant diversity in woodland invaded by *Solidago canadensis* (Asteraceae) in Lushan Nature Reserve. Plant Sci J 30(4):366–373. (in Chinese with English abstract)

Wei YL, Ye D, Wen YG, Men YY (2006) Invasive plant species in Shiwandashan Mountain National Natural Reserve, Guangxi, China. China For Sci Technol 20(6):23–26. (in Chinese with English abstract)

Wu XS, Wang LX, Qu B (2010) An investigation of exotic invasive plants in nature reserves in Liaoning province. Ecol Environ 3:71–75. (in Chinese with English abstract)

Xu XR, Chen J, Xu P, Yao ZS (2012) Analysis of invasive plants of Taohongling Nature Reserve in Jiangxi. Jiangxi Sci 30(6):753–756. (in Chinese with English abstract)

Ye D, Li XK, Wen YG, Tang SC (2008) Alien plant species in Shiwandashan Mountain Natural Reserve, Guangxi, China. J Guangxi Agric Biol Sci 27(4):445–450. (in Chinese with English abstract)

Zeng ZY, Wu HP, Jiang Y, Peng GH (2013) Dynamics and impacts of invasion by exotic species to Poyang Lake National Nature Reserve. Energy Res Manag 4(15–18):35. (in Chinese with English abstract)

Zhang RJ, Xing FW (2011) The composition and origin of alien plants in Ganshiling Nature Reserve in Hainan, China. Guizhou Agric Sci 39(7):31–33. (in Chinese with English abstract)

Zheng JM, Xu M, Sun Y, Wan HL, Liang TJ (2011) Comparison of roadside alien plant composition inside and outside Lushan Nature Reserve, Jiangxi Province. J Beijing For Univ 33(3):51–56. (in Chinese with English abstract)

Zhu KJ, Wang B, Sun L, Chen JR (2006) Xiamen Sci Technol 3:38–41. (in Chinese)

Index

A

Acerophagus coccois, parasitoid, 60
Achatina fulica, giant African snail, 22
Achnatherum inebrians, 118
Aconitum carmichaeli, 118
Adaptation, ecological adaptability
  epigenetic regulatory system, 35
  generalists, *C. odorata*, 36
  heat/thermal stress, insects, 168, 246
  host plants, insects, 163–164, 247–248
  hybridization, 87
  immune priming, pathogens, 248
  insecticides, resistance, 162, 167, 203, 270, 293
  multi-species synergy, 60
  physical/ambient stresses, 33–34, 261, 271, 286, 304
  research topics, 8, 152–154 (*see also* Evolution)
Agasicles hygrophila, flea beetle, 39, 40
Ageratina adenophora, crofton weed, 7, 9, 11, 37, 41, 42, 150
Ageratum conyzoides, tropic ageratum, 32, 34, 42, 132, 305
Agreements, international, 154
Agriculture
  arthropods and, 11
  characteristics of, 100–101
  facility agriculture, 116
  impacts of invasions, 117–119, 150
  international cooperation and, 64
  pests and, 86, 160
  research activities, 153
Agroecosystems
  biological invasions and, 5, 21–44
  ecosystem invasibility, 10
  research and, 8
Algae, 70, 72, 81, 89
Allee effect, 142–143
Allelochemicals, 34, 36
Allelopathy /allelopathic effects, 34, 35, 150
Alligator weed. *See Alternanthera philoxeroides*, alligator weed
Allotropa oracellae, 60
Alternanthera philoxeroides, alligator weed, 10, 11, 22, 28, 30, 31, 130
Alternanthera sessilis, 32
Amaranthus spinosus, 130
Ambrosia artemisiifolia, common ragweed, 7, 11, 22, 31, 35, 37, 39, 41, 43, 118, 119
Amorpha fruticosa, indigobush, 42
Amphibia, 83
Amphiesma vibakari, Asian keelback, 130
Amylostereum areolatum, white rot fungus, 57
Animals, 5, 7, 22, 30, 71, 102, 129–131
Aquaculture, 68, 71, 81–86, 89–90
Aquarium, 68, 71, 82–84, 90
Aquatic ecosystems, 67–92
  damage/influence, 7, 84–87
  global changes, 142
  nitrogen deposition, 128
  non-native species
    distribution, 70–71
    diversity, 69–71
    introduction/spread, 71–84
    management, 88–92
  research topics, 8
Aquatic organisms, aquatic (IAS), 8
Arbuscular mycorrhizal fungi (AMF), 37

F. Wan et al. (eds.), *Biological Invasions and Its Management in China*,
Invading Nature - Springer Series in Invasion Ecology 11,
DOI 10.1007/978-94-024-0948-2

Aristichthys nobilis, bighead carp, 87
Arsenophonus, 166
Arthropod, 172, 300, 305
Aschersonia aleyrodis, entomopathogenic fungi, 174
Asecodes hispinarum, 236–238
Asymmetric mating interactions, 33, 162
Attractant, 39, 61, 213, 221, 223, 274, 275, 309
Axinoscymnus cardilobus, lady beetle, 174

B
Bacillus spp., 37, 197, 236, 274
Bacteria
  biocontrol agents, engineered, 211
  gut bacteria, 252
  horizontal gene transfers, 166
  host immune responses, 261
  invasion species, 101
  multitrophic interactions, 225
  pathogens, 38, 248, 251, 294
  symbiotic species, 9, 190, 274
Bactrocera dorsalis, oriental fruit fly, 41, 267–278, 307
Ballast water, 71, 88, 89
Beauveria bassiana, 174, 189, 276, 309
Bemisia tabaci, sweet potato whitefly, 9, 13, 28, 31, 33, 41, 43, 116, 150, 159–176
Benthic species, benthic community, 7, 86
Bidens pilosa, 130
Biocontrol. *See* Biological control
Biodiversity, biological diversity, 7
Biological control, 11, 39–41
Birds, 7
Body size, 185, 272
Brontispa longissima, coconut leaf beetle, 56, 230, 231, 233, 234, 236–240
Bursaphelenchus xylophilus, pinewood nematode, 4, 9

C
Carbon dioxide (CO_2), 142
Carbon sequestration, 220
Cardinium, 166
Carpomyia vesuviana, 56, 110
Carps, 86, 87
Cenchrus pauciflorus, 119
Ceratocystis fimbriata, 112, 120
Chouioia cunea, 57
Chromolaena odorata, Siam weed, 9, 22, 32, 35, 37, 130
Cladosporium dadosporioides, 60
Clavibacter michiganense subsp. *Sepedonicus*, potato ring rot, 38
Clavibacter michiganensis subsp. *Michiganensis*, tomato bacterial canker, 38
Climate /climate change /climatic conditions/ climatic zone, 68, 99, 114
  aquatic system, 142
  IAS distribution/invasions, 28, 112, 116
  nature reserve, invasions, 135
  oases/grassland, desert, 100, 117
Clonal integration/integrity, 37
Cocos nucifera, coconut palm, 233, 247
Common ragweed. *See Ambrosia artemisiifolia*, common ragweed
Competition, competitive ability, 34, 36
  Ageratina adenophora, 9, 35, 150
  Alternanthera philoxeroides, 32
  Chromolaena odorata, 9
  community and, 85
  competitive displacement/interactions, 162, 167
  increased competitive ability (EICA), 35
  Mikania micrantha, 32
  Solenopsis invicta, red imported fire ant (RIFA), 301–303
Conservation district/area, 126–128, 131–140
Control. *See* Managements
Coronilla varia, crownvetch, 42
Crayfish, 73, 130
Crustaceans, 81
Cryptic species, 160, 162
Ctenopharyngodon idellus, grass carp, 87
Cydia pomonella, codling moth, 11, 39, 285–295

D
Databases, 11–12, 63, 128, 141, 155
Deladenus siricidicola, 58
Dendroctonus valens, red turpentine beetle, 13, 54, 61, 65, 219–226
Desert, 8, 14, 97–121
Detection, 10, 38–39
  alien species in natural reserves, 120–121
  aquatic species, 90
  Bactrocera dorsalis, 275
  Brontispa longissima, 230
  database and, 11–12
  DNA barcoding, 10, 38
  forest pest, 64
  Hyphantria cunea, Fall webworm, 54

research, 154
Rhynchophorus ferrugineus, 246, 248, 250, 252, 253
Solenopsis invicta, red imported fire ant, 309
Dispersal, 71, 115, 143
Displacement, invasion impact, 87, 162, 164, 167
Distribution, invasive species, 6
Disturbances, 10, 31, 33, 37, 112, 128, 143
Diversity. *See* Biodiversity

E
Ecological resistance, of ecosystems. *See* Invasibility
Economic factors, economic development/growth, 16, 28, 114, 128
Economic loss, invasion impacts, 6, 31, 117
Eichhornia crassipes, water hyacinth, 22, 36
Encarsia bimaculata, 172
Encarsia formosa, 39–41, 172, 173
Encarsia sophia, *En. transvena*, 40, 172, 173
Endophytes, 37
Enemy release, 8, 33, 35–36
Energy-use capacity/strategy, 9, 36
Epiblema strenuana, ragweed borer, 39, 40
Eretmocerus hayati, 172
Eretmocerus sp. nr. *furuhashii*, 172, 173
Erigeron annuus, 130
Eriosoma lanigerum, woolly apple aphid, 42, 55
Erwinia amylovora, 38
Eupatorium adenophorum, crofton weed, 10, 54
Eupatorium catarium, praxelis, 32, 42
Euphorbia hirta, 130
Evolution, 9 (*see also* Adaptation, ecological adaptability)
allelopathic effects, allelochemicals, 34, 36
defensive mechanisms, 309
immune protection, 248
increased competitive ability, 9, 35–36
insecticide resistance, 162, 270
interactions, virus, *B.tabaci*, 9
microevolution in invasive, *D. valens*, 61, 62
nitrogen allocation, 36
nitrogen allocation , tradeoff, 9
tolerating freezing, 32, 33
Evolution of increased competitive ability (EICA), 9, 35

F
Flaveria bidentis, yellowtop, 42
Fopius arisanus, 40, 275
Forests, forest ecosystem, 6, 11, 14, 15, 53–65
Frankliniella occidentalis, western flower thrips, 28, 41, 43
Freshwater habitats, ecosystem, and organisms
aquaculture, 73
invasive species diversity, 68, 70, 77–80
Penaeus vannamei., Pacific white shrimp, 73
vectors, introduction, 71–84
Fungi
arbuscular mycorrhizal fungi, 34, 37
biocontrol agents, 172, 189, 236–240, 251, 262, 275, 309
invasive species, 5, 102
pathogens, 62
symbiotic species, 9, 57, 61–63
Fusarium oxysporum f. sp. *cubense*, 11, 38

G
Gastropoda, 83
Genetic control. *See* Managements
Genetic diversity, genetic variation, genetic differentiation, 8, 118, 142, 154, 220, 246, 248–250, 269, 288–290
Genetic pollution/extinction, 87
Geographic origins, of invasive species. *See* Origin, geographic
Giant African snail. *See Achatina fulica*, giant African snail
Global warming, 135, 222, 302
Glomus spp., 34
Government, 4 (*see also* Laws and regulations)
Grazing, 100–102, 114

H
Hamiltonella, 166
Harmonia axyridis, 173, 174
Helianthus tuberosus, Jerusalem artichoke, 42
Hemiberlesia pitysophila, 130
Hippophae rhamnoides, common sea-buckthorn, 42
Human activities, 22
human disturbance, 32, 115, 127, 131
Hyphantria cunea, fall webworm, 54–57
IAS, distribution, 6
IAS, introduction/dispersal, vector, 7, 29, 62, 68, 71–84
research, 154

Hybridization, 87, 118
Hylastes parallelus, 222, 223
Hyoscyamus niger, black henbane, 39, 207
Hyphantria cunea, fall webworm, 4, 54, 130
Hypophthalmichthys molitrix, silver carp, 87

I

Ibalialeu cospoides, 58
Impacts (of invasions)
  biodiversity, ecosystems/habitat damage, 86–87
  economic loss, 117
  genetic pollution/extinction/introgression, 87, 118
  human health, 118–119
  species extinction, 87
Insecticides
  adaptation/resistance, invasive trait, 8, 32
  resistance in *Bemisia tabaci*, 162, 167–169, 174, 176
  resistance mechanisms, 8, 203
  resistance, research topics, 8
Insects
  control, 42
  detection, 10, 38
  invasions, 31–37, 115
  invasive species, 4, 6, 102–114
  research topics, 8, 129
  vectors, 9, 60
Interactions, interspecies. *See* Mutualism, symbiosis, competition, allelopathy, asymmetric mating interactions
Introductions, intentional, 6, 28, 70–73
Introgression, 87, 118
Invasibility, affecting factors, 32
  biodiversity, 31
  competitors, 32
  disturbance, 31
  empty niches, 134
  natural enemies, 32
  resources, 34
Invasibility, of ecosystems, 10, 31–32, 134, 135, 140, 143–144
Invasion mechanisms, 8, 32–37
  asymmetric mating interactions, 162–163
  multi-species synergy, multispecies interactions, 60–61, 221–222
  parthenogenesis, 32
  vector-virus mutualism, 9, 33
Invasive species, of China, 5–7
Invertebrate, 70, 305
Ipomoea purpurea, 130
Isaria fumosorosea, 174

L

Lakes, 68, 86, 87, 114
Land use, 22, 30
Laws and regulations, 89, 90, 141, 155
Legislations. *See* Laws and regulations
Leptinotarsa decemlineata, Colorado potato beetle (CPB), 11, 31, 117, 195–213
Leptocybe invasa, 56, 58–59
Leptographium procerum, 61, 221
Life form, 103–107, 133
Light-saturated photosynthetic rate (Pmax), 36
Liriomyza sativae, leafminer, 31, 43, 116
Lissorhoptrus oryzophilus, rice water weevil, 33, 186–188, 190–191
Locusta migratoria, 116, 117

M

Malas baccata, Siberian crabapple, 42
Mammals, Mammalia, 8
Managements
  genetic control, 294
  integrated use, 42–43
  mechanical control/removal, 141, 234
  prevention/intercepting belt, 11, 39, 205, 207, 208
  replacement control, 11, 41–42
  resistant cultivars/plants, 11, 42, 171, 176, 189, 209, 234
  sterile insect technique, 276, 292, 294
Marine ecosystems/habitat, 5, 71, 72, 85
Marine species/organisms, 68, 70, 74–76
Mechanical control/removal. *See* Managements
Megarhyssa nortoni, 58
Metarhizium anisopliae, 190, 210, 236, 251, 263
Microbes, microorganisms, microbiota, microbial community. *See also* Bacteria, fungi, viruses
  beneficials, 61
  feedback (invasion mechanism), 36
  gut microbiota, 252
  IAS groups, 22, 102, 112
  rhizospheric microorganisms, 7
  soil biota, 32
  symbiotic bacteria, 274
Mikania micrantha mile-a-minute, 32
Ministry of Science & Technology (MST), 4, 13, 44, 83, 152
Molluscs (mollusks), 81
Monochamus, 60
Mutualism/ mutualistic symbiotic relationship, 9, 57, 162, 164, 220, 226, 306

Mylopharyngodon piceus, black carp, 86
Mytilopsis sallei, black-striped mussel, 86

N
Natural enemies
biocontrol agents, 11, 39, 41, 43, 60
natural enemy release, 8, 35, 36, 143
Nature reserves, 8, 14, 128–130, 133, 134, 139–143, 145
invasions, researches, 128, 129
invasions, risk assessment, 140–142
invasive patterns, 131, 133–135, 139, 140
invasive species, 130
Neochetina bruchi, water hyacinth weevils, 39–41
Neochetina eichhorniae, water hyacinth weevils, 39–41
Neovison vison, American mink, 130
Nitrogen
allocation (to plant tissues), 9
deposition, 128
nutrient in soils, 31
use (by plants), 36
Noctiluca scintillans, sea sparkle, 85
North America, invasion origin, 6, 28, 112
Nutrients (in soil), 31, 32, 35–37, 128

O
Octodonta nipae, Nipa palm hispid beetle, 56, 257–264
Oncorhynchus mykiss, steelhead trout, 130
Ondatra zibethica, 130
Ondatra zibethicus, 119
One Belt One Road, 43
Ophiostoma ips, 221
Ophraella communa, ragweed leaf beetle, 39, 40
Oracella acuta, pine mealybug, 59, 60, 130
Origin, geographic, 6, 28
Orius sauteri, 41, 173, 174
Ornamental trades, 71, 82, 84, 90
Oxytropis glabra, 118

P
Paecilomyces lilacinus, 309
Pedicularis abrotanifolia, 118
Penaeus vannamei., Pacific white shrimp, 73
Phenacoccus solenopsis, cotton mealybug, 28
Phenotypic plasticity. *See* Plasticity
Pheromones, 39, 223, 251, 292, 293
Phyllostachys edulis, 129
Phytophthora sojae, 9, 11, 13, 33, 38
Pine wilt disease (PWD), 60
Plants, invasive species, 34–35
adaptation, 33
allelopathic effects, 33, 34
clonal integration, 37
competition, 34
dispersal, 30
diversity, 7, 101, 128
energy-use capacity, 36
intentionally introduced, 28
microbial community, associations, 36–37
natural enemy release, 35
phenotypic plasticity, 34–35
Plasticity, 34, 38
Poa pratensis, Kentucky bluegrass, 42
Policy (Policy-making), 43
Pomacea canaliculata, golden apple snail, 7, 86, 130
Portiera, 166
Prevention, control method. *See* Managements
Procambarus clarkia, red swamp crayfish, 73, 130
Propylea japonica, 173–174
Public awareness, 12, 44, 121, 129, 144
Public education, 12, 90

R
Radopholus similis, burrowing nematode, 11, 39, 42
Rana catesbeiana, bullfrog, 75, 130
Rattus norvegicus, 130
Red imported fire ant. *See Solenopsis invicta*, red imported fire ant (RIFA)
Regulations. *See* Laws and regulations
Reproduction, 162
allelopathic chemicals and, 34
asymmetric mating interactions and, 9, 33, 162
habitats and, 112
invasion mechanisms, 32, 33
reproduction capacity, 150, 162, 176
reproductive interference (*see* Asymmetric mating interactions)
research topics, 8
symbiosis and, 60, 166
Resistance to invasion. *See* Invasibility
Resource (in ecosystems)
capture/ acquisition of, 36
invasibility and, 31
nitrogen allocation, 9, 35
resource availability, 10
soil, 32, 100
Restoration, 71, 84, 120, 144, 151, 155

Rhynchophorus ferrugineus , red palm weevil, 56, 246, 248, 250, 251, 253
Rhyssa persuasoria, 58
Risk assessment/analysis, 10, 37, 89, 91, 92, 140–142, 154
RNA interference (RNAi), 271, 273, 276, 277

S
Seeds
  A.adenophora, of, 22
  adaptation, 32
  Agrostemma githago, impacts, 119
  germination, 34, 118
  dispersal, 28
Selitrichodes neseri, 59
Semiochemical, 222–224, 226
Shrubs, 130, 133
Sirex noctilio, Sirex woodwasp, 57–58
Snail, 22, 86
Soils, 32
  contamination, pollution, 33, 100
  disturbance, 141
  erosion, 100
  microbes, microbial community, 36–38
  nutrient, resource, 7, 31, 32, 34, 100
  pathogen, 32, 34
  soil biota, 10
  type, 99
Solanum rostratum, buffalobur, 34, 117
Solenopsis invicta, red imported fire ant (RIFA), 7, 11, 28, 56, 299–310
Solidago canadensis, Canada Goldenrod, 33, 34, 37, 129
Spalangia endius, 276
Spartina alterniflora, smooth cordgrass, 7, 11, 74, 84, 129
Species complex, 28, 162, 163, 175
Species richness. *See* Biodiversity
Sporothrix sp., 61
Spread. *See* Dispersal
Steinernema feltiae, 190
Sterile insect technique (SIT). *See* Managements
Susceptibility to invasion. *See* Invasibility
Sweet potato whitefly. *See Bemisia tabaci*, sweet potato whitefly
Symbiosis, symbionts, 8, 9, 57, 60, 162, 166, 190, 220, 274

T
Tapinoma melanocephalum, ghost ant, 305
Tetrastichus brontispae, 236, 238, 262, 264
Tilletia controversa, 38
Tomato yellow leaf curl virus (TYLCV), 165, 171
Trachemys scripta, red-eared turtle, 84
Trade, international, 4, 53, 68, 71, 72, 83, 84, 90, 112, 115
Transport (air, land, sea), 30, 43, 58, 64, 81
Transport (arrival) (stage of invasion), 30
Trapping, traps, *control method*
  B. dorsalis, 274
  Cydia pomonella, 289
  Dendroctonus valens, 221, 224
  kairomones and, 221, 223
  light trapping, 186
  pheromone trapping, 57
  Rhynchophorus ferrugineus, 252
  S. invicta, 307
  trap crops, 172, 211
  whiteflies, 172
  yellow stick traps, 39, 41
Trialeurodes vaporariorum, 31, 41, 168
Trichoderma sp., 210
Trichogramma platneri, 294
Trout, 130
Turtles, 82, 84

U
Unintentional introductions, 28, 29, 112

V
Vectors, 61, 71–84, 86, 88 (*see also* Dispersal)
Vegetation. *See* Plants
Veriticillium sp., 210
Vertebrates, 70
Verticillium dahliae, 117
Verticillium lecanii, 174
Viruses, 5, 9, 33, 102, 116, 160, 162, 164–165, 171

W
Wedelia trilobata, 36, 37
Weeds, 30, 38, 118, 307
Wetlands, 7, 68, 71, 84, 126
Wind, 28, 112, 186, 202, 231

X
Xanthium italicum, 117
Xanthium spinosum, 104, 117
Xanthomonas axonopodis, 38

Z
Zarhopalus debarri, 60

Kolibri 360 GmbH
Ferdinand-Jühlke-Straße 7
99095 Erfurt, Deutschland
produktsicherheit@kolibri360.de